T0258068

Handbook of Biomaterials

Handbook of Biomaterials

Edited by **Ralph Seguin**

New York

Published by NY Research Press,
23 West, 55th Street, Suite 816,
New York, NY 10019, USA
www.nyresearchpress.com

Handbook of Biomaterials
Edited by Ralph Seguin

International Standard Book Number: 978-1-63238-229-0 (Hardback)

Printed in the United States of America.

Contents

Preface

This book has been a concerted effort by a group of academicians, researchers and scientists, who have contributed their research works for the realization of the book. This book has materialized in the wake of emerging advancements and innovations in this field. Therefore, the need of the hour was to compile all the required researches and disseminate the knowledge to a broad spectrum of people comprising of students, researchers and specialists of the field.

This book consists of reviews and original researches conducted by experts and scientists working in the field of biomaterials, its development and applications. It offers readers the potentials of distinct synthetic and engineered biomaterials. This book gives a comprehensive summary of the applications of various biomaterials, along with the techniques required for designing, developing and classifying these biomaterials without any intervention by any industrial source. It also elucidates the various techniques used to produce biomaterials with the required physical and biological features for medical and clinical applications.

At the end of the preface, I would like to thank the authors for their brilliant chapters and the publisher for guiding us all-through the making of the book till its final stage. Also, I would like to thank my family for providing the support and encouragement throughout my academic career and research projects.

Editor

Part 1

Biomechanical and Physical Studies

Biomechanical Properties of Synovial Fluid in/Between Peripheral Zones of Articular Cartilage

Miroslav Petrtyl, Jaroslav Lisal and Jana Danesova
Laboratory of Biomechanics and Biomaterial Engineering, Faculty of Civ. Engineering,
Czech Technical University in Prague
Czech Republic

1. Introduction

The properties and behaviour of articular cartilage (**AC**) have been studied from numerous aspects. A number of biomechanical models of the properties and behaviour of AC are available today. The traditional model presents cartilage as homogeneous, isotropic and biphase material (Armstrong et al., 1984). There also exist models of transversally isotropic biphase cartilage material (Cohen et al., 1992; Cohen et al., 1993), non-linear poroelastic cartilage material (Li et al., 1999), models of poroviscoelastic (Wilson et al., 2005) and hyperelastic cartilage material (Garcia & Cortes, 2006), models of triphase cartilage material (Lai et al., 1991; Ateshian et al., 2004), and other models (Wilson et al., 2004; Jurvelin et al., 1990). The published models differ, more or less, by the angle of their authors' view of the properties and behaviour of articular cartilage during its loading.

The authors base their theories on various assumptions concerning the mutual links between the structural components of the cartilage matrix and their interactions on the molecular level.

The system behaviour of AC very depend on nonlinear properties of synovial fluid (**SF**). Certain volumes of SF are moveable components during the mechanical loading in the peripheral zone of AC. Biomechanical properties of peripheral zone of AC are significantly influenced by change of SF viscosity due to mechanical loading.

The hydrodynamic lubrication systems and influences of residual strains on the initial presupplementation of articular plateaus by synovial fluid were not sufficiently analyzed up to now.

Our research has been focused on analyses of residual strains arising in AC at cyclic loading and on the viscous properties of SF. Residual strains in articular cartilage contribute the preaccumulation of articular surfaces by synovial fluid.

SF reacts very sensitively to the magnitude of shear stress and to the velocity of the rotation of the femoral and tibial part of the knee joint round their relative centre of rotation when the limb shifts from flexion to extension and vice versa. Shear stresses decrease aggregations of macromolecules of hyaluronic acid in SF.

Articular cartilage (AC) is a viscohyperelastic composite biomaterial whose biomechanical functions consist

1. in transferring physiological loads into the subchondral bone and further to the spongious bone,
2. in ensuring the lubrication of articular plateaus of joints and
3. in protecting the structural components of cartilage from higher physiological forces.

The macromolecular structure of AC in the peripheral zone (Fig. 1.) has two fundamental biomechanical safety functions, i.e. to regulate the lubrication of articular surfaces and to protect the chondrocytes and extracellular matrix from high loading.

The rheological properties of SF play the key role in the achievement of the optimum hyaluronan concentration.

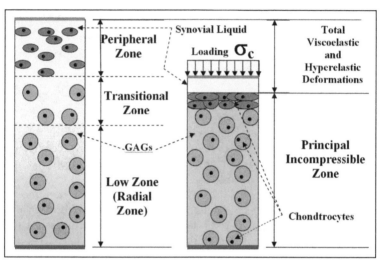

Fig. 1. Complex structural system of articular cartilage (collagen fibres of 2nd type are not drawn)

The properties of SF in the gap between the opposite surfaces of articulate cartilage are not homogeneous during loading. The properties of SF change not only during biomechanical loading, but also during each individual's life time. The viscous properties of this fluid undergo changes (in time) due to mechanical loading. As a consequence of its very specific rheological characteristics, SF very efficiently adapts to external biomechanical effects. Exact knowledge of the rheological properties of synovial fluid is a key tool for the preservation and treatment of AC. The significance of the specific role of SF viscosity and viscosity deviations from predetermined physiological values were first pointed out as early as the 1950s to 1990s (Johnson et al., 1955; Bloch et al., 1963; Ferguson et al., 1968; Anadere et al., 1979; Schurz & Ribitsch, 1987; Safari et al., 1990 etc.). The defects of concentrations of the dispersion rate components were noticed by Mori (Mori et al., 2002). In this respect, it cannot be overlooked that mechanical properties of SF very strongly depend on the molecular weight of the dispersion rate (Sundblad et al., 1953; Scott & Heatley, 1999; Yanaki et al., 1990; Lapcik et al., 1998) and also on changes in the aggregations of macromolecular complexes in SF during mechanical effects (Myers et al., 1966; Ferguson et al., 1968; Nuki & Ferguson, 1971; Anadere et al., 1979 and Schurz & Ribitsch, 1987).

Synovial fluid is a viscous liquid characterized by the apparent viscosity η. This viscosity depends on stress and the time during which the stress acts. SF is found in the pores of the peripheral zone of AC and on its surface (in the gap between the opposite AC surfaces). The viscosity of synovial fluid is caused by the forces of attraction among its molecules being fully manifested during its flow. In other words, viscosity is a measure of its internal resistance during the SF flow. In the space between the opposite AC surfaces, its flow behaves like a non-Newtonian fluid.

As was pointed out above, biomechanical effects play a non-negligible and frequently a primary role in regulating rheological properties.

The principal components of synovial fluid are water, hyaluronic acid **HA**, roughly 3-4 mg/ml, D-glucuronic acid and D-N-acetylglucosamine (Saari et al., 1993 and others). By its structure, hyaluronic acid is a long polymer, which very substantially predetermines the viscous properties of synovial fluid. Its molecular structure is evident from Fig. 2. *Synovial fluid* also contains an essential growth hormone *prolactin* (PRL) and *glycoprotein lubricin*.

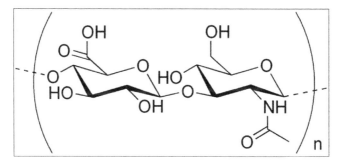

Fig. 2. Molecular complex of hyaluronic acid (HA)

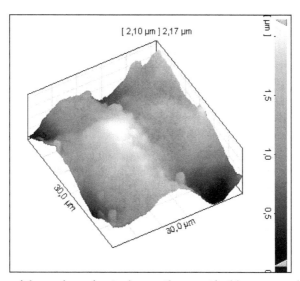

Fig. 3. Topography of the surface of articular cartilage verified by means of FAM (Force Atomic Microscope). The height differences of surface points range up to ca 200 nm - 2,4 μm. In unloaded condition, they are flooded by synovial fluid

Prolactin induces the synthesis of proteoglycans and, in combination with glucocorticoids, it contributes to the configuration of chondrocytes inside AC and to the syntheses of type II collagen. The average molecular weight of human SF is 3 – 4 MDa.

Important components of SF are lubricin and some proteins from blood plasma (γ-globulin and albumin), which enhance the lubricating properties of SF (Oates, 2006). The importance of HA and proteins for the lubricating properties of SF was also described (Swann et al., 1985; Rinaudo et al., 2009).

In the gap between AC surfaces, synovial fluid forms a micro-layer with a thickness of ca 50 μm. It fills up all surface micro-depressions (Fig. 3. and 4., Petrtýl et al., 2010) and in accessible places its molecules are in contact with the macromolecules of residual SF localized in the pores of the femoral and tibial peripheral zone of AC.

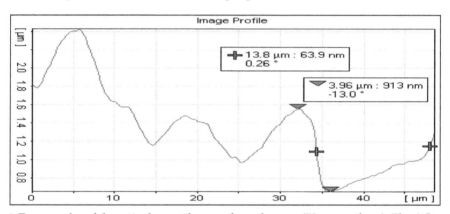

Fig. 4. Topography of the articular cartilage surface of a man (58 years of age). The AC surface oscillates to relative heights of 2.5 μm. During fast shifts of the AC surface (due to the effect of dynamic shifting forces/dynamic bending moments or shear stresses), the AC surface is filled up with generated synovial gel (with less associated NaHA macromolecules) with *low viscosity*

SF is a rheological material whose properties change in time (Scott, 1999 and others). As a consequence of loading, associations of polymer chains of HA (and some proteins) arise and rheopexic properties of SF are manifested (Oates et al., 2006). Due to its specific rheological properties, SF ensures the lubrication of AC surfaces. The key component contributing to lubrication is HA/NaHA. In healthy young individuals, the endogenous production of hyaluronic acid (HA) reaches the peak values during adolescence. It declines with age. It also decreases during arthritis and rheumatic arthritis (Bloch et al., 1963; Anadere et al., 1979; Davies & Palfrey, 1968; Schurz & Ribitsch, 1987 and numerous other authors). Some AC diseases originate from the disturbance of SF lubrication mechanisms and from the defects of genetically predetermined SF properties. Therefore, the lubrication mechanisms of AC surfaces must be characterized with respect to the rheological properties of SF.

2. Contents

The objectives of our research has been aimed on the definition of the biomechanical properties of SF which contribute to the lubrication of the opposite surfaces of articular cartilage, on the analysis of the effects of shear stresses on changes in SF viscosity and on the analysis of the residual strains arising in AC at cyclic loading.

2.1 Rheological properties of synovial fluid

With respect to the project objectives, the focus of interest was on the confirmation of the rheological properties of hyaluronic acid with sodium anions (sodium hyaluronan, NaHA) in an amount of 3.5 mg ml^{-1} in distilled water without any other additives. The use of only NaHA was based on the verification of the association of HA macromolecules and on the manifestation of highly specific rheological properties of SF, which regulate its lubrication function. The rheological properties were verified using the rotation viscometer Rheolab QC (Anton Paar, Austria). Viscosity values were measured continuously within 8 minutes.

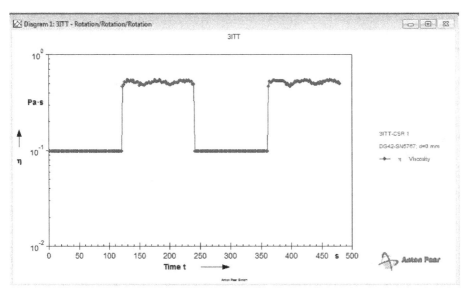

Fig. 5. SF apparent viscosity as related to time (velocity gradient 100s^{-1})

Samples were subjected to the effect of constant velocity gradient (100s^{-1} – 500s^{-1} – 1000s^{-1} – 2000s^{-1}) in time 0 – 120s and 240 – 360s. Samples were subjected in the tranquility state in time 120 – 240s and 360 – 480s. The measurements were performed at the temperature of human body (37°C).

Fig. 5. clearly shows that at the constant SF flow velocity gradient 100s^{-1} there is a distinct time-related constant values in viscosity. The verified synthetic synovial fluid possesses *pseudoplastic properties*. It is evident that the *macromolecules of hyaluronic acid (NaHA/HA) in a water dispersion environment principally contribute to the pseudoplastic behaviour of the fluid. This property is of key importance for controlling the quality of the AC surface protection.*

Fig. 6. also shows that at the constant SF flow velocity gradient 2000s^{-1} there is a distinct time-related constant values in viscosity. The viscosity of SF after unloading always returns to the same values (ca 0.8 Pa s).

Fig. 7. shows that viscosity values of SF with increasing rate of flow velocity gradient 0 – 2000 s^{-1} (in time 0 – 60s) decrease. Viscosity values of SF are constant with constant rate of velocity gradient 2000 s^{-1} (in time 60 – 180s).

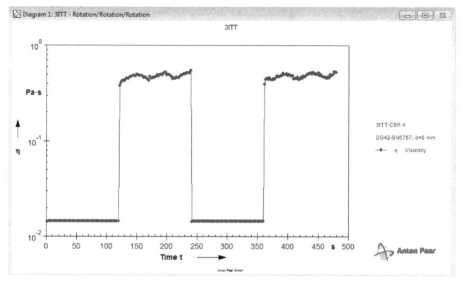

Fig. 6. SF apparent viscosity as related to time (velocity gradient 2000s⁻¹)

Due to the fact that the lubrication abilities of SF strongly depend on the magnitude of viscosity, and SF viscosity depends on the SF flow velocities, the effects of the magnitudes and directions of shifting forces or shear stresses respectively on the distributions of the magnitudes and directions of SF flow velocity vectors in the space between the opposite AC surfaces had to be analyzed.

The kinematics of the limb motion (within one cycle) shows that during a step the leg continuously passes through the phases of flexion – extension – flexion (Fig. 8.). The effect of shifting forces (or shear stresses respectively) is predominantly manifested in the phases of flexion, while normal forces representing the effects of the gravity (weight) of each individual mostly apply in the phases of extension, Fig. 8. The distributions of the magnitudes of SF flow velocity vectors depend on the shifts of the tibial and femoral part of the knee joint, Fig. 9., reaching their peaks in places on the interface of SF with the upper and lower AC surface, Fig. 10. The velocities of SF flows very substantially affect the SF behavior contributing to the lubrication of AC surfaces and their protection.

At rest the bonds are created among the macromolecules of hyaluronic acid (HA) leading to the creation of associates. *By associating molecular chains of HA (at rest) into a continuous structure, a spatial macromolecular grid is created in SF which contributes to the growth in viscosity and also to the growth in elastic properties.*

The associations of HA molecules are the manifestation of cohesive forces among HA macromolecular chains. SF represents a dispersion system (White, 1963) in which the dispersion rate is dominantly formed by snakelike HA macromolecules. The dispersion environment is formed by water. Cohesive forces among NaHA polymer chains in SF are of physical nature. The density (number) of bonds among HA macromolecules is dominantly controlled by mechanical effects. Fig. 9. In relation to the magnitudes of velocity gradients, NaHA macromolecules are able to form "thick" synovial gel which possesses elastic properties characteristic of solid elastic materials, even though the dispersion environment of synovial gel is liquid.

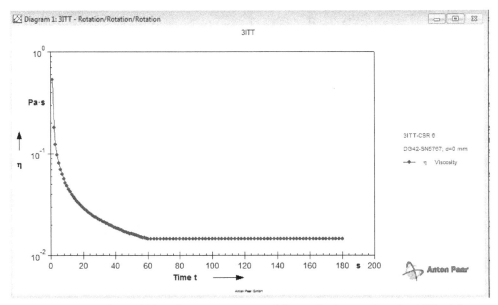

Fig. 7. Viscosity values of SF with increasing rate of flow velocity gradient 0 – 2000 s^{-1} (in time 0 – 60s) decrease. Viscosity values of SF are constant with constant rate of velocity gradient 2000 s^{-1} (in time 60 – 180s)

SF represents a mobile dispersion system in which *synovial gel is generated due to non-Newtonian properties of SF.* Within this system, the macromolecules of hyaluronic acid can be intertwined into a three-dimensional grid, which continuously penetrates through the dispersion environment formed by water. The pseudoplastic properties of SF are manifested through mechanical effects (for example while walking or running), Fig. 8., Fig. 9. *Physical netting occurs, which is characterized by the interconnection of sections of polymer chains into knots or knot areas.* Generally speaking, the association of individual molecules of hyaluronic acid (HA/NaHA) occurs in cases of reduced affinity of its macromolecular chains to the solvent. In other words, the *macromolecules of hyaluronic acid (HA) form a spatial grid structure in a water solution* (Fig. 9.).

Mutually inverse shifts and inverse rotations of the opposite AC surfaces cause inverse flows of SF on its interface with the AC surface (Fig. 10.).

The greatest magnitudes of SF velocity vectors due to the effect of shear stresses τ_{xy}, (or the effects of shifting forces respectively) are found near the upper and lower AC surface. They are, however, mutually inversely oriented. Fig. 10. displays the right-oriented velocity vector direction near the upper surface, and the left-oriented one near the lower AC surface. The magnitudes of velocity vectors decrease in the direction towards the central SF zone. In this thin neutral zone, the velocity vector is theoretically zero in value. A very thin layer (zone) of SF in the vicinity of the central zone, with very small to zero velocities, can be appointed **neutral SF zone.**

At very small velocities of SF flows, the *viscosity of the neutral central zone is higher than the viscosity in the vicinity of AC surfaces.* Under the conditions of very low viscosity, the SF material in the vicinity of AC surfaces is characterized by a low friction coefficient. Friction reaches values of ca 0.024 – 0.047 (Radin et al., 1971).

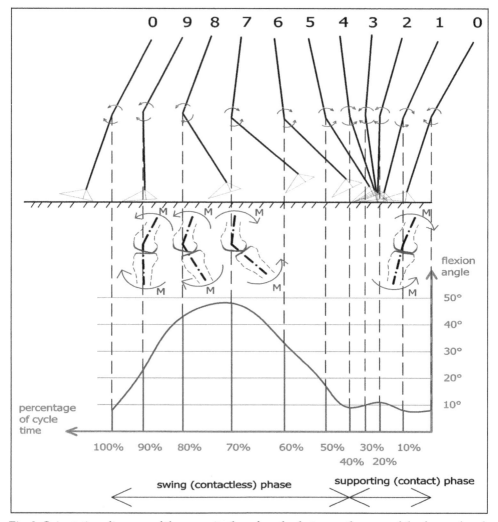

Fig. 8. Orientation diagram of the magnitudes of angles between the axes of the femoral and tibial diaphysis during the "flexion – extension – flexion" cycle of the lower limb in relation to the time percentage of the cycle

The total *thickness of the gap* between the opposite AC surfaces is only ca 50 μm, including height roughness of the surfaces near both peripheral layers 2 x 2.5 μm, Fig. 4., Fig. 9. (Petrtýl et al., 2010). In quiescent state, the AC surfaces are flooded with SF (synovial gel) while during the leg motion (from flexion to extension and vice versa) synovial sol with the relatively low viscosity is generated in SF in peripheral zones of AC. In other words, *due to the effect of shear stresses* τ_{xy} the viscosity η of SF decreases and synovial sol is generated. Aggregations of macromolecules of hyaluronic acid decrease. *The most intense aggregations* are in places of the smallest SF velocities, i.e. in neutral (central) zone of SF between the AC surfaces.

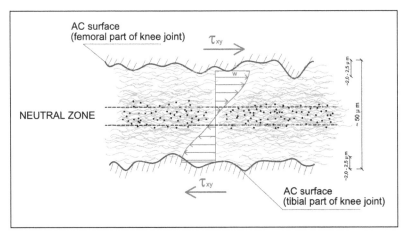

Fig. 9. Diagram of the distribution of magnitudes and directions of SF flow velocity vectors in the gap between AC surfaces. Associations of NaHA/HA macromolecules decline in places with the greatest SF flow velocity gradient, i.e. in zones adjoining each AC surface. The SF flow velocity gradient decreases in the direction towards the neutral zone

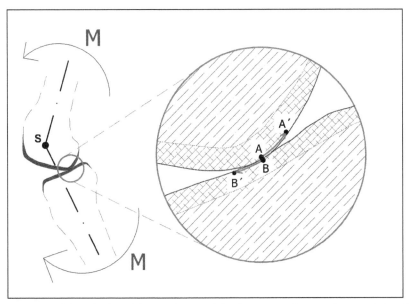

Fig. 10. Rotation of the tibial and femoral part of the knee joint during the transition from flexion to extension. During the rotation of the femoral part of the knee joint (due to the effect of the left-hand rotation moment M) round the current (relative) centre of rotation (which is the intersection of longitudinal axes of the femur and tibia), point **A** moves to position **A´**. During a simultaneous rotation of the tibial part of the knee joint (due to the effect of the right-hand rotation moment M) round the same current (relative) centre of rotation, point **B** moves to position **B´**

Due to pseudoplastic properties of SF in the space between the opposite AC surfaces (Fig. 9.), non-physiological *abrasive wear* of the surfaces of AC peripheral zones is efficiently prevented.

The SF solution process in the gap between the AC surfaces is not an isolated phenomenon. It is interconnected (during walking, running etc.) with residual SF in the pores of the intercellular matrix in peripheral zones of the tibial and femoral part of AC. Under high loads, an *integrated unit* is generated which, after the *formation of an incompressible "cushion"*, is able to transfer extreme loads thus protecting the peripheral and internal AC structures from their destructions.

2.2 Residual strains during the cyclic loading in the articular cartilage

In agreement with our analyses, the properties and behaviors of articular cartilage in the biomechanical perspective may be described by means of a complex viscohyperelastic model (Fig. 11.). The biomechanical compartment is composed of the Kelvin Voigt viscoelastic model (in the peripheral and partially in the transitional zone of AC) and of the hyperelastic model (in the middle transitional zone and the low zone of AC). The peripheral zone is histologically limited by oval (disk shaped) chondrocytes. The viscohyperelastic properties of AC are predetermined by the specific molecular structures.

The mechanical/biomechanical properties of articular cartilage are topographically non homogeneous. The material variability and non homogeneity depends on the type and the size of physiological loading effects (Akizuki et al., 1986; Petrtyl et al., 2008).

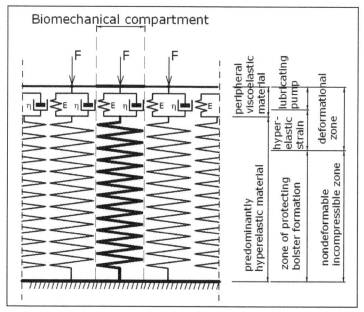

Fig. 11. Mechanical diagram of the complex viscohyperelastic model of articular cartilage. The mechanical compartment is composed of the Kelvin Voigt viscoelastic model (in the peripheral and transitional zone of AC) and of the hyperelastic model (in the middle transitional zone and the low zone of AC)

AC is composed of cells (chondrocytes), of extracellular composite material representing a reinforcing component – collagen 2nd type (Benninghoff, 1925) and of a non reinforcing, molecularly complex matrix (Bjelle, 1975). A matrix is dominantly composed of glycoprotein molecules and firmly bonded water. In the peripheral zone, there is synovial fluid unbound by ions.

The principal construction components of the matrix are glycoproteins. They possess a saccharide component (80-90 %) and a protein component (ca 20 - 10 %). Polysaccharides are composed of molecules of chondroitin-4-sulphate, chondtroitin-6-sulphate and keratansulphate. They are bonded onto the bearing protein, which is further bonded onto the hyaluronic acid macromolecule by means of two binding proteins. Keratansulphates and chondroitinsulphates are proteoglycans which, through bearing and binding proteins and together with the supporting macromolecule of hyaluronic acid, constitute the proteoglycan (or glycosaminoglycan) aggregate. As the saccharide part contains spatial polyanion fields, the presence of a large number of sulphate, carboxyl and hydroxyl groups results in the creation of *extensive fields of ionic bonds with water molecules.*

The proteoglycan aggregate, together with *bonded water*, creates an amorphous extracellular material (matrix) of cartilage, which is bonded onto the reinforcing component – collagen 2nd type. Glycosaminoglycans are connected onto the supporting fibres of collagen type II by means of electrostatic bonds. In articular cartilage, nature took special efforts in safeguarding the biomechanical protection of chondrocytes in the peripheral zone. In the biomechanical perspective, chondrocytes are protected by glycocalix (i.e. a spherical saccharide envelope with firmly bonded water). Glycocalix is composed of a saccharide envelope bonded onto chondrocytes via transmembrane proteoglycans, transmembrane glycoproteins and adsorbed glycoproteins. *The glycocalix envelopes create gradually the incompressible continuous layer during the loading in peripheral zone of AC* (Fig. 12.).

Our research has been focussed on analyses of viscoelastic strains of the upper peripheral cartilage zone, on the residual strains arising at cyclic loading, on the analyses of strain rate and on the creation of a peripheral incompressible cartilage cushion.

The peripheral cartilage zone consists of chondrocytes packaged in proteoglycans (GAGs) with firmly bonded molecules of water. In the intercellular space, there is unbound synovial fluid which contains water, hyaluronic acid, lubricin, proteinases and collagenases. Synovial fluid exhibits non-Newtonian flow characteristics. As was pointed out above, under a load the synovial fluid is relocated on the surfaces of AC.

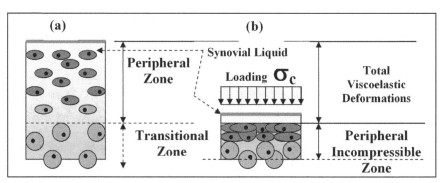

Fig. 12. Peripheral zone of articular cartilage without (a)/with (b) loads. The peripheral incompressible zone is integrated with the incompressible zone in the middle (transitional) zone and low (radial) zone

During loading, the chondrocytes with GAGs encapsulation (in the peripheral zone) create a continuous incompressible mezzo layer with protected chondrocytes. Simultaneously, an incompressible peripheral zone arises in the middle of the transitional zone and in the low (radial) zone of AC. There are dominantly hyperelastic properties in the transitional and the low radial zone (Fig. 11.). Stress states can be simulated by the modified Cauchy stress tensor for incompressible hyperelastic material.

Viscous properties in the peripheral zone of articular cartilage result from the interaction between the molecules of the extracellular matrix and the molecules of free (unbound) synovial fluid. The transport of SF molecules through the extracellular space and the lack of bonding of these molecules onto glycosaminoglycans create the basic condition for the viscous behaviors of cartilage. High dynamic forces are dominantly undertaken by the AC matrix with firmly bonded water in its low and middle zone with a simultaneous creation of an incompressible tissue, a cushion (Fig. 1.).

The articular cartilage matrix with viscoelastic properties functions dominantly as a *protective pump* and a regulator of the amount of SF permanently maintained (during cyclic loading) between articular plateaus. The importance of the *protective pump* is evident from the function of retention of AC strains during cyclic loading. Due to slow down viscoelastic strain, part of accumulated (i.e. previously discharged) SF from the preceding loading cycle is *retained in articular cartilage* (Fig. 13.).

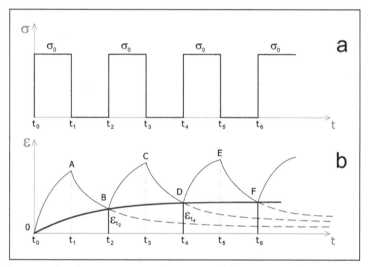

Fig. 13. Application of Kelvin Voigt viscoelastic model for the expression of step by step increments of strains ε_{ti} in the peripheral zone of AC during cyclic loading (e.g. while walking or running)

Fig. 13. in its upper part (a) shows the loading cycles e.g. during walking, while in the lower part (b) strains during the strain time growth and during strain relaxation are visible. The strain time growth occurs during the first loading (see the first concave curve OA of the strain growth). At the time t_1 after unloading strain relaxation occurs (see the convex shape of the second curve AB). At the time t_2 the successive (second) loading cycle starts. The strain time growth during the successive loading cycle, however, does not start at a zero value (as was the case during the initial, first loading cycle), but at point B, or at the value of the residual strain ε_{t2}. *The first residual strain provides the initial presupplementation of articular plateaus with synovial fluid.* Fig. 13. manifests that the envelope curve OBDF slightly grows during cyclic loading to stabilize after a certain time *at a steady value characterizing long-term strain (during the time of cyclic loading) and long-term presupplementation of articular space with synovial fluid.* After cyclic loading stops (i.e. after AC unloading) during the last loading cycle, as seen in Fig. 1., the strain relaxation follows the convex curve, and strains asymptotically approach to the time axis t (or zero). After the termination of the last loading cycle, SF (in the form of synovial sol) is sucked back into the peripheral layer of AC. The mechanism of viscous strain time growth and viscous strain relaxation creates a highly efficient *protective pump* functioning not only to discharge and suck back synovial fluid, but also to pump (accumulate) it into the articular space.

Stresses in the peripheral zone may be expressed for the Kelvin Voigt model by the constitutive equation:

$$\sigma(t) = \eta \frac{d\varepsilon(t)}{dt} + E\varepsilon(t) \tag{1}$$

where η is the coefficient of viscosity, E is the modulus of elasticity, $\varepsilon(t)$ is the strain of AC and $\dfrac{d\varepsilon(t)}{dt}$ is the strain rate of cartilage tissue in the peripheral zone.

Equation (1) is a first order linear differential equation for an unknown function $\varepsilon(t)$. The solution to the non-homogeneous equation (1) under the given initial conditions determines the time related strain of articular cartilage. In our case, it is in the form:

$$\varepsilon(t) = e^{\frac{-1}{\eta}Et} \left[\frac{1}{\eta} \int_{t_0}^{t} \sigma(\tau) e^{\frac{1}{\eta}E\tau} \, d\tau \right] \tag{2}$$

Let us further consider the case where articular cartilage is loaded by a constant load $\sigma(\tau) = \sigma_c = const$ (Fig. 13.) :

$$\varepsilon(t) = \frac{\sigma_c}{E} \left[1 - e^{\frac{-1}{\eta}E(t-t_0)} \right] \tag{3}$$

Equation (3) implies that the strain of AC is a function of time depending on the magnitude of the constant stress σ_c also (for example by shifting an individual's weight onto one foot).

The presence of residual strain (marked by a thick line in Fig. 13.) ensures the accumulation of synovial fluid between articular plateaus. It means that during each step (during cyclic loading) articular plateaus are presupplemented with the lubrication medium – synovial fluid. The magnitudes of residual strains of AC play a key role in the presupplementation of AC surface plateaus with synovial fluid. The magnitudes of residual strains may be determined from the functions expressing strain during the strain time growth and from the functions expressing strain during the strain relaxation of AC, this may be performed separately for each loading cycle of cartilage (Fig. 13.).

For the 1st phase of the first loading cycle, for $t \in < t_0; t_1 >$, (Fig. 13.) the concave curve is defined by function (3) for the articular cartilage strain.

Discrete strain at the time t_0 is $\varepsilon_{t_0} = 0$, at the time t_1 discrete strain is:

$$\varepsilon_{t_1} = \frac{\sigma_c}{E}\left[1 - e^{\frac{-1}{\eta}E(t_1 - t_0)}\right] \tag{4}$$

For the 2nd phase of the first loading cycle (for $t \in < t_1 ; t_2 >$) (Fig. 13.), the convex curve AB is defined by the function for articular cartilage strain:

$$\varepsilon(t) = \varepsilon_{t_1} e^{\frac{-1}{\eta}E(t - t_1)} \tag{5}$$

Discrete strain at the time t_0 is $\varepsilon_{t_0} = 0$, at the time t_1 discrete strain is:

$$\varepsilon_{t_2} = \varepsilon_{t_1} e^{\frac{-1}{\eta}E(t_2 - t_1)} \tag{6}$$

The magnitudes of strains during cyclic loading at the starting points of loading and unloading of articular cartilage may be expressed by recurrent relations. For the time t_i with an odd index, the strain at the respective nodal points is:

$$\varepsilon_{t_{(2k+1)}} = \frac{\sigma_c}{E}\left[1 - e^{\frac{-E}{\eta}(k+1)l}\right]_{k=0,1,2...} \tag{7}$$

where l is the length of the time interval $<t_i ; t_{i+1}>$. For the time t_i with an even index, the strain is :

$$\varepsilon_{t_{2k}} = \frac{\sigma_c}{E}\left[e^{\frac{-E_l}{\eta}} - e^{\frac{-E}{\eta}(k+1)l}\right]_{k=0,1,2...} \tag{8}$$

where l is the length of the time interval $<t_i ; t_{i+1}>$, $i = 0, 1, 2, \ldots$ During long-term cyclic loading and unloading, for $k \to \infty$ the strain $\varepsilon_{t(2k+1)}$ asymptotically approaches the steady state σ_c/E ; for $k \to \infty$ the strain ε_{t2k} asymptotically approaches the steady state $\frac{\sigma_c}{E}e^{\frac{-E_l}{\eta}}$. It is evident that for $k \to \infty$ it holds true that:

$$\varepsilon_{t_{(2k+1)}} = \frac{\sigma_c}{E} > \varepsilon_{t_{2k}} = \frac{\sigma_c}{E}e^{\frac{-E_l}{\eta}} \tag{9}$$

2.2.1 The strain rate of articular cartilage in peripheral zone during the strain time growth

Strain $\varepsilon(t)$ of AC during the strain time growth in the interval of $t \in \langle t_0; t_1 \rangle$ is given by equation (3). Because $\dfrac{d\varepsilon(t)}{dt} = \dot{\varepsilon}(t) > 0$ (in the indicated interval) the function $\varepsilon(t)$ is increasing. The strain rate of AC during the strain-time growth in interval of $t \in \langle t_0; t_1 \rangle$ is given by equation (10):

$$\frac{d\varepsilon(t)}{dt} = \frac{\sigma_c}{\eta} e^{\frac{-1}{\eta}E(t-t_0)} \tag{10}$$

From equation (10) is evident, that the strain rate of articular cartilage in interval of $t \in \langle t_0; t_1\rangle$ decelerates. The strain rate shortly after the load is the highest.

2.2.2 The strain rate of articular cartilage in peripheral zone during the strain relaxation

The strain of peripheral zone in time t_1 during unloading is given by equation (11):

$$\varepsilon(t) = \varepsilon_{t_1} e^{\frac{-1}{\eta}E(t-t_1)} \tag{11}$$

The strain rate $\dfrac{d\varepsilon(t)}{dt} = \dot{\varepsilon}(t) < 0$. It means that the strain function $\varepsilon(t)$ in interval of $t \in \langle t_1; t_2\rangle$ is decreasing. The strain rate in the same interval of $\langle t_1; t_2\rangle$ is decreasing also. Strain rate during the strain relaxation in interval of $\langle t_1; t_2\rangle$ is given by equation (12):

$$\frac{d\varepsilon(t)}{dt} = \varepsilon_{t_1} e^{\frac{-1}{\eta}E(t-t_1)}\left[-\frac{E}{\eta}\right] \tag{12}$$

The strain rate of articular cartilage shortly after the unloading (during the strain relaxation) is distinctly higher than to the end of interval of $\langle t_1; t_2\rangle$. Strain rate $\left|\dfrac{d\varepsilon(t)}{dt}\right|$ with increasing time in interval of $t \in \langle t_1; t_2\rangle$ is decreasing.

3. Conclusions

The above described analyses lead to the formulation of the following key conclusions:
Synovial fluid is a viscous pseudoplastic non-Newtonian fluid. Apparent viscosity of SF decreases with increasing rate of flow velocity gradient. SF does not display a decrease in viscosity *over time at a constant flow velocity gradient* (as it is typical for thixotropic material). The rheological properties of synovial fluid essentially affect the biomechanical behaviour of SF between the opposite AC surfaces and in the peripheral AC zone also. During the shifts of the femoral and tibial part of AC in opposite directions the velocities of SF flows decrease in the direction towards the neutral central zone of the gap between the AC surfaces. *Non-linear abatement in viscosity* in the direction from the *neutral ("quiescent") layer of* SF towards the opposite AC surfaces contributes to the lubrication quality and very efficiently protect the uneven micro-surfaces of AC.
The viscoelastic properties of the peripheral zone of AC and its molecular structure ensure the regulation of the transport and accumulation of SF between articular plateaus. The hydrodynamic lubrication biomechanism adapts with high sensitivity to biomechanical stresses. The viscoelastic properties of AC in the peripheral zone ensure that during cyclic loading some amount of SF is always retained accumulated between articular plateaus, which were presupplemented with it in the previous loading cycle. During long-term harmonic cyclic loading and unloading, the strains stabilize at limit values.
The limit strain value of AC during loading is always greater than its limit strain value after unloading. Shortly after loading, the strain rate is always greater than before unloading. In

this way, the hydrodynamic biomechanism quickly presupplements the surface localities with lubrication material. Shortly after unloading, the strain rate is high. During strain relaxation, it slows down. This is the way how the articular cartilage tissue attempts to retain the lubrication material between the articular plateaus of synovial joints as long as possible during cyclic loading.

Analogically to the low and the middle zone of AC where an incompressible zone arises under high loads whose dominant function is to bear high loads and protect chondrocytes with the intercellular matrix from destruction, in the peripheral zone as well a partial incompressible zone arises whose function is to bear high loads and protect the peripheral tissue from mechanical failure. The appearance of the incompressible tissue in all zones is synchronized aiming at the creation of a single (integrated) *incompressible cushion*. The existence of an incompressible zone secures the protection of chondrocytes and extracellular material from potential destruction.

3.1 Significance of results for clinical practice

Metabolic processes during the HA synthesis are very dynamic. The chondrocytes in AC actively synthesize and catabolise HA so that its optimal "usability" is achieved (in a relatively short time). The HA synthesis is usually in equilibrium with catabolic processes. These processes result in the achievement of the optimum HA concentration. The studies of metabolic processes (Schurz et al., 1987) implied that the half-life of the functional existence of HA molecules are mere 2-3 weeks. The solved project makes it evident that the "short lifecycle" of HA is dominantly caused by dynamic (biomechanical) effects. During leg movements, long snakelike NaHA/HA macromolecules are exposed to fast changes in shape accompanied by permanently arising and vanishing physical (non-covalent) bonds. To avoid the shortage of HA/NaHA, old polymer chains are replaced with new chains. The disturbance of HA new formation processes may lead to initiations of pathological processes. Mechanical effects during movements continuously initiate new groupings of HA macromolecules and newly arising (and vanishing) bonds among them. Frequented variations of kinetic energy transfers into HA molecular structures contribute to HA fragmentations in the biophysical perspective. These fragmentations may also be biochemically accelerated by hyaluronisades (Saari et al., 1993). HA fragments may initiate the formation of macrophages and extensive inflammations of AC.

The above examples of the interrelated nature of the causes of some AC defects show the key role of the rheological properties of non-Newtonian synovial fluid.

4. Acknowledgment

The contents presented in this chapter was supported by the Research Grant from MSMT No.6840770012.

5. References

Akizuki, S.; Mow, V.C.; Muller, F.; Pita, J.C; Howell, D.S. & Manicourt, D.H. (1986). Tensile properties of human knee joint cartilage: I. Influence of ionic conditions, weight bearing, and fibrillation on the tensile modulus, *J. of orthopaedic research*, Vol. 4, No. 4, pp. 379-392

Anadere, I.; Chmiel, H. & Laschner, W. (1979). Viscoelasticity of "normal" and pathological synovial fluid. *Biorheology*, Vol. 16, No. 3, pp. 179-184

Armstrong, C.G.; Lai, W.M. & Mow, V.C. (1984). An analysis of the unconfined compression of articular cartilage. *J. Biomech. Eng.*, Vol. 106, No. 2, (May 1984), pp. 165-173, ISSN 0148-0731

Ateshian, G.A.; Chahine, N.O.; Basalo, I.M. & Hung, C.T. (2004). The correspondence between equilibrium biphasic and triphasic material properties in mixture models of articular cartilage. *J. of Biomechanics*, Vol. 37, No. 3, (March 2004), pp. 391-400, ISSN 0021-9290

Benninghoff, A. (1925). Form und Bau der Gelenkknorpel in ihren Beziehungen zur Funktion, *Zeitschrift für Zellforschung und Mikroskopische Anatomie*, Vol. 2, No. 5, pp. 783-862

Bjelle, A. (1975). Content and Composition of Glycosaminoglycans in Human Knee Joint Cartilage: Variation with Site and Age in Adults. *Connective tissue research*, Vol. 3, No. 2-3, (January 1975), pp. 141-147, ISSN 0300-8207

Bloch, B. & Dintenfass, L. (1963). Rheological study of human synovial fluid. *Australian and New Zealand Journal of Surgery*, Vol. 33, No. 2, (November 1963), pp. 108-113

Cohen, B.; Gardner, T.R. & Ateshian, G.A. (1993). The influence of transverse isotropy on cartilage indentation behavior - A study of the human humeral head. In: *Transactions Orthopaedic Research Society*, Orthopaedic Research Society, pp. 185, Chicago, IL

Cohen, B.; Lai, W.M.; Chorney, G.S.; Dick, H.M.; Mow, V.C. (1992). Unconfined compression of transversely-isotropic biphasic tissue, In: *Advances in Bioengineering*, American Society of Mechanical Engineers, pp. 187-190, ISBN 0791811166

Davies, D.V. & Palfey, A.J. (1968). Some of the physical properties of normal and pathological synovial fluids. *J. of Biomechanics*, Vol. 1, No. 2, (July 1968), pp. 79-88, ISSN 0021-9290

Ferguson, J.; Boyle, J.A.; McSween, R.N.; Jasani, M.K. (1968). Observations on the flow properties of the synovial fluid from patients with rheumatoid arthritis. *Biorheology*, Vol. 5, No. 2, (July 1968), pp. 119-131

Garcia, J.J.; Cortes, D.H. (2006). A nonlinear biphasic viscohyperelastic model for articular cartilage, *J. of Biomechanics*, Vol. 39, No. 16, pp. 2991-2998, ISSN 0021-9290

Johnson, J.P. (1955). The viscosity of normal and pathological human synovial fluids. *J. Biochem*, Vol. 59, No. 3, (April 1955), pp. 633-637

Jurveli, J.; Kiviranta, I.; Saamanen, A.M.; Tammi, M. & Helminen, H.J. (1990). Indentation stiffness of young canine knee articular cartilage—Influence of strenuous joint loading. *J. of Biomechanics*, Vol. 23, No. 12, pp. 1239-1246, ISSN 0021-9290

Lai, W.M.; Hou, J.S. & Mow, V.C. (1991). A Triphasic Theory for the Swelling and Deformation Behaviors of Articular Cartilage. *J. Biomech. Eng.*, Vol. 113, No. 3, (August 1991), pp. 245-351, ISSN 0148-0731

Lapcik, L. Jr.; Lapcik, L.; De Smedt, S.; Demeester, J. & Chabrecek, P. (1998). Hyaluronan: Preparation, Structure, Properties, and Applications. *Chemical reviews*, Vol. 98, No. 8, (December 1998), pp. 2663-2684, ISSN 0009-2665

Li, L.P.; Soulhat, J.; Buschmann, M.D.; Shirazi-Adl, A. (1999). Nonlinear analysis of cartilage in unconfined ramp compression using a fibril reinforced poroelastic model. *Clinical Biomechanics*, Vol. 14, No. 9, (November 1999), pp. 673-682, ISSN 0268-0033

Mori, S.; Naito, M. & Moriyama, S. (2002). Highly viscous sodium hyaluronate and joint lubrication. *International Orthopaedics*, Vol. 26, No. 2, (April 2002), pp. 116-121, ISSN 0341-2695

Myers, R.R.; Negami, S. & White, R.K. (1966). Dynamic mechanical properties of synovial fluid. *Biorheology*, Vol. 3, pp. 197-209

Nuki, G. & Ferguson, J. (1971). Studies on the nature and significance of macromolecular complexes in the rheology of synovial fluid from normal and diseased human joints. *Rheologica acta*, Vol. 10, No. 1, pp. 8-14

Oates, K.M.N.; Krause, ,W.E.; Jones, R.L. & Colby, R.H. (2006). Rheopexy of synovial fluid and protein aggregation. *J. R. Soc. Interface*, Vol. 3, No. 6, (February 2006), pp. 167-174, ISSN 1742-5689

Petrtyl, M.; Bastl, Z.; Krulis, Z.; Hulejova, H.; Polanska, M.; Lisal, J.; Danesova, J. & Cerny, P. (2010). Cycloolefin-Copolymer/Polyethylene (COC/PE) Blend Assists with the Creation of New Articular Cartilage. *Macromolecular Symposia Special Issue: Layered Nanostructures – Polymers with Improved Properties*, Vol. 294, No. 1, (August 2010), pp. 120-132

Petrtyl, M.; Lisal, J. & Danesova, J. (2008). The states of compressibility and incompressibility of articular cartilage during the physiological loading (in czech), *Locomotor Systems, Advances in Research, Diagnostics and Therapy*, Vol. 15, No. 3-4, (October 2008), pp. 173-183, ISSN 1212-4575

Radin, E.L.; Paul, I.L.; Swann, D.A. & Schottstaedt, E.S. (1971). Lubrication of synovial membrane. *Ann. Rheum. Dis.*, Vol. 30, No. 3, (May 1963), pp. 322-352, ISSN 0003-4967

Rinaudo, M.; Rozand, Y.; Mathieu, P. & Conrozier, T. (2009). Role of different pre-treatments on composition and rheology of synovial fluids. *Polymers*, Vol. 1, No. 1, pp. 16-34, ISSN 2073-4360

Saari, H.; Konttinen, Y.T.; Friman, C. & Sorsa, T. (1993). Differential effects of reactive oxygen species on native synovial fluid and purified human umbilical cord hyaluronate. *Inflammation*, Vol. 17, No. 4, (August 1993), pp.403-415, ISSN 0360-3997

Safari, M.; Bjelle, A.; Gudmundsson, M.; Högfors, C. & Granhed, H. (1990). Clinical assessment of rheumatic diseases using viscoelastic parameters for synovial fluid. *Biorheology*, Vol. 27, No. 5, pp. 659-674

Schurz, J. & Ribitsch, V. (1987). Rheology of synovial fluid. *Biorheology*, Vol. 24, No. 4, pp. 385-399

Scott, J.E. & Heatly, F. (1999). Hyaluronan forms specific stable tertiary structures in aqueous solution: a 13C NMR study., *Proc. Natl. Acad. Sci. USA*, Vol. 96, No. 9, (April 1999), pp. 4850-4855, ISSN 0027-8424

Sundblad, L. (1953). Studies on hyaluronic acid in synovial fluids. *Acta Soc. Med. Ups.*, Vol. 58, No. 3-4, (April 1953), pp. 113-238

Swann, D.A.; Silver, F.H.; Slyater, H.S., Stafford, W. & Shore, E. (1985). The molecular structure and lubricating activity of lubricin isolated from bovine and human synovial fluids. *The Biochemical Journal*, Vol. 225, No. 1, (January 1985), pp. 195-201

White, R.K. (1963). The rheology of synovial fluid. *The journal of bone and joint surgery*, Vol. 45, No. 5, (July 1963), pp. 1084-1090, ISSN 1535-1386

Wilson, W.; van Donkelaar, C.C.; van Rietbergen, B.; Ito, K. & Huiskes, R. (2005). Erratum to "Stresses in the local collagen network of articular cartilage: a poroviscoelastic fibril-reinforced finite element study" [Journal of Biomechanics 37 (2004) 357–366] and "A fibril-reinforced poroviscoelastic swelling model for articular cartilage" [Journal of Biomechanics 38 (2005) 1195–1204]. *J. of Biomechanics*, Vol. 38, No. 10, (October 2005), pp. 2138-2140, ISSN 0021-9290

Wilson, W.; van Donkelaar, C.C.; van Rietbergen, B.; Ito, K. & Huiskes, R. (2004). Stresses in the local collagen network of articular cartilage: a poroviscoelastic fibril-reinforced finite element study. *J. of Biomechanics*, Vol. 37, No. 3, (March 2004), pp. 357-366, ISSN 0021-9290

Yanaki, T. & Yamaguchi, T. (1990). Temporary network formation of hyaluronate under a physiological condition. 1. Molecular-weight dependence. *Biopolymers*, Vol. 30, No. 3-4, pp. 415-425, ISSN 0006-3525

2

Biomimetic Materials as Potential Medical Adhesives – Composition and Adhesive Properties of the Material Coating the Cuverian Tubules Expelled by *Holothuria dofleinii*

Yong Y. Peng[1], Veronica Glattauer[1], Timothy D. Skewes[2],
Jacinta F. White[1], Kate M. Nairn[1], Andrew N. McDevitt[3],
Christopher M. Elvin[3], Jerome A. Werkmeister[1],
Lloyd D. Graham[4] and John A.M. Ramshaw[1]

1. Introduction

Novel, distinct adhesive systems have been described for a wide range of marine species (Kamino, 2008). These highly effective, natural materials provide a link between biological science and material science, and can serve as models on which new, bioinspired synthetic materials could be based. These various adhesive systems have developed independently, on many occasions, and provide a wide range of opportunities for the development of new, biologically-inspired adhesives. The natural adhesives include, for example, the marine mussel (*Mytilus sp.*) (Lin et al., 2007), barnacle (Nakano et al., 2007) and stickleback (Jones et al., 2001) adhesives, which are protein-based, as well as sponge, certain algal and marine bacterial adhesives (Mancuso-Nichols et al., 2009) that are polysaccharide-based.

In the present paper, we examine the adhesive system found associated with the Cuverian tubules of a holothurian species (sea cucumber), *Holothuria dofleinii*. This is an example of the particularly rapid marine adhesive that is found on the surface of Cuverian tubules when they are expelled (DeMoor et al., 2003; Müller et al., 1972; VandenSpiegel & Jangoux, 1987). The unique nature of this natural adhesive system, especially its rapid action under water, has suggested that if the mechanism can be understood, then it may prove to be possible to mimic the adhesive through biotechnology and/or synthetic chemistry. An adhesive that functions readily in an aqueous environment would be particularly valuable, especially in medical applications, as the majority of existing adhesives bind to dry surfaces more strongly than the same surfaces when wet.

Cuverian tubules provide a host defence mechanism for certain species of holothurians (Lawrence, 2001; VandenSpiegel & Jangoux, 1987). It has long been known that, on expulsion, the Cuverian tubules fill with liquid and lengthen, become sticky and rapidly

[1]*CSIRO Materials Science and Engineering, Bayview Avenue, Clayton, VIC 3169, Australia*
[2]*CSIRO Marine and Atmospheric Research, Middle Street, Cleveland, QLD 4163, Australia*
[3]*CSIRO Livestock Industries, Carmody Road, St Lucia, QLD 4067, Australia*
[4]*CSIRO Food and Nutritional Sciences, Julius Ave, North Ryde, NSW 2113, Australia*

immobilise most organisms with which they come into contact (VandenSpiegel & Jangoux, 1987). The tubules, once expelled, are immediately adhesive on contact with a solid surface (VandenSpiegel & Jangoux, 1987), such as the exoskeleton or skin of a predator. Crabs, molluscs and sea stars can stimulate tubule expulsion, and the tubules stick to these species. This adhesion happens entirely under water, and does not need the mixed environment of the intertidal zone where many of the other potential adhesives are sourced.

Sticky tubules are found only within the family Holothuridae within the order Aspidochirotida, and mostly in the genus Bohadschia and the genus Holothuria. Various authors have described the ultrastructure of the tubules, especially for *H. forskåli* (Lawrence, 2001; VandenSpiegel & Jangoux, 1987; VandenSpiegel et al., 2000), as well as their expulsion and release (Flammang et al., 2002) and the timeframe of regeneration (Flammang et al., 2002; VandenSpiegel et al., 2000).

Flammang and Jangoux (2004) suggested, from the differences in the surface (adhesive) protein types and compositions in *H. forskåli* and *H. maculosa*, that the adhesion proteins and mechanism may differ between species. Other studies showed that adhesive strengths varied between species, with the adhesion in *H. leucospilota* being several times greater than for six other species (Flammang et al., 2002). A limited number of studies have probed the mechanism of adhesion, focusing on *H. forskåli* and *H. leucospilota* (De Moor et al., 2003; Müller et al., 1972; Zahn et al., 1973). These studies have shown that best adhesion is found at temperatures, salinity and pH similar to those found in the marine environment in which the organism flourishes, and is most effective with hydrophilic surfaces (Flammang et al., 2002; Müller et al., 1972; Zahn et al., 1973). Increasing concentrations of urea led to a loss of adhesion, suggesting that native protein structure(s) or interctions(s) may be required for effective bonding (Müller et al., 1972). Later biochemical studies have also suggested that the adhesive mechanism involves protein components (DeMoor et al., 2003).

In the present study, we have extended the information on Cuvierian tubule adhesion. In this study we examined the tubules of a different species, *H. dofleinii* Augustin, 1908. We have examined the distribution of the adhesive substance on the surface of expelled tubules, along with the molecular weights and amino acid compositions of its main protein components. We have estimated the strength of adhesion of *H. dofleinii* tubules to different substrata, and examined the effects of salinity, pH, ionic strength and denaturants on the adhesive properties.

2. Materials and methods

2.1 Collection of materials

Individual *H. dofleinii* were obtained from shallow subtidal seagrass banks in Moreton Bay, Queensland, at a depth of about 1-2 metres at low tide, close to the western side of Stradbroke Island (153° 26.4' E 27° 25.13' S to 27° 25.68' S), and were held for up to 5 days prior to use in filtered, recirculating seawater tanks at 21.5 – 22 °C. The identification of the animals was based on morphology, spicule shape and size and 18S-RNA sequencing (Peng & Skewes, unpublished data).

2.2 Sample preparation

To collect expelled Cuvierian tubules, *H. dofleinii* individuals were held and gently stimulated underwater until tubules were expelled. Immediately after expulsion, a tubule was individually collected using polytetrafluoroethylene-tipped forceps, and was allowed

Biomimetic Materials as Potential Medical Adhesives – Composition and Adhesive Properties
of the Material Coating the Cuvierian Tubules Expelled by Holothuria dofleinii

23

to drain briefly (<20 sec), but not by squeezing as had been proposed by others (Zahn et al., 1973). Intact tubules prior to expulsion were obtained by dissection of animals that had been euthanised by freezing at minus 20 °C.

2.3 Microscopy

To look for the presence of glycoprotein on the surface of expelled tubules, samples were treated with fluorescently-labelled lectins; fluorescein isothiocyanate (FITC)-labelled concanavilin A (ConA), FITC-labelled *Datura stramonium* agglutinin (DSA), and FITC-labelled *Lycopersicon esculentum* agglutinin (LEA) (all from Sigma, St Louis). All FITC-labelled lectins were applied as 20 µg/mL solutions in Tris-buffered saline (TBS) for 60 min, followed by 3 × 5 min washes in TBS. Samples were examined using appropriate narrow pass filters on an Olympus BX61 fluorescence microscope.

To examine the distribution of adhesiveness on tubules, individual freshly expelled tubules after draining (see above) were transferred to a wash solution in a plastic trough which contained a suspension of 0.5% w/v Bio-Gel P2 (45-90 µm particle size) in 3.5% w/v NaCl, 10 mM sodium phosphate, pH 7.6. After 5 sec immersion, the tubules were washed 3 times in 3.5% NaCl, 10 mM sodium phosphate, pH 7.6 and were then drained and placed onto glass slides. After air drying, the tubules were examined by microscopy.

For scanning electron microscopy (SEM) expelled tubules were examined using a Philips XL30 FESEM microscope at an accelerating voltage of 2 kV.

2.4 Gel electrophoresis analysis

Freshly expelled and drained tubules were allowed to adhere to a glass plate and were air dried. The tubules on glass plates were removed by peeling, leaving the layer of adhesive, and potentially other components of the tubule wall as a print on the glass (DeMoor et al., 2003). This material was collected by removal with a sharp razor blade and was then extracted in electrophoresis sample buffer, containing 2-mercaptoethanol. SDS-polyacrylamide gel electrophoresis (SDS-PAGE) was based on the method of Laemmli (1970) using Invitrogen NuPAGE Novex 4-14% Bis-Tris Gel with MES running gel buffer, at 180V for 60 min. Molecular weights were determined by comparison to globular protein standards (BioRad) using BioRad Quantity One v.4.4.0 software. For protein identification, gels were stained by Coomassie Blue R-250. Samples that had not been dried completely, but only sufficient to remove excess liquid, appeared to give samples that contained less insoluble material, although the yield of adhesive proteins was less.

2.5 Amino acid analysis

Protein extracts were separated by SDS-PAGE, followed by transfer of the protein bands to PVDF membrane using Invitrogen NuPAGE Transfer buffer (NP0006-1). Amino acid analysis of PVDF membrane pieces used vapour-phase hydrolysis (5.8 M HCl at 108 °C for 18 h), followed by precolumn derivatisation with 6-aminoquinolyl-N-hydroxysuccinimidyl carbamate (Cohen & DeAntonis, 1994). Derivatives were separated and quantified by reversed phase (C18 Waters AccQTag) HPLC at 37 °C (Cohen, 2001) (Australian Proteome Analysis Facility), using a Waters Alliance 2695 Separation Module, a Waters 474 Fluorescence Detector and a Waters 2487 Dual λ Absorbance Detector in series.

2.6 Adhesion properties

Adhesion properties were measured using a 90 Degree Adhesive Peel Strength Test (Dimas et al., 2000), adapted to the rapid evaluation of tubules from a single animal under different experimental conditions. Individual expelled tubules, as above, were transferred to a wash solution in a plastic trough containing a wash solution determined by the particular test (see below). The numbers of samples tested in a given experiment are given in the results Tables; in each case, tubules from a minimum of 3 separate animals were used. After 60 ± 2 sec, the tubule was removed from the wash solution and allowed to drain for 5 sec. It was then laid across the width of a 25 mm wide strip of substratum, selected for the particular test. The tubule was allowed to adhere to the test substratum under its own weight for 60 ± 2 sec. This differs from previous studies where a load was applied during adhesion (Flammang et al., 2002). During the adhesion period the tubule was trimmed to leave <10 mm overhanging one side of the substratum and about 50 mm on the other side. The flat width of the tubule was measured and also recorded photographically, with a ruler placed adjacent, for subsequent verification. At the end of the 60 sec adhesion period, each tubule-substratum assembly was then transferred to a frame that allowed the substratum to be held horizontally with the adhered tubule on the underside, i.e. with the free c. 50 mm length of tubule hanging below. The load was then increased stepwise (2.5 g/5 sec) to the overhang of the tubule until the tubule-substratum adhesion failed by peeling. The total load at failure was recorded. The maximum force tested was 0.2 N (approximately 20 g load) which equated to about 0.05 N/mm for an average tubule, because higher loads typically took too long to add and the tubule could have begun to desiccate at that stage, potentially changing the adhesive strength. A minimum of six determinations was made for each test condition. Data are presented as the total force at failure (N) divided by tubule width (mm). Although we did not test values above 0.05 N/mm, our conservative approach did not hinder examination of conditions that led to reduction of adhesive strength.

Experiments to test adhesion to different substrata used a wash solution of simulated sea-water comprising 3.5% NaCl, 10 mM sodium phosphate buffer, pH 7.6. Various substrata were tested, including clean glass (microscope slide), aluminium, polyvinyl chloride, chitin (from crab), polycarbonate, poly(methyl methacrylate) (PMMA) and polytetrafluoroethylene (PTFE), all cut to a similar size. As the chitin substratum lacked stiffness, the samples were first glued with cyanoacrylate onto a glass microscope slide. The chitin sample also had an irregular surface and was not uniform like the other materials.

The effects on tubule-glass adhesion of various chloride or sodium salts (50 mM) were examined by supplementing the 3.5% NaCl, 10 mM sodium phosphate buffer, pH 7.6, before washing the tubules. The effect of NaCl concentration on tubule-glass adhesion was examined using different NaCl concentrations in 10 mM sodium phosphate buffer, at final pH 7.6. The effect of pH on tubule-glass adhesion was examined using solutions prepared using three salts: Tris/chloride, sodium citrate and sodium acetate, each at 50 mM in 3.5% NaCl, 10 mM sodium phosphate. Similarly, the effect of urea on tubule-glass adhesion was examined using different urea concentrations in 3.5% NaCl, 10 mM sodium phosphate at a final pH of 7.6. Glass was used as the standard substratum as it was readily available in uniform quality, and had previously been shown to be an excellent material for adhesion of *H. forskåli* tubules (Flammang et al., 2002).

Biomimetic Materials as Potential Medical Adhesives – Composition and Adhesive Properties
of the Material Coating the Cuvierian Tubules Expelled by Holothuria dofleinii

25

3. Results and discussion

3.1 Tubule structure and microscopy

Dissection of euthanised *H. dofleinii* showed that the body cavity contained a large number, several dozen, Cuvierian tubules in their compressed form (Figure 1A). In their compressed form the tubules were not sticky, but they rapidly became sticky on mechanical extension, even from the dead animals. The compressed, individual tubules showed a corrugated and folded surface (Figure 1B), which would allow extension when required, like a concertina bellows. In cross-section (Figure 1C) a three-lobed channel could be seen that would allow fluid insertion for expansion of the tubules. When expelled and fully extended *in vivo*, these tubules became instantly sticky and changed from the 25-35 mm compressed length up to around 350-400 mm. The fully extended tubules (Figure 1D) were typically about 4mm flat width when fully inflated, and still showed some patterning from the folding that was present in the compressed state. Individual animals contained several dozen non-inflated tubules, but when the animals were stimulated only a small number were expelled, normally around 8-12.

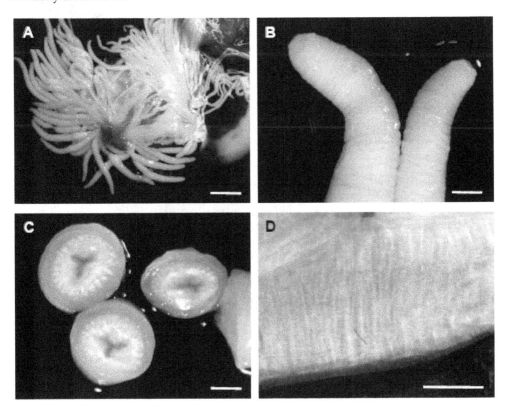

Fig. 1. Cuvierian tubules for *H. dofleinii*. (A, B, C) After dissection of a euthanised animal, showing, (A) the total mass of tubules, (B) the tips of compressed tubules, and (C) the cross-section of compressed tubules. (D) The surface of an *in vivo* expelled tubule. Bar (A) = 10 mm, Bars (B, C, D) = 1 mm.

The surfaces of naturally extended tubules showed strong binding of FITC-DSA (Figure 2A) and FITC-LEA lectins to the tubule surface, with a series of bands that resemble the original folding of the un-extended tubule. FITC-ConA lectin binding was weak, suggesting that the DSA and LEA binding was specific for carbohydrate or glycosylated protein on the surface, rather than non-specific binding. These 2 lectins recognise very similar carbohydrate entities; (N-acetyl glucosamine)$_2$ by DSA and (N-acetyl glucosamine)$_3$ by LEA, which are distinct from the α-mannose or α-glucose recognised by ConA. SEM of the surface of expelled tubules (Figure 2B) showed that the surface had fibrous-like structures, suggesting aggregates of the adhesive material overlaying a further fibrous, collagenous layer of the wall of the tubule.

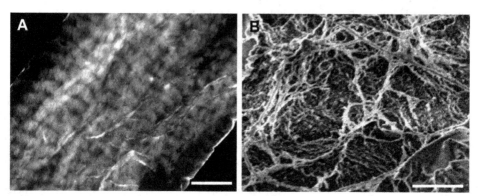

Fig. 2. Cuvierian tubules from *H. dofleinii*. (A) Fluorescence microscopy of FITC-labelled DSA bound to expelled Cuvierian tubules. Bar = 0.5 mm. (B) SEM of expelled Cuvierian tubules. Bar = 2.5 μm

When freshly expelled tubules were briefly immersed in 3.5% NaCl containing Bio-Gel P2 particles, and then washed in 3.5% NaCl, particles bound to the tubule (Figure 3). These data showed that the surface of the tubule was generally adhesive, and there was no specific localisation of the particles, for example to the fibrous-like patches seen by SEM.

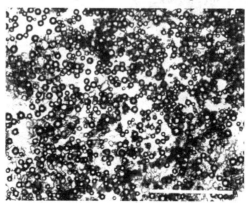

Fig. 3. Cuvierian tubules from *H. dofleinii*. Adhesion of Bio-Gel P2 beads to freshly expelled Cuvierian tubules from *H. dofleinii*. Bar = 0.5 mm

3.2 Gel electrophoresis analysis

The adhesion print isolated from glass showed a range of proteins of well-defined molecular weights when analysed under reducing conditions by SDS-PAGE (Figure 4). A similar pattern, except as noted below, was observed from >10 separate samples from different animals. Seven bands, H2 at 89 kDa, H3 at 70 kDa, H4 at 61 kDa, H6 at 44 kDa, H7 at 37 kDa, H8 26 kDa and H9 at 17 kDa, were consistently present in all tubule samples that were examined (n > 12). Band H4 has been examined as a single entity, but in some gels (Figure 4) it appeared that it may comprise 2 components. For all other bands, although each appeared to be a single component, it is also possible that more than one component could be present, migrating similarly. In many samples, but not all, an additional band, H5 at 53 kDa, was present. As it was not consistently present it was assumed that it may not be a key component of the adhesive system and hence was not examined further. In a few samples, an additional band H1 was observed at 150-170 kDa. This band seemed to be more prevalent in samples where tubule fragments were present in the adhesion print, and could possibly be related to the collagen that is the main structural component of the tubule (Watson and Silvester, 1959). The collected material sometimes contained a proportion of material that remained insoluble in the sample buffer. Examination of the adhesive prints under a microscope suggested that the samples that subsequently contained more insoluble material also contained more fibrous material that could be from the collagenous wall of the tubule. Samples with little wall material typically had little, if any insoluble material.

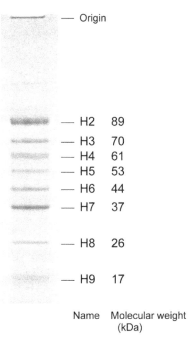

— Origin

— H2   89
— H3   70
— H4   61
— H5   53
— H6   44
— H7   37

— H8   26

— H9   17

Name   Molecular weight
(kDa)

Fig. 4. SDS-PAGE of the reduced proteins from the surface of a freshly expelled Cuvierian tubule from *H. dofleinii*. The gel was stained with Coomassie Blue R-250. Key bands are labelled and their estimated molecular weights, interpolated from a standard curve using globular protein standards (BioRad), are given.

Previously, Flammang and colleagues (DeMoor et al., 2003) have shown a gel electrophoresis pattern for the tubule print from *H. forskali* samples. In this case, a high background staining was present, and the bands were generally less well defined and more poorly resolved. In some cases the apparent *H. forskali* bands had comparable molecular weights to those observed in the present study. Thus the sharp bands at 95 kDa and 45 kDa may be similar to the H2 (89 kDa) and H6 (44 kDa) bands, while the diffuse bands at 63 kDa and 33 kDa may be similar to the H4 (63 kDa) and the H7 (37 kDa) bands, respectively.

3.3 Amino acid analysis
The amino acid compositions determined from amino acid analyses of the 6 principal bands are given in Table 1. No data were collected for Band H4 as it appeared to be a doublet (Figure 3). Deamidation during acid hydrolysis means that Asn cannot be distinguished from Asp, nor can Gln be distinguished from Glu; the two pairs are given as Asx and Glx respectively. This prevents an estimation of pI for each of these proteins. Hydroxyproline was not observed in any of the 6 principal bands.

Comparison of the analyses for the various bands did not show signature features for any particular band, and the compositions were broadly similar for all bands. All of the bands had high contents of Gly (7.8-16.8 mol%) and Glx (11.6-16.2 mol%) relative to the average for eukaryotic proteins (6.9 and 9.7 mol%, respectively) (Doolittle, 1986); similarly, DeMoor et al. (2003) observed high Gly contents (16-22 mol%) for the proteins extracted from *H. forskåli*. As noted above, the SDS-PAGE molecular weight data suggests that it is possible that some bands could be related between the species - H2 and 95 kDa, H6 and 45 kDa and H7 and 33 kDa. Comparison of the amino acid composition data, however, did not show strong similarities. However, it has been suggested (Flammang and Jangoux, 2004), that the protein components present in the adhesives differ between species.

	H2	H3	H6	H7	H8	H9
Asx	9.7	8.7	12.1	13.1	9.7	15.0
Ser	9.8	11.6	7.0	6.5	10.2	8.0
Glx	16.2	13.3	13.2	12.7	15.0	11.6
Gly	16.1	16.8	7.8	8.1	13.6	10.5
His	1.1	1.0	1.3	ND	0.7	1.3
Arg	3.6	5.2	4.9	4.1	5.5	4.4
Thr	3.9	5.2	5.1	5.0	4.9	4.6
Ala	7.8	7.0	9.2	9.4	6.8	8.0
Pro	3.2	4.9	3.9	3.2	6.3	3.1
Tyr	1.8	2.5	2.4	1.9	3.0	2.4
Val	4.6	5.5	6.4	6.4	5.5	5.7
Met	ND	0.4	3.2	3.8	0.4	5.0
Lys	8.8	4.9	8.3	9.4	5.0	7.0
Ile	4.1	3.8	5.6	5.3	3.5	4.8
Leu	7.0	7.0	7.3	8.3	7.5	5.8
Phe	2.4	2.4	2.5	3.0	2.5	2.6

Table 1. Amino acid analysis of individual protein bands after separation by SDS-PAGE. Results are given as Mol %. ND = Not detected. Trp was not determined.

3.4 Adhesion characteristics

In the present study, a 90 Degree Peel Test was used to evaluate the adhesion of freshly expelled Cuvierian tubules. This method was chosen as we had encountered problems when tensile testing the *H. dofleinii* tubules following the approach used by Flammang and colleagues (Flammang et al., 2002). Specifically, when *H. dofleinii* tubules were sandwiched between two materials to which there was good adhesion, e.g. glass or metals, testing could lead to strength values which reflected the structural failure of the Cuvierian tubule rather than the failure of the adhesive, especially if some drying had occurred (data not shown). This method would only allow the determination of a minimum value for the adhesive strength as the latter exceeded the break strength of the tubule material itself.

The present test was suitable for rapidly examining numerous, freshly expelled samples, thus allowing ready comparison between the effects of various treatment solutions. The various treatments (i.e., incubations of tubules in the appropriate wash solutions) prior to adhesive testing were rapid (1 min) as it appeared that the adhesion could decline if tubules were left soaking for lengthy periods (data not shown). With *H. forskåli*, a lag period of about 60 min at 16 °C was recorded before adhesion started to decline, decreasing to about 15 min at 26 °C (Müller et al., 1972). In another study (Flammang et al., 2002) a longer lag phase was observed, and an initial increase in adhesive strength was reported. Yet others have reported adhesive strength to fall after 20 min (Zahn et al., 1973). The present approach, therefore, used short incubations in order to minimise time-based variations and to mimic the timescale over which tubules would be required to act in the natural environment.

Previous studies (Flammang et al., 2002) have shown that a compressive force of 2–10 N during adhesion led to a 6- to 8-fold increase in the resulting bond strength. In the present case, no compressive load was added so as to better simulate the natural process of ensnaring a predator.

Tubule widths showed little variation between individual samples, the average size being 4.0 mm. Tubules that were not fully expelled, and which therefore had a lesser diameter, were discarded. The observed width is larger than that found for *H. forskåli* (Flammang et al., 2002; Zahn et al., 1973) and *H. leucospilota* (Flammang et al., 2002), the species previously studied in detail, and also larger than for *H. impatiens* and *H. maculosa;* these other species generally have tubule diameters of 1–2 mm (Flammang et al., 2002). Although there are many potential tubules within the body cavity (Figure 1A) *H. dofleinii* expels only a few, typically 8 - 12 for organisms stimulated in the holding tanks compared with the more numerous thin tubules expelled by *H. leucospilota* or *H. forskali* (Flammang et al., 2002).

Adhesive strength was also found to vary when different substrata were examined, all after washing the tubules in 3.5% NaCl 10 mM Na/PO$_4$, pH 7.6. There was a trend for strongest adhesion to be observed with hydrophilic substrata, glass and aluminium (Table 2). Adhesion to polycarbonate, PMMA, and PTFE was very poor; indeed, for PMMA and PTFE, the load required for peel was barely more than the weight of the 50 mm of tubule overhang. Intermediate adhesion values were observed with polyvinyl chloride and crab chitin surfaces (Table 2). The chitin samples were unusual in having a textured surface rather than a smooth one. Previously, Zahn, Flammang and colleagues had shown strong adhesion to hydrophilic surfaces such as glass and stainless steel, and poor adhesion to hydrophobic ones such as paraffin wax, polystyrene and polyethylene (Zhan et al., 1973; Flammang et al, 2002). In general our results are consistent with this trend: the best adhesion was observed with glass whilst the poorest was observed with PTFE.

| Solution | Force/width | S.D. | n |

	(N/mm)		
Glass	> 0.050	-	10
Aluminium	> 0.050	-	6
Polyvinyl chloride	0.024	0.006	11
Chitin	0.021	0.005	8
Polycarbonate	0.010	0.002	8
PMMA	0.009	0.003	7
PTFE	0.008	0.002	6

Table 2. Force required to peel Cuvierian tubules off various substrata to determine adhesive strength.

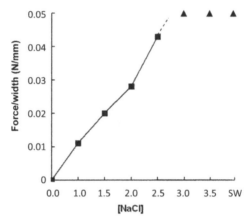

Fig. 5. The effect of different washing solutions on the adhesiveness of *H. dofleinii* Cuvierian tubules for glass. The effect of NaCl concentration; where ▲ indicates conditions where the force per unit width exceeded 0.05 N/mm. SW is natural sea water.

Adhesive strength decreased with decreasing NaCl concentration (Figure 5). At ≥3% NaCl the adhesion exceeded 0.05 N/mm. Reducing the NaCl concentration incrementally from 2.5% to 1.0% NaCl led to a steady decline in adhesive strength (Figure 5). The adhesive strength at 1% NaCl, which is comparable in concentration to physiological saline, was significantly weaker than in 3.5% NaCl simulated seawater. This is consistent with the previous observations on *H. forskåli* tubules (Flammang et al., 2002). It suggests that hydrophobic interactions may be important in the adhesive mechanism.

The effects on tubule-glass adhesion of other chloride or sodium salts (50 mM) (Table 3) showed that in all cases there was a loss of adhesive strength. For chloride salts, the loss was smaller when Tris rather than ammonium was the cation (Table 3). The other salts examined were all sodium salts of carboxylic acids, for whose action no simple mechanism could be proposed. Thus while formate (a monocarboxylate) and oxalate (a dicarboxylate) both showed similar adhesion, that observed with acetate (another monocaboxylate) was ~35% below the value observed for formate. However, the values presented in Table 3 show only a trend as the errors in measurement are such that the different systems are not necessarily distinguishable. Supplementation with EDTA (a tetracarboxylate) was the most effective at disrupting bond strength, and essentially led to complete loss of adhesion (Table 3). It is not clear whether this is due to the multiple carboxyl groups of this salt, to its strong metal ion chelating capability, or to some other property. However, certain other marine adhesives, such as that from *Mytilus*, do require metal activity (Hwang et al., 2010). A previous study which tested 15 different amino acids at 0.5% w/v solutions on adhesion by *H. forskåli* tubules (Müller et al., 1972) showed that most had little, if any, effect. The exceptions were the hydrophobic amino acids leucine (20% loss) and phenylalanine (57% loss). For phenylalanine, the loss was slow to develop (taking several minutes) and could not be reversed by washing (Zahn et al., 1973).

Solution	Force/width (N/mm)	S.D.	n
3.5% NaCl	> 0.050		>8
Tris/chloride	0.050	0.008	6
Sodium formate	0.047	0.012	6
Sodium oxalate	0.046	0.011	8
Ammonium chloride	0.036	0.008	8
Sodium citrate	0.035	0.009	7
Sodium acetate	0.030	0.006	8
Sodium EDTA	<0.003		8

Table 3. Comparison of the effects of different salt solutions on adhesion of Cuvierian tubules onto glass. All salts were 50mM in 3.5% NaCl, 10 mM sodium phosphate, pH 7.6

The effect of pH on the adhesive strength of the *H. dofleinii* adhesive showed that for Tris/chloride buffer, the best observed strength of those tested was at pH 7.6, and that the observed strength decreased at both lower and higher pH values (Figure 6). For citrate and acetate buffers, adhesive strength declined progressively as the pH was lowered from pH 7.6, with little adhesion remaining at pH 5.0 (Figure 6). A loss of adhesive strength at acidic pH values was also observed by Müller et al. (1972), who used paraffin wax as a (poor) substratum for tubule adhesion.

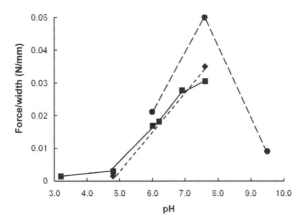

Fig. 6. The effect of different washing solutions on the adhesiveness of *H. dofleinii* Cuvierian tubules for glass. The effect of changes in pH, where ■ indicates acetate buffers, ♦ indicates citrate buffers and ● indicates Tris/Chloride buffers.

Some reports suggest that proteins may play an important role in the adhesion of Cuvierian tubules from *H. forskåli* (DeMoor et al., 2003; Flammang & Jangoux, 2004; Müller et al., 1972). For example, the adhesive residue left when tubules are peeled from a surface consists mainly of protein (DeMoor et al., 2003), and the treatment of tubules with proteases causes loss of adhesion (Müller et al., 1972). However, it has been reported that the proteins most likely differ between species (Flamman & Jangoux, 2004), making further comparative biochemical surveys important for elucidating the mechanism.

The effect of urea on tubule-glass adhesion showed that bond strength decreased progressively with increasing urea concentration until it was completely lost at 2 M urea (Figure 7). However, if tubules that had been incubated for about 60 sec in 2 M urea were then rinsed for about 60 sec in simulated sea water, some adhesion was restored, although the extent was rather variable. Urea disrupts hydrogen bonding, and its effect on adhesion may reflect some partially reversible protein unfolding (Zahn et al., 1972). The rapidity and partial reversibility of the effect indicates that there is not a complete urea-mediated release of proteins from the tubule surface.

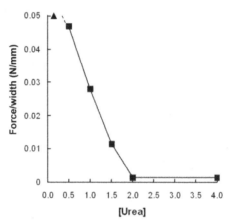

Fig. 7. The effect of different washing solutions on the adhesiveness of *H. dofleinii* Cuvierian tubules for glass. (C) The effect of urea concentration, where ▲ indicates conditions where the force per unit width exceeded 0.05 N/mm.

4. Conclusion

The distinct features of the Cuvierian tubule adhesion mechanism, especially its rapid action under water, are unique. If the mechanism can be understood, then it may be possible to design a synthetic system with analogous properties. An adhesive that provided instant grip in an aqueous environment would be very valuable, especially in medical applications, as the majority of existing adhesives bind well only to dry surfaces.

It appears that although the tubules of *H.dofleinii* are distinct from those of other species, especially in their size and the number expelled, the adhesive properties of these Cuvierian tubules (including preferences for hydrophilic surfaces, pH optima, etc.) are similar to those found in other species, even if mechanistic details may differ between species as has been proposed previously (Flammang and Jangoux, 2004). It is thought that during expulsion and tubule elongation, granular cells that are internal in the pre-release tubule become located

Biomimetic Materials as Potential Medical Adhesives – Composition and Adhesive Properties
of the Material Coating the Cuvierian Tubules Expelled by Holothuria dofleinii

33

on the tubule surface and release their contents on contact with a surface (VandenSpiegel and Jangoux, 1987) leading (in whole or in part) to the observed adhesion. Histology has shown that these granules contain protein and lipid, but lack polysaccharide (VandenSpiegel and Jangoux, 1987). Biochemical studies have indicated that the granules contain a protein of around 10 kDa, and it has been suggested that polymers of this protein account for the higher molecular weight proteins that are seen in the adhesive prints (Flammang and Jangoux, 2004), but this seems highly unlikely in *H. dofleinii* as the protein bands are very well resolved by gel electrophoresis and the calculated molecular weights of these bands do not conform to such a regular series of increases.

Our present study emphasises that the adhesives of natural systems are optimised for the specific environments in which they have evolved, such as the present marine environment. An analogue intended for medical use would need to be optimised to yield maximum adhesion in the physiological conditions that prevail in mammalian tissues. In the present case, the adhesion works better at higher NaCl concentrations that found in medical applications so understanding more about the mechanism and the protein structures and properties will be needed in order to adapt this system for applications where lower NaCl concentrations are present.

5. Acknowledgments

We wish to thank Nicole Murphy for assistance with Holothurian collection and Dr Anita Hill for helpful discussions. This study was facilitated by access to the Australian Proteome Analysis Facility supported under the Australian Government's National Collaborative Research Infrastructure Strategy (NCRIS). The project received support from the CSIRO Wealth from Oceans National Research Flagship.

6. References

Cohen, S.A. (2001). Amino acid analysis using precolumn derivatisation with 6-aminoquinolyl-*N*-hydroxysuccinimidyl carbamate. *Methods in Molecular Biology*, Vol.159, pp. 39-47, ISSN 1064-3745

Cohen, S.A. & DeAntonis, K.M. (1994). Applications of amino acid analysis derivatisation with 6-aminoquinolyl-*N*-hydroxysuccinimidyl carbamate: Analysis of feed grains, intravenous solutions and glycoproteins. *Journal of Chromatography*, Vol.661, No.1-2, (February), pp. 25-34, ISSN 0021-9673

DeMoor, S.; Waite, J.H.; Jangoux, M. & Flammang, P. (2003). Characterization of the adhesive from cuvierian tubules of the sea cucumber *Holothuria forskali* (Echinodermata, Holothuroidea). *Marine Biotechnology*, Vol.5, No.1, (January), pp. 45-57, ISSN 1436-2228

Dimas, D.A.; Dallas, P.P.; Rekkas, D.D. & Choulis, N.H. (2000) Effect of several factors on the mechanical properties of pressure-sensitive adhesives used in transdermal therapeutic systems. *AAPS PharmSciTech*, Vol.1, No.2, (June), pp. 80-87, ISSN 1530-9932

Doolittle, R.F. (1986). *Of URFs and ORFs: a primer on how to analyse derived amino acid sequences*, University Science Books, ISBN 0-935702-54-7, Mill Valley, CA, USA

Flammang, P. & Jangoux, M. (2004). *Final Report ONR Grant* N00014-99-1-0853

Flammang, P.; Ribesse, J. & Jangoux, M. (2002) Biomechanics of adhesion in sea cucumber Cuvierian tubules (Echinodermata, Holothuroidea). *Integrative and Comparative Biology*, Vol.42, No.6, (December), pp. 1107-1115, ISSN 1540-7063

Hamel, J.-F. & Mercier, A. (2000). Cuvierian tubules in tropical holothurians; usefulness and efficiency as a defence mechanism. *Marine and Freshwater Behaviour and Physiology*, Vol.33, pp. 115-139, ISSN 1023-6244

Hwang, D.S.; Zeng, H.; Masic, A.; Harrington, M.J.; Israelachvili, J.N. & Waite, J.H. (2010). Protein- and metal-dependent interactions of a prominent protein in mussel adhesive plaques. *Journal of Biological Chemistry*, Vol.285, No.33, (August), pp. 25850-25858, ISSN 0021-9258

Jones, I.; Lindberg, C.; Jakobsson, S.; Hellqvist, A.; Hellman, U.; Borg, B. & Olsson, P.E. (2001). Molecular cloning and characterization of spiggin. *Journal of Biological Chemistry*, Vol.276, No.21, (May), pp. 17857-17863, ISSN 0021-9258

Kamino, K. (2008). Underwater adhesive of marine organisms as the vital link between biological science and material science. *Marine Biotechnology*, Vol.10. No.2, (March), pp. 111-121, ISSN 1436-2228

Laemmli, U.K. (1970). Cleavage of structural proteins during the assembly of the head of bacteriophage T4. *Nature*, Vol.227, No.5259, (August), pp. 680-685, ISSN 0028-0836

Lawrence, J.M. (2001). Function of eponymous structures in echinoderms; a review. *Canadian Journal of Zoology*, Vol.79, No.7, (July), pp. 1251-1264, ISSN 0008-4301

Lin, Q.; Gourdon, D.; Sun, C.; Holten-Andersen, N.; Anderson, T.H.; Waite, J.H. & Israelachvili, J.N. (2007). Adhesion mechanisms of the mussel foot proteins mfp-1 and mfp-3. *Proceeding of the National Academy of Sciences of the United States of America*, Vol.104, No.10, (March), pp. 3782-3786, ISSN 0027-8424

Mancuso-Nichols, C.A.; Nairn, K.M.; Glattauer, V.; Blackburn, S.I.; Ramshaw, J.A.M. & Graham, L.D. (2009) Screening microalgal cultures in search of microbial exopolysaccharides with potential as adhesives *Journal of Adhesion*, Vol.85, No.2-3, pp. 97-125, ISSN 0021-8464

Müller, W.E.G.; Zahn, R.K. & Schmid, K. (1972). The adhesive behaviour in Cuvierian tubules of *Holothuria forskåli*. Biochemical and biophysical investigations. *Cytobiology*, Vol.5, No.3, pp. 335-351, ISSN 0070-2463

Nakano, M.; Shen, J.R. & Kamino, K. (2007). Self-assembling peptide inspired by a barnacle underwater adhesive protein. *Biomacromolecules*, Vol.8, No.6, (June), pp. 1830-1835, ISSN 1525-7797

VandenSpiegel, D. & Jangoux, M. (1987). Cuvierian tubules of the holothuroid *Holothuria forskali* (Echinodermata): a morphofunctional study. *Marine Biology*, Vol.96, No.2, pp. 263-275, ISSN 0025-3162

VandenSpiegel, D.; Jangoux, M. & Flammang, P. (2000). Maintaining the line of defence: regeneration of Cuvierian tubules in the sea cucumber *Holothuria forskali* (Echinodermata, Holothuroidea). *Biological Bulletin*, Vol.198, No.1, (February), pp. 34-49, ISSN 0006-3185

Watson, M.R. & Silvester, N.R. (1959). Studies of invertebrate collagen preparations. *Biochemical Journal*, Vol.71, No.3, (March), pp. 578-584, ISSN 0264-6021

Zahn, R.K.; Müller, W.E.G. & Michaelis, M. (1973). Sticking mechanisms in adhesive organs from a *Holothuria*. *Research in Molecular Biology*, Vol.2, pp. 47-88, ISSN 0340-5400

3

Mechanical and Biological Properties of Bio-Inspired Nano-Fibrous Elastic Materials from Collagen

Nobuhiro Nagai[1,2], Ryosuke Kubota[2],
Ryohei Okahashi[2] and Masanobu Munekata[2]
[1]*Division of Clinical Cell Therapy, Center for Advanced Medical Research and
Development, ART, Tohoku University, Graduate School of Medicine*
[2]*Division of Biotechnology and Macromolecular Chemistry, Graduate
School of Engineering, Hokkaido University*
Japan

1. Introduction

Collagen-based biomaterials have been widely used in medical applications, because of its many advantages, including low antigenicity, abundant availability, biodegradability, and biocompatibility [1]. Collagen represents the major structural protein, accounting for nearly 30% of all vertebrate body protein. The collagen molecule comprises three polypeptide chains (α-chains) which form a unique triple-helical structure (Fig. 1) [2]. Each of the three chains has the repeating structure glycine–X–Y, where X and Y are frequently the imino acids proline and hydroxyproline (Fig. 1b). The collagen molecules self-aggregate through fibrillogenesis into microfibrils forming extracellular matrix (ECM) in the body [3-5]. The fibrils provide the major biomechanical matrix for cell growth (Fig. 1a), allowing the shape of tissues to be defined and maintained. The main application of collagen for biomaterials is as a scaffold for tissue engineering and a carrier for drug delivery [2, 6-9]. Many different forms of collagen biomaterials, such as film [10, 11], gel [12-17], sponge [18-20], micro-/nano-particle [21, 22], and fiber [23], have been fabricated and used in practice. However, most collagen biomaterials become brittle and fail under quite low strains, which limit their application to biomedical engineering fields that need larger mechanical properties, especially elasticity [16].

Recently, we reported a novel crosslinking method of improving the mechanical properties and thermal stability of collagen [24]. The method mimics actual biological events to form collagen matrix in the body; monomeric collagens extruded from cells into extracellular environment initially form microfibrillar aggregates, then lysyl oxidase crosslinking during their assembly to form fibrils (Fig. 1). The *in vitro* crosslinking during collagen fibrillogenesis, namely "bio-inspired crosslinking", creates a crosslinked collagen fibrillar gel with high mechanical properties at certain crosslinking agent concentrations [25, 26]. Fibril formation involves the aggregation and alignment of collagen molecules, and helps increase the collagen's thermal stability. The introduction of crosslinking during fibril formation further increases the thermal stability of collagen. The synergistic effects of crosslinking and fibril formation are found to enable an increase in the thermal stability of

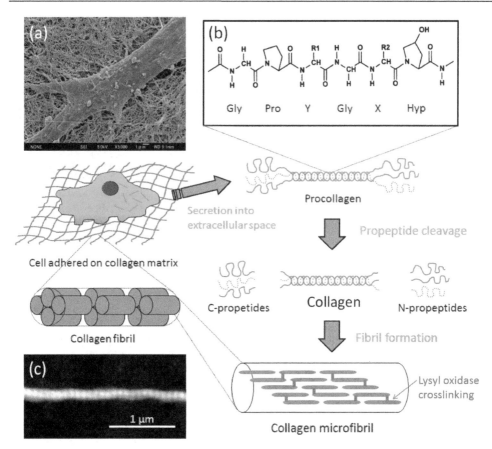

Fig. 1. Schematic representation of collagen synthesis process. Procollagen consists of a 300 nm long triple helical domain (comprised of three alpha-chains each of approximately 1000 residues) flanked by a trimeric globular C-propeptide domain and a trimeric N-propeptide domain. Procollagen is secreted from cells and is converted into collagen by the removal of the N- and C-propeptides by proteases. The collagen spontaneously self-assembles into cross-striated fibrils that occur in the extracellular matrix of connective tissues. The fibrils are stabilized by covalent crosslinking, which is initiated by oxidative deamination of specific lysine and hydroxylysine residues in collagen by lysyl oxidase. (a) Scanning electron microscopy image of a human fibroblast adhered on reconstituted salmon collagen fibrillar matrix [17]. (b) Chemical structure of a collagen alpha-chain. (c) Atomic force microscopy image of reconstituted salmon collagen fibril [17]. The repeating broad dark and light zones (where D=67 nm, the characteristic axial periodicity of collagen) can be seen in the fibril.

salmon-derived collagen (SC) [24]. Additionally, heat denaturation of the bio-inspired crosslinked SC gel provides elastic materials, which has the elongation of over 200% at break point [24]. So far, the elasticity of the collagens that had been crosslinked "after" fibrillogenesis is not as high as that crosslinked "during" fibrillogenesis (bio-inspired crosslinking) [25]. This may indicate that the bio-inspired crosslinking confer intrafibrillar crosslinking as well as interfibrillar one, allowing for homogenous crosslinking sites and larger mechanical properties. Therefore, the bio-inspired crosslinking may provide more wide range of applications of collagen biomaterials. However, the bio-inspired crosslinking has not yet been applied to bovine collagen (BC), which is widely used for medical applications for past decades, because the bio-inspired crosslinking has been developed for thermal stabilization of "fish collagens" with low-denaturation temperature to develop marine-derived collagen biomaterials [13-17, 24-26]. In this study, we studied the crosslinking condition to create bio-inspired crosslinked BC gel and prepared the elastic material from the BC gel (e-BC gel) by heat treatment. The mechanical properties (tensile strength and elongation rate) and biological properties (biodegradability, cell culture, and blood compatibility) of the e-BC gel were evaluated. Herein, we report the fabrication of the bio-inspired elastic material from BC and demonstrate its applicability for biomaterials, especially in vascular tissue engineering.

2. The bio-inspired crosslinking conditions for BC

Acid-soluble collagen molecules self-assemble and form fibrils under physiological conditions. The pH, NaCl concentraton, and temperature are important factors to provide a successful reconstituted collagen fibrillar gel. First, we evaluated the effect of NaCl concentrations on fibril formation of BC at constant pH of 7.4 and temperature of 37°C. The fibril formation of BC was monitored by a turbidity change observed at 310 nm [26, 27]. Figure 2 shows that a rapid rise in turbidity was observed in the mixture of BC solution and 30 mM Na-phosphate buffer at NaCl concentration from 50mM to 100 mM. Then, the rise in turbidity increase was gradually decreased at over 140 mM NaCl. The fibril formation rate of collagen is known to be reduced by addition of salts [4], which appears to reduce electrostatic interaction among collagen molecules. The bio-inspired crosslinking needs active fibrillogenesis during crosslinking (see below) [17, 26]. Therefore, the optimum range of NaCl concentration for BC fibril formation was determined to be 50-100 mM.

Cross-linking generally reinforces the biomaterials composed of collagen fibrils for further improvement of mechanical properties. Various techniques for stabilizing collagen have been developed and reported. These techniques are divided into chemical treatments and physical treatments. Glutaraldehyde is one of the most widely used chemical agents [28, 29]; it is known, however, that there are side effects to its use in cross-linking [30], for example, cytotoxicity, enhancement of calcification, and a mild inflammatory response compared with using other reagents. The water soluble condensign agent 1-ethyl-3-(3-dimethylaminopropyl) carbodiimide hydrochloride (EDC) has been reported to be significantly less cytotoxic than glutaraldehyde because EDC reagents do not remain in the linkage and are simply washed away during the cross-linking process [28, 29]. On the other hand, physical cross-linking methods such as UV irradiation [31, 32] and dehydrothermal treatment [33, 34] do not introduce any additional chemical units. These methods may therefore be more biocompatible than chemical treatments. However, the mechanical properties of materials cross-linked by physical treatments are lower than those cross-linked by chemical treatments. Therefore, EDC was used for a crosslinking agent in this study.

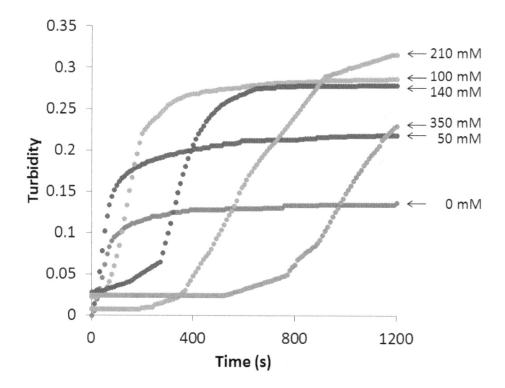

Fig. 2. Effect of NaCl concentration on fibril formation curve. The values in the graphs indicate the final NaCl concentrations in the gels.

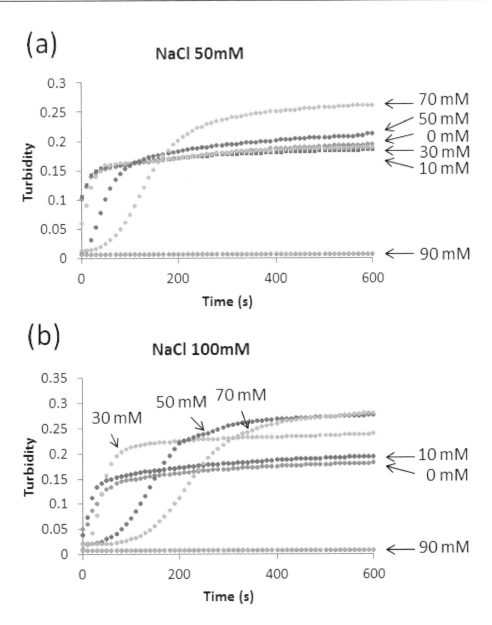

Fig. 3. Effect of EDC concentrations on fibril formation curve in the presence of 50 mM NaCl (a) and 100 mM NaCl (b). The values in the graphs indicate the final EDC concentrations in the gels.

The turbidity changes in BC solution mixed with Na-phosphate buffer including various EDC concentration at 50 mM or 100 mM NaCl were monitored using a spectrophotometer

(Fig. 3a, b). The rate of fibril formation decreased with increasing EDC concentration, which indicates EDC exhibited an inhibitory effect on collagen fibril formation. These inhibitory effects were probably the results of the rapid reaction of EDC to monomeric collagens upon mixing of the EDC containing buffer and acidic collagen solution, by which the ability to form fibrils was reduced or lost through random and nonfibrous aggregation of monomeric collagens. Further increased EDC concentration above 90 mM completely suppressed fibril formation irrespective of NaCl concentrations. Based on these results, it appeared that a buffer that would enable a faster fibril formation rate would be desirable. According to a previous report, EDC is sufficiently stable and active under such conditions [35].

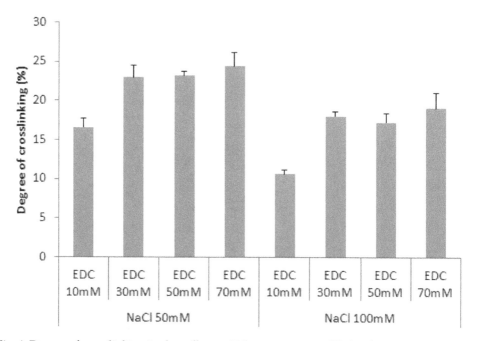

Fig. 4. Degree of crosslinking in the collagen. Values are mean ± SD (n=4).

The degree of crosslinking was determined as the decrease in the free amino group content of the collagen molecules [19, 26]. The free amino group content was measured spectrophotometrically after the reaction of the free amino groups with 2,4,6-trinitro-benzensulfonic acid and was expressed as the decrease in the ratio of the free amino group content of the crosslinked sample to that of the uncrosslinked sample. An increase in the degree of crosslinking with increase in the EDC concentration was observed (Fig. 4). The degree of crosslinking was slight lower at NaCl concentration of 100 mM compared to the NaCl concentration of 50 mM. This may be attributed to lower ability of fibril formation observed at NaCl concentration of 50 mM. Fig 2 shows that the plateau level in turbidity at 50 mM NaCl was lower than that at 100 mM NaCl. Lower fibril formation may result in an increase of nonfibrous aggregates, which can lead to high degree of crosslinking due to increased sites of crosslinking in monomeric collagens. The synergistic effects of crosslinking and fibril formation were thought to be complete at EDC and NaCl concentrations of 70 mM and 100 mM, respectively.

3. Fibril formation of BC

To produce an elastic material from bio-inspired crosslinked BC gel (e-BC gel), heat denaturation process is needed. By heat treatment, the cross-linked collagen fibrils shrink, maintaining the cross-linkage among the collagen molecules and fibrils through the denaturation of triple-helical collagen molecules to the random-coil form [24]. At the same time, uncross-linked collagen molecules and fibrils may be lost through dissolution to water. In fact, the original BC gels crosslinked with the EDC concentrations of 30-70 mM showed drastic shrinkage (Fig. 5) and rubber-like elasticity after heat treatment at 60°C for 5 min. The BC gels crosslinked with EDC concentrations of 0-10 mM dissolved away after heat treatment due to incomplete crosslinking.

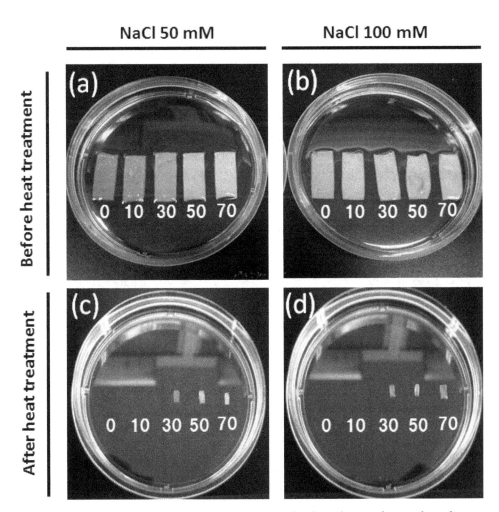

Fig. 5. Appearances of BC gels (a, b) and e-BC gels (c, d). The values in the graphs indicate the final EDC concentrations (mM) in the gels.

The collagen fibrils were observed by high-resolution scanning electron microscopy (SEM). The preparation of the specimen was performed according to a previous report [16, 24]. Figure 6 shows the well-developed networks of nano-fibrils on the BC gel and e-BC gel. The width of the fibrils on the BC-gel was in the range of 50–100 nm (Fig. 6a). However, a wider (width; >200 nm) and winding fibril-like structure was observed on the e-BC gel (Fig. 6b), indicating that the fibril structure of collagen was deformed through the heat treatment. The wide and winding fibril-like structure of the e-BC gel should be directly derived from the collagen nano-fibrils of the original BC gel through swelling of the fibrils by comparison of both surface structures.

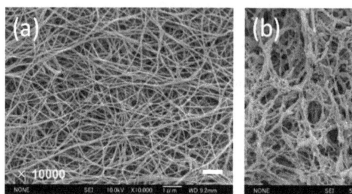
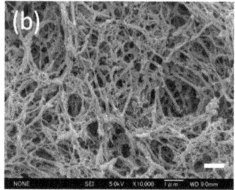

Fig. 6. SEM images of BC gels (a) and e-BC gels (c). Bars: 1 μm.

4. Mechanical properties of e-BC gel

The mechanical properties of the e-BC gel were evaluated by tensile tests. The original BC gel rarely had elasticity and stretchability similar to the usual collagen materials. However, the e-BC gel showed rubberlike elasticity and high stretchability. Figure 7 shows the stress–strain curves to the breaking point obtained in the strain rate of 0.1 mm/s (n = 5). Salmon-derived elastic collagen gels (e-SC gel) were used as controls [24]. The mean values ± standard deviation (SD) of elongation at the break of the e-BC gel and e-SC gel were 201 ± 47% and 260 ± 59%, respectively (Fig. 7c). At the early stage of loading, stress was almost linearly increased depending on the strain. Above a strain of ca. 100%, an increase in strain hardening was observed. There was no significant difference in elongation between the two e-gels. According to the report by Koide and Daito [31], collagen films reinforced by traditional cross-linking reagents, glutaraldehyde and tannic acid, showed only small elongation at the breaking point (6.6% and 12.4%, respectively). Weadock et al. showed small ultimate strains (approximately 40% and 30%) of collagen fibers cross-linked by UV irradiation and dehydrothermal treatment, respectively [36]. Even a purified skin with an intact fibrous collagen network gives elongation at a breaking point of 125% [36]. Recently, it was reported that a chemically cross-linked collagen-elastin-glycosaminoglycan scaffold, which are the contents analogous to the actual tissue/organs, demonstrated good stretchability (150% strain) [37]. To the best of our knowledge, this is the first report of a material from bovine collagen with elongation at a breaking point over 200%. Although the mechanism of the high stretchability was not well understood, the denaturation of the collagen molecules probably plays an important role in the elongation, i.e., the bend structure of denatured collagen fibrils observed in Fig. 6b is considered to provide its rubber-like stretchability.

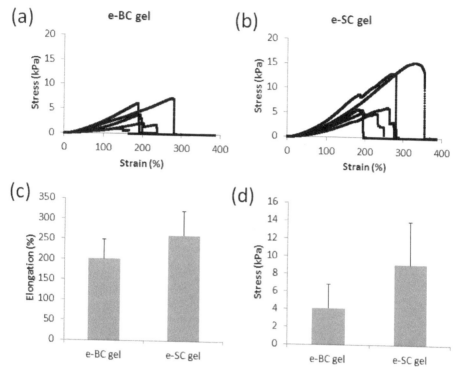

Fig. 7. Stress–strain curves generated from tensile testing of e-BC gels (a) and e-SC gels (b). The specimen (5 × 3 × 12 mm) was gripped to achieve a gauge length of 8 mm and stretched in the strain rate of 1.25%/s. (c) Elongation at the break of the e-BC gel and e-SC gel. (d) Ultimate strength at the break of the e-BC gel and e-SC gel. Values are mean ± SD (n=5).

The mean values ± SD of ultimate strength at the break of the e-BC gel and e-SC gel were 4.1 ± 2.6 kPa and 9.0 ± 4.8 kPa, respectively (Fig. 7d). The strength of the e-BC gels was lower than that of e-SC gels, although there was no significant difference between the two. Collagen exhibits a limited mechanical resistance so that collagen requires an additional skeletal material such as inorganic materials [38]. Bio-inspired crosslinking can provide a collagen scaffold with high mechanical strength as well as stretchability. It is useful without the addition of any other material. Because a collagen solution is a precursor for fabrication of various collagen forms, bio-inspired crosslinking would be a useful fabrication process of widely various collagen biomaterials.

5. Biological properties of e-BC gel

To investigate the potential of the e-BC gel for use as a cell culture scaffold, we measured the growth rates of human umbilical vein endothelial cells (HUVECs) cultured on the e-BC gel.

The cell number was evaluated by the 3-[4,5-dimethylthiazol-2-yl]-2,5-diphenyl tetrazolium bromide (MTT) test [13]. The MTT test is an established method to determine viable cell number by measuring the metabolic activity of cellular enzymes. Figure 8 shows steady increases in cell number with culture time on both the e-BC gel and tissue culture plate (TCP). There was a lag time before steady increase in cell number on the e-BC gel. Semler et al. reported that cells shows high growth rate on matrices with high mechanical compliance such as TCP, whereas the cells aggregate three-dimensionally on matrices with low mechanical compliance such as collagen gel [39]. Therefore, the difference in the mechanical properties of the surface of the culture substrate might affect the initial rate of cell growth.

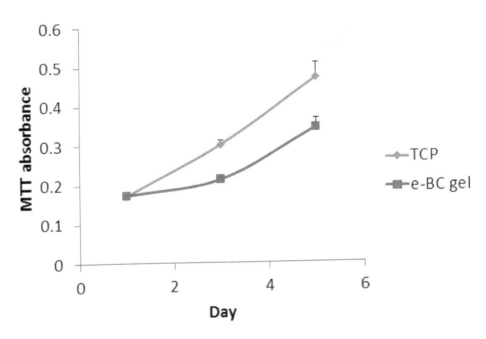

Fig. 8. Growth curves of HUVECs cultured on e-BC gels and plastic tissue culture plate. Values are mean ± SD (n=4).

The SEM images show the distribution and spreading morphology of the HUVECs cultured on the e-BC gel at day 1 and day 5 after cultivation (Fig. 9). Apparently, the cells at 5d cultivation were confluent and showed cobblestone-like morphologies (Fig 9b). HUVECs also grow better on the e-SC gel and shows confluent at day 6 after cultivation [13]. Therefore, the e-BC gel could be used as a cell culture scaffold as well as the e-SC gel. EDC cross-links collagen molecules by the formation of isopeptides without being incorporated itself, thus precluding depolymerization and the possible release of potentially cytotoxic reagents. Furthermore, the by-product of the cross-linking reaction and un-reacted EDC in the e-BC gel should be completely removed by the drastic shrinkage in hot water. It is expected that the e-BC gel has good biocompatibility and no cytotoxicity.

To assess the degradability of the e-BC gel in collagenase solution (50 U/ml), protein content measurement was performed using a bicinchoninic acid protein assay kit as previously described [40]. Figure 10 shows that e-BC gel was completely digested for 24 h as well as the e-SC gel. The e-BC gel degraded slightly later than the e-SC gel. This may be due to the low denaturation temperature of SC ($19°C$). The physiological concentration range of collagenase is approximately 1 U/ml [41], therefore, the e-BC gels might show the slow degradation profile *in vivo* than the *in vitro* results. Additionally, e-SC gel gradually degrades 1 month after implantation in rat subcutaneous pouches [16], e-BC gel, therefore, would show the same or later biodegradation profile *in vivo*.

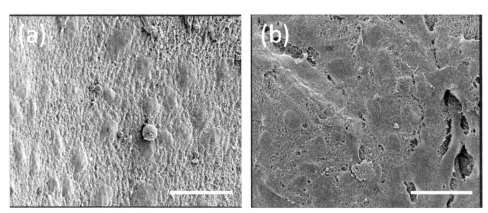

Fig. 9. SEM images of HUVECs cultured on e-BC gel for 1d (a) and 5d (b). Bars: 50 µm.

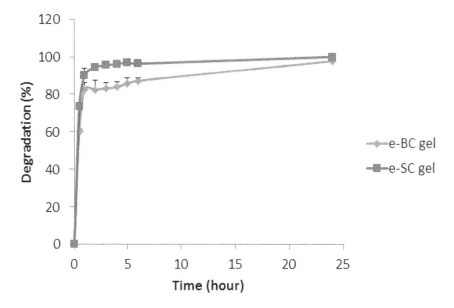

Fig. 10. In vitro degradation rate of e-BC gel and e-SC gel in collagenase solution. Values are mean ± SD (n=4).

Owing to the mechanical, biological, and biodegradable properties of the e-BC gel, it could potentially be used to engineer blood vessels *in vivo*. Synthetic materials such as polyethylene terephthalate (Dacron) and expanded polytetrafluoroethylene (ePTFE) have been clinically applied as vascular grafts for a long time to replace or bypass large-diameter blood vessels. However, when used in small-diameter blood vessels (inner diameter < 6 mm), the patency rates are poor compared to autologous vein grafts. These failures are due to early thrombosis and gradual neointimal hyperplasia, and the pathological changes occurred due to the lack of blood or mechanical compatibility of the synthetic grafts [42]. To address this problem, tissue engineering approach is promising. A variety of biodegradable polymers and scaffolds have been evaluated to develop a tissue-engineered vascular graft [43-46]. These approaches depend on either the *in vitro* or *in vivo* cellular remodeling of a polymeric scaffold. For successful *in vivo* cellular remodeling, the biocompatibility, biodegradability, and mechanical properties of the scaffold must be suitable to the dynamic environment of the blood vessel. Therefore, the ideal scaffold should employ a biocompatible and biodegradable polymer with elastic properties that interact favorably with cells and blood. Therefore, the interaction of the e-BC gel with rat whole blood and plasma was investigated to assess their blood compatibility for use in vascular-tissue engineering.

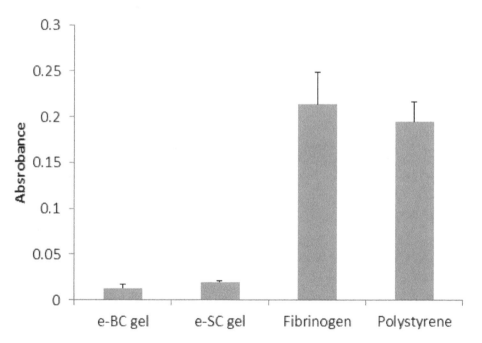

Fig. 11. Platelet adhesion rates on the e-BC gel, e-SC gel, and the control samples. Values are mean ± SD (n=4).

After incubation of the e-BC gel with platelet-rich plasma (PRP) collected from rat blood, the colored *p*-nitrophenol produced by the acid-phosphatase reaction of the platelets was measured with a microplate reader at an absorbance of 405 nm. The percentage of adherent platelets was calculated according to the methods reported previously [14]. Figure 11 shows the number of platelets that adhered to the samples. The platelet number was estimated from the acid phosphatase reaction [47]. There was a linear relationship between the PRP concentration and the absorbance values at 405 nm, indicating that the acid phosphatase reaction of the platelets may be considered a reliable indicator of platelet number (data not shown). The results demonstrate that the platelet adhesion rates were markedly low on the e-BC gel when compared to the fibrinogen-coated or polystyrene surfaces. The e-SC gel also showed an adhesion rate as low as the e-BC gel. We are separately studying the fabrication of a vascular graft using the e-SC gel [16]. Consequently, the e-BC gel can potentially be used for the fabrication of tissue-engineered vascular graft.

Considering that the platelets adhered better to the collagen-coated than to the gelatin-coated surface [14], the anticoagulant ability of the e-BC gel may have been due to heat denaturation. The e-BC gel was prepared by heat treatment at 60°C resulting in collagen denaturation (gelatinization). Polanowska-Grabowska and coworkers reported that the platelet adhesion rate on a gelatin-coated surface was lower than on collagen- coated or fibrinogen-coated surfaces [48]. However, blood coagulation is known to depend on material properties, such as surface-free energy, surface charge, and wettability; these properties govern protein adsorption involving platelet adhesion [49, 50]. Experiments using human whole blood are needed to test the clinical applications of the gels. Further examinations are necessary to ensure the blood compatibility of the e-BC gel.

6. Conclusion

In conclusion, we successfully fabricated an elastic collagen material from EDC cross-linked BC fibrillar gel by heat treatment. "Bio-inspired crosslinking" used in this study involves collagen fibril formation in the presence of EDC as a crosslinking reagent, which was developed in an attempt to mimic the in vivo simultaneous occurrence of fibril formation and crosslinking. We successfully prepared the bio-inspired crosslinked BC gels by adjusting the NaCl and EDC concentrations during collagen fibril formation. An advantage of bio-inspired crosslinking is the achievement of homogenous intrafibrillar crosslinking as well as interfibrillar one, providing higher mechanical properties compared to the traditional sequential crosslinking in which monomeric collagen initially forms fibril, then subsequently crosslinked using chemical or physical methods. Another advantage is the elastic properties of bio-inspired crosslinked BC gels after heat treatment. Although common collagen materials dissolved in water at a temperature above their denaturation temperature, we found that the bio-inspired cross-linked BC gel drastically shrank at a high temperature without remarkable dissolution. The collagen gel obtained interestingly showed rubber-like elasticity and high stretchability. The human cells showed good attachment and proliferation on this elastic material, suggesting its potential to be utilized in biomaterials for tissue engineering. Additionally, the elastic material demonstrated excellent blood compatibility. Our future work will focus on fabrication of small-caliber tubes (inner diameter < 6 mm) for small-caliber vascular grafts and preclinical animal studies to further assess the safety and effectiveness of the collagen-based vascular grafts.

7. Acknowledgment

This study was supported by Grants-in-Aid for Young Scientists (B) (20700393) from the Ministry of Education, Science, and Culture, Japan.

8. References

[1] Lee CH, Singla A, Lee Y. Biomedical applications of collagen. Int J Pharm. 2001;221:1-22.

[2] Friess W. Collagen--biomaterial for drug delivery. Eur J Pharm Biopharm. 1998;45:113-36.

[3] Na GC, Phillips LJ, Freire EI. In vitro collagen fibril assembly: thermodynamic studies. Biochemistry-Us. 1989;28:7153-61.

[4] Williams BR, Gelman RA, Poppke DC, Piez KA. Collagen fibril formation. Optimal in vitro conditions and preliminary kinetic results. J Biol Chem. 1978;253:6578-85.

[5] Kadler KE, Holmes DF, Trotter JA, Chapman JA. Collagen fibril formation. Biochem J. 1996;316 (Pt 1):1-11.

[6] Cen L, Liu W, Cui L, Zhang W, Cao Y. Collagen tissue engineering: development of novel biomaterials and applications. Pediatr Res. 2008;63:492-6.

[7] Glowacki J, Mizuno S. Collagen scaffolds for tissue engineering. Biopolymers. 2008;89:338-44.

[8] Sano A, Maeda M, Nagahara S, Ochiya T, Honma K, Itoh H, et al. Atelocollagen for protein and gene delivery. Adv Drug Deliv Rev. 2003;55:1651-77.

[9] Ramshaw JA, Peng YY, Glattauer V, Werkmeister JA. Collagens as biomaterials. J Mater Sci Mater Med. 2009;20 Suppl 1:S3-8.

[10] Schlegel AK, Mohler H, Busch F, Mehl A. Preclinical and clinical studies of a collagen membrane (Bio-Gide). Biomaterials. 1997;18:535-8.

[11] Tiller JC, Bonner G, Pan LC, Klibanov AM. Improving biomaterial properties of collagen films by chemical modification. Biotechnol Bioeng. 2001;73:246-52.

[12] Hao W, Hu YY, Wei YY, Pang L, Lv R, Bai JP, et al. Collagen I gel can facilitate homogenous bone formation of adipose-derived stem cells in PLGA-beta-TCP scaffold. Cells Tissues Organs. 2008;187:89-102.

[13] Kanayama T, Nagai N, Mori K, Munekata M. Application of elastic salmon collagen gel to uniaxial stretching culture of human umbilical vein endothelial cells. J Biosci Bioeng. 2008;105:554-7.

[14] Nagai N, Kubota R, Okahashi R, Munekata M. Blood compatibility evaluation of elastic gelatin gel from salmon collagen. J Biosci Bioeng. 2008;106:412-5.

[15] Nagai N, Mori K, Munekata M. Biological properties of crosslinked salmon collagen fibrillar gel as a scaffold for human umbilical vein endothelial cells. J Biomater Appl. 2008;23:275-87.

[16] Nagai N, Nakayama Y, Zhou YM, Takamizawa K, Mori K, Munekata M. Development of salmon collagen vascular graft: mechanical and biological properties and preliminary implantation study. J Biomed Mater Res B Appl Biomater. 2008;87:432-9.

[17] Nagai N, Mori K, Satoh Y, Takahashi N, Yunoki S, Tajima K, et al. In vitro growth and differentiated activities of human periodontal ligament fibroblasts cultured on salmon collagen gel. J Biomed Mater Res A. 2007;82:395-402.

[18] Hosseinkhani H, Hosseinkhani M, Tian F, Kobayashi H, Tabata Y. Bone regeneration on a collagen sponge self-assembled peptide-amphiphile nanofiber hybrid scaffold. Tissue Eng. 2007;13:11-9.

[19] Nagai N, Yunoki S, Suzuki T, Sakata M, Tajima K, Munekata M. Application of cross-linked salmon atelocollagen to the scaffold of human periodontal ligament cells. J Biosci Bioeng. 2004;97:389-94.

[20] Shen X, Nagai N, Murata M, Nishimura D, Sugi M, Munekata M. Development of salmon milt DNA/salmon collagen composite for wound dressing. J Mater Sci Mater Med. 2008;19:3473-9.

[21] Nagai N, Kumasaka N, Kawashima T, Kaji H, Nishizawa M, Abe T. Preparation and characterization of collagen microspheres for sustained release of VEGF. J Mater Sci Mater Med. 2010;21:1891-8.

[22] Glattauer V, White JF, Tsai WB, Tsai CC, Tebb TA, Danon SJ, et al. Preparation of resorbable collagen-based beads for direct use in tissue engineering and cell therapy applications. J Biomed Mater Res A. 2010;92:1301-9.

[23] Srinivasan A, Sehgal PK. Characterization of biocompatible collagen fibers--a promising candidate for cardiac patch. Tissue Eng Part C Methods. 2010;16:895-903.

[24] Yunoki S, Mori K, Suzuki T, Nagai N, Munekata M. Novel elastic material from collagen for tissue engineering. J Mater Sci Mater Med. 2007;18:1369-75.

[25] Yunoki S, Matsuda T. Simultaneous processing of fibril formation and cross-linking improves mechanical properties of collagen. Biomacromolecules. 2008;9:879-85.

[26] Yunoki S, Nagai N, Suzuki T, Munekata M. Novel biomaterial from reinforced salmon collagen gel prepared by fibril formation and cross-linking. J Biosci Bioeng. 2004;98:40-7.

[27] Nagai N, Kobayashi H, Katayama S, Munekata M. Preparation and characterization of collagen from soft-shelled turtle (Pelodiscus sinensis) skin for biomaterial applications. J Biomater Sci Polym Ed. 2009;20:567-76.

[28] Lee CR, Grodzinsky AJ, Spector M. The effects of cross-linking of collagen-glycosaminoglycan scaffolds on compressive stiffness, chondrocyte-mediated contraction, proliferation and biosynthesis. Biomaterials. 2001;22:3145-54.

[29] Olde Damink LH, Dijkstra PJ, van Luyn MJ, van Wachem PB, Nieuwenhuis P, Feijen J. In vitro degradation of dermal sheep collagen cross-linked using a water-soluble carbodiimide. Biomaterials. 1996;17:679-84.

[30] Goissis G, Marcantonio E, Jr., Marcantonio RA, Lia RC, Cancian DC, de Carvalho WM. Biocompatibility studies of anionic collagen membranes with different degree of glutaraldehyde cross-linking. Biomaterials. 1999;20:27-34.

[31] Koide T, Daito M. Effects of various collagen crosslinking techniques on mechanical properties of collagen film. Dent Mater J. 1997;16:1-9.

[32] Suh H, Lee WK, Park JC, Cho BK. Evaluation of the degree of cross-linking in UV irradiated porcine valves. Yonsei Med J. 1999;40:159-65.

[33] Weadock KS, Miller EJ, Bellincampi LD, Zawadsky JP, Dunn MG. Physical crosslinking of collagen fibers: comparison of ultraviolet irradiation and dehydrothermal treatment. J Biomed Mater Res. 1995;29:1373-9.

[34] Toba T, Nakamura T, Matsumoto K, Fukuda S, Yoshitani M, Ueda H, et al. Influence of dehydrothermal crosslinking on the growth of PC-12 cells cultured on laminin coated collagen. ASAIO J. 2002;48:17-20.

[35] Nakajima N, Ikada Y. Mechanism of amide formation by carbodiimide for bioconjugation in aqueous media. Bioconjug Chem. 1995;6:123-30.

[36] Olde Damink LH, Dijkstra PJ, Van Luyn MJ, Van Wachem PB, Nieuwenhuis P, Feijen J. Changes in the mechanical properties of dermal sheep collagen during in vitro degradation. J Biomed Mater Res. 1995;29:139-47.

[37] Daamen WF, van Moerkerk HT, Hafmans T, Buttafoco L, Poot AA, Veerkamp JH, et al. Preparation and evaluation of molecularly-defined collagen-elastin-glycosaminoglycan scaffolds for tissue engineering. Biomaterials. 2003;24:4001-9.

[38] Wahl DA, Czernuszka JT. Collagen-hydroxyapatite composites for hard tissue repair. Eur Cell Mater. 2006;11:43-56.

[39] Semler EJ, Ranucci CS, Moghe PV. Mechanochemical manipulation of hepatocyte aggregation can selectively induce or repress liver-specific function. Biotechnol Bioeng. 2000;69:359-69.

[40] Nagai N, Yunoki S, Satoh Y, Tajima K, Munekata M. A method of cell-sheet preparation using collagenase digestion of salmon atelocollagen fibrillar gel. J Biosci Bioeng. 2004;98:493-6.

[41] Yao C, Roderfeld M, Rath T, Roeb E, Bernhagen J, Steffens G. The impact of proteinase-induced matrix degradation on the release of VEGF from heparinized collagen matrices. Biomaterials. 2006;27:1608-16.

[42] Isenberg BC, Williams C, Tranquillo RT. Small-diameter artificial arteries engineered in vitro. Circ Res. 2006;98:25-35.

[43] Iwai S, Sawa Y, Ichikawa H, Taketani S, Uchimura E, Chen G, et al. Biodegradable polymer with collagen microsponge serves as a new bioengineered cardiovascular prosthesis. J Thorac Cardiovasc Surg. 2004;128:472-9.

[44] Lepidi S, Grego F, Vindigni V, Zavan B, Tonello C, Deriu GP, et al. Hyaluronan biodegradable scaffold for small-caliber artery grafting: preliminary results in an animal model. Eur J Vasc Endovasc Surg. 2006;32:411-7.

[45] Shum-Tim D, Stock U, Hrkach J, Shinoka T, Lien J, Moses MA, et al. Tissue engineering of autologous aorta using a new biodegradable polymer. Ann Thorac Surg. 1999;68:2298-304; discussion 305.

[46] Wildevuur CR, van der Lei B, Schakenraad JM. Basic aspects of the regeneration of small-calibre neoarteries in biodegradable vascular grafts in rats. Biomaterials. 1987;8:418-22.

[47] Eriksson AC, Whiss PA. Measurement of adhesion of human platelets in plasma to protein surfaces in microplates. J Pharmacol Toxicol Methods. 2005;52:356-65.

[48] Polanowska-Grabowska R, Simon CG, Jr., Gear AR. Platelet adhesion to collagen type I, collagen type IV, von Willebrand factor, fibronectin, laminin and fibrinogen: rapid kinetics under shear. Thromb Haemost. 1999;81:118-23.

[49] Kulik E, Ikada Y. In vitro platelet adhesion to nonionic and ionic hydrogels with different water contents. J Biomed Mater Res. 1996;30:295-304.

[50] Lee JH, Khang G, Lee JW, Lee HB. Platelet adhesion onto chargeable functional group gradient surfaces. J Biomed Mater Res. 1998;40:180-6.

4

Charge Transport and Electrical Switching in Composite Biopolymers

Gabriel Katana[1] and Wycliffe Kipnusu[2]

[1]Physics Department, Pwani University College, P.O. Box 195 Kilifi,
[2]Institute of Experimental Physics I, University of Leipzig, 04103 Leipzig,
[1]Kenya
[2]Germany

1. Introduction

Polymers are long chain macromolecules made up of many repeating units called monomers. They are found in nature and can also be made synthetically. Natural/bio polymers are considered to be environmental benign materials as opposed to synthetic polymers. Research geared toward producing innocuous products from biopolymers has intensified. Improved understanding of properties of biopolymers allows for the design of new eco-friendly materials that have enhanced physical properties and that make more efficient use of resources. Biopolymers also have the advantage of being biodegradable and biocompatible. They are therefore of interest for application in advanced biomedical materials, for instance tissue engineering, artificial bones or gene therapy (Eduardo et al., 2005). Other possible fields of applications are related to electrical properties, making this class of materials attractive for potential uses in electronic switches, gates, storage devices, biosensors and biological transistors (Finkenstadt & Willett 2004). Plant biopolymers constitute the largest pool of living organic matter most of which can be attributed to four distinct classes of organic compounds; lignin, cellulose, hemicellulose and cuticles. Cuticles are mainly made up of polymethylenic biopolymers which include cutin and suberin. Cutin-containing layers are found on the surfaces of all primary parts of aerial plants, such as stems, petioles, leaves, flower parts, fruits and some seed coats. In addition, cutin may be found on some internal parts of plants such as the juice-sacs of citrus fruits (Heredia, 2003). Composition, structure and biophysical data of plant cuticles have recently been reviewed (Jeffree, 2003; Pollard et al. 2008; and Dom ínguez et al., 2011) and will only be mentioned briefly. The main constituents of cutin are esterified fatty acids hydroxylated and epoxy hydroxylated with chain lengths mostly of 16 and 18 carbon atoms. It also contains some fraction of phenolic and fluvanoid compounds.

2. Charge transfer mechanism in biopolymers

Several biopolymers have well documented properties as organic semiconductors (Eley et al., 1977; Leszek et al., 2002; Radha & Rosen, 2003; Mallick & Sakar, 2000; Lewis & Bowen, 2007; Ashutosh & Singh 2008). DNA-based biopolymer material possesses unique optical and electromagnetic properties, including low and tunable electrical resistivity, ultralow optical and microwave low loss, organic field effect transistors, organic light emitting diodes (LED) (Hagen et al., 2006). Nonlinear optical polymer electro-optic modulators fabricated from biopolymers have demonstrated better performance compared to those made from other materials (Piel et al., 2006).Naturally occurring Guar gum biopolymer chemically modified with polyaniline exhibits electrical conductivity in the range of 1.6×10^{-2} S/cm at room temperature (Ashutosh & Singh 2008). Mallick & Sakar (2000) investigated electrical conductivity of gum arabica found in different species of Acacia babul [*Acacia Arabica*] (Boutelje, 1980) and found that its electrical properties are similar to that of synthetic conducting polymer doped with inorganic salt and are proton conducting in nature.

Charge transport in biomolecular materials takes place mainly through two processes: Super exchange transport and hopping transport. Super exchange is a chain mediated tunneling transport. In this process electrons or holes are indirectly transferred from a donor to acceptor group through an energetically well- isolated bridge, where the bridge orbitals are only utilized as coupling media. In the hopping mechanism, the electron temporarily resides on the bridge for a short time during its passing from one redox center to the other, but in the super-exchange, the conjugated bridge only serves as a medium to pass the electron between the donor and acceptor (Tao et al., 2005). Tunnelling is a process process that decays exponentially with the length of the molecule. A simple tunneling model assumes a finite potential barrier at the metal-insulator interface. It describes free electron flow for a short distance into the sample from the metal contact. At low voltages the charge transfer is described by Simmons relations (DiBenedetto et al., 2009) but at higher applied voltages the tunneling is determined by Fowler-Nordheim process. Superexchange process can either be coherent or non-coherent. Coherent tunneling process is whereby a charge carrier moves from a donor to acceptor fast enough such that there is no dephasing by nuclear motions of the bridge (Weiss et al., 2007). Consequently, charges do not exchange energy with the molecules. However this process does not take place at significantly long distances. Incoherent superexchange on the other hand is a multi-step process in which a localized charge carrier interacts with phonons generated by thermal motion of the molecules (Singh et al., 2010)

As opposed to superexchange, hopping transport in biopolymers is a weakly distance dependent incoherent process. Generally superexchange is a short range transfer of charges in a spatial scale of a few Å while hopping transport takes place over a longer distance greater than 1nm. The exact mechanism of tunneling and hopping is not fully understood but it is known to be influenced by several factors. First, type of charge carriers in biopolymers which can either be holes, electrons or even polarons influences charge transport. Hole transfer is initiated by photo-oxidation of the donor groups attached to the terminus of the molecule whereas electron transfer occurs by chemical reduction of the acceptor group. A direct reduction of the molecule in contact with the metal electrode occurs when the voltage is applied. The interplay between donor acceptor and coupling fluctuation in biological electron transfer has also been observed (Skourtis et al., 2010). Secondly, band structure and hoping sites also influence charge transport. Although there are no band structures in biomolecules, energy gaps exist due to different hybridized electronic states. These energy gaps provide hopping sites through which charges propagate. Finally, conformation and spatial changes for the conducting state may overlap and hence create hopping sites as described by variable range hopping (Shinwari et al., 2010). Variable range hopping mainly describes transport mechanism in solid-state materials, but has also been observed in biopolymers (Mei Li et al., 2010). Similar to superexchange, electron hopping in biopolymers is a multi-step process. According to this mechanism, the overall distance between primary electron donor and final electron acceptor is split into a series of short electron transfer steps. The essential difference is the existence of bridge units (oxidized or reduced species) that function as relays system and the fact the hopping process has a weak distance dependence (Cordes & Giese, 2009).Both of the two processes can take place at the same time and have been observed by several experimental groups. Treadwaya et al. (2002) noted that DNA assemblies of different lengths, sequence, and conformation may allow tunneling, hopping, or some mixture of the two mechanisms to actually dominate.

From measurements that probed changes in oxidized guanine damage yield with response to base perturbations, Armitage et al. (2004) noted that charge transfer through base-base of DNA molecules takes place through hopping via the π-π bond overlap. Tao et al. (2005) reported electron hopping and bridge-assisted superexchange charge transfer between donor and an acceptor groups in peptide systems. The charge and dipole of the peptide play an important role in the electron transfer (Amit et al., 2008). Galoppini & Fox (1996) demonstrated the effect of electric field generated by the helix dipole on electron transfer in Aib-rich α helical peptides and found out that other than the effect from secondary structure (α helix and β sheet), dipole and hydrogen bonding, the solvent also has a marked influence on the study of the electron transfer. Due to complexity of peptides, the importance of individual amino acids in controlling electron transfer is not yet understood in detail.

Similar studies in proteins have concluded that electron transfer can occur across hydrogen bonds and that the rate of such transfer is greatly increased when the electron motions are strongly coupled with those of the protons (Ronald et al., 1981). While studying energy transport in biopolymers, Radha & Rossen (2003) suggested, based on the experimental results, that a soliton in biopolymers is an energy packet (similar to the "conformon" which is the packet of conformational strain on mitochondria) associated with a conformational strain localized in region much shorter than the length of a molecule. It was also noted by the same group that as the soliton (localized curvature) moves on the polymer, it could trap an electron and drag it along. This mechanism may be important in understanding charge transport in biological molecules, where curvatures abound. Studies on charge transport in ethyl cellulose- chloranil systems have also been done, (Khare et al., 2000) where the space charge limited current (SCLC) was found to be the dominant mode of electrical conduction at high field in these systems.

Mechanisms leading to charge conduction in metal-polymer-metal configuration have been the subject of intensive study in the past two decades. Much of these studies have focused on doped and undoped synthetic polymers where the commonly discussed high-field electronic conduction mechanism for various films are Fowler-Nordheim tunneling, Poole-Frenkel (P-F), Richardson-Schottky (R-S) thermionic emissions, space charge limited conduction and variable range hopping. Based on the same sample geometry it is reasoned that the mechanism mentioned above could also contribute to detectable current flow in biopolymers sandwiched between metal electrodes. These mechanisms are discussed hereunder and in section 5, experimental results based on cutin biopolymer are presented and discussed in reference to the charge transport mechanism mentioned above.

2.1 Fowler-Nordheim tunneling

In Fowler-Nordheim tunneling the basic idea is that quantum mechanical tunneling from the adjacent conductor into the insulator limits the current through the structure. Once the carriers have tunneled into the insulator they are free to move within the material.

Determination of the current is based on the Wentzel, Kramers and Brillouin (WKB) approximation from which Eq. (1) is obtained.

$$J_{FN} = \chi_{FN} E^2 \exp\left[-\frac{4}{3} \frac{\sqrt{2m^*}}{q\hbar} \frac{(q\varphi_B)^{3/2}}{E} \right] \qquad (1)$$

where J_{FN} is the current density according to Fowler -Nordheim, χ_{FN} is the Fowler Nordheim constant, E is the electric field, m^* is the effective mass of the tunneling charge, $\hbar$ is a reduced plancks constant, q is the electron charge and ϕ_B is the potential barrier height at the conductor/insulator interface. To check for this current mechanism, experimental I-V characteristics are typically plotted as $\ln(J/E^2)$ vs $1/E$, a so-called Fowler-Nordheim plot. Provided the effective mass of the insulator is known, one can fit the experimental data to a straight line yielding a value for the barrier height.

2.2 Field emission process

Whereas Fowler-Nordheim tunneling implies that carriers are free to move through the insulator, it cannot be the case where defects or traps are present in an insulator. The traps restrict the current flow because of a capture and emission process. The two field emission charge transport process that occur when insulators are sandwiched between metal electrodes are Poole-Frenkel and Schottky emission process. Thermionic (schottky) emission assumes that an electron from the contact can be injected into the dielectric once it has acquired sufficient thermal energy to cross into the maximum potential (resulting from the superposition of the external and the image-charge potential). If the sample has structural defects, the defects act as trapping sites for the electrons. Thermally traped charges will contribute to current density according to Poole-Frenkel emission. They are generally observed in both organic and inorganic semiconducting materials. Poole-Frenkel effect is due to thermal excitation of trapped charges via field assisted lowering of trap depth while Schottky effect is a field lowering of interfacial barrier at the blocking electrode. Expression for Poole-Frenkel and Schottky effects are given in Eq. (2) and (3) respectively.

$$J_{PF} = J_{PFO} \exp[(\beta_{PF} E^{1/2})/kT] \qquad (2)$$

$$J_S = J_{SO} \exp[(\beta_S E^{1/2})/kT] \qquad (3)$$

J_{SO} and J_{PFO} are pre-exponential factors, β_S is the Schottky coefficient, β_{PF} is the Poole-Frenkel coefficient, and E is the electric field. The theoretical values of Schottky and Poole-Frenkel coefficient are related by Eq.(4):

$$\beta_S = \left(e^3 / 4\pi\varepsilon\varepsilon_0 \right) = \beta_{PF} / 2 \qquad (4)$$

where q is electron charge, ε is relative permittivity, ε_0 permitibity in free space

2.3 Space charge limited current

For structures where carriers can easily enter the insulator and freely move through the insulator, the resulting charge flow densities are much higher than predicted by Fowler-Nordheim tunneling and Poole-Frenkel mechanism. The high density of these charged carriers causes a field gradient, which limits the current density and the mechanism is then referred to as space charge limited current. Starting from the basic Gauss's law in one-dimension, assuming that the insulator contains no free carriers if no current flows the expression for the space charge limited current can be obtained as shown in Eq. (5)

$$J = \frac{9\varepsilon\mu V^2}{8d^3} \tag{5}$$

where J is current density, ε is relative permittivity, μ is charge mobility, V is applied voltage and d is electrode spacing. Space charge limited current results from the fact that when the injected carrier concentration exceeds the thermal carrier concentration, the electric field in the sample becomes very non-uniform, and the current no longer follows Ohm's law.

2.4 Variable-range hopping mechanism

Charge conduction in semiconducting polymers is thought to take place by hopping of charge carriers in an energetically disordered landscape of hopping sites (Meisel et al., 2006). The variable-range hopping (VRH) conduction mechanism originally proposed by Mott for amorphous semiconductors (Mott & Davis, 1979) assuming a phonon-assisted hopping process has also been observed in conducting polymers and their composites at low temperature (Ghosh et al., 2001; Luthra et al., 2003; Singh et al., 2003). Bulk conductivity of conducting polymers depends upon several factors, such as the structure, number and nature of charge carriers, and their transport along and between the polymer chains and across the morphological barriers (Long et al., 2003). When the phonon energy is insufficient (low temperature), carriers will tend to hop larger distances in order to locate in sites which are energetically closer than their nearest neighbours. Eq. (6) gives the DC conductivity based on the VRH conduction model.

$$\sigma = \frac{\sigma_0}{\sqrt{T}} \exp\left[-\left(\frac{T_d}{T}\right)^{1/4}\right] \tag{6}$$

where the pre-exponential factor σ_0 is given by Eq.(7)

$$\sigma_0 = \frac{q^2 v_{ph}}{2(8\pi k)^{1/2}} \left[\frac{N_{(EF)}}{\gamma}\right]^{1/2} \tag{7}$$

and q is the electron charge, k is the Boltzmann's constant, v_{ph} is the typical phonon frequency obtained from the Debye temperature ($\approx 10^{13}$ Hz), γ is the decay length of the localized wave function near the Fermi level and $N_{(EF)}$ is the density of states at the Fermi level. The characteristic Mott temperature T_d, as shown in Eq.(8) corresponds to the hopping barrier for charge carriers (also known as the pseudo-activation energy) and measures the degree of disorder present in the system.

$$T_d = 18.11 \frac{\gamma^3}{k N_{(EF)}} \tag{8}$$

Two other Mott parameters, the variable range hopping distance (R_{VRH}) and hopping activation energy (W) are given by Eq. (9) and (10) respectively

$$R_{VRH} = \left[\frac{9}{8\pi\gamma kTN_{(EF)}} \right]^{1/4}$$ (9)

$$W = \frac{3}{4\pi R^3 N_{(EF)}}$$ (10)

3. Electrical switching mechanism in bioploymers

Biomolecules often have sensitive bio-active sites that can change under external stimuli such as temperature, light, electrical signals, PH and chemical/biochemical reactions of their environs. Such switchable biomolecules are of tremendous usefulness in diverse areas including biological, medical and bio-electronic technology. Most research groups in this field are interested in investigating new class of switchable biological systems albeit the field is still at its infancy stage. Chu et al. (2010) reported electro-switchable oligopetides as a function of surface potential. Oligolysine peptides exhibit protonated amino side chain at PH-7 providing the basis of switching "ON" and "OFF" of the biological activity on the surface upon application of negative potential. Switching initiated by PH changes has been observed in other biomolecules and biopolymers (Zimmermann et al., 2006). Biomolecular motors of actomyosin experience rapid and reversible on-off switching by thermal activation (Mihajlovi et al., 2004). The most optimistic approach of integrating photo switchable biomolecules into opto-electronic devices is provided by highly photo sensitive bacteriorhodopsin. This molecule has shown remarkable photo sensitive switching of its electrical properties that mimic conventional Gate transistors (Roy et al., 2010; Pandey, 2006; Qun et al., 2004). Bottom-up approach toward building optical nano-electronic devices is also feasible with the discoveries of switchable photoconductivity in even the smallest structures such as quantum dots of cross-linked ligands (Lilly et al., 2011).

Electrical switching in biomolecules has wider applications in electronic industry. However, a lucid understanding of microscopic switching mechanism in these biological systems is still an outstanding challenge. In most investigations probing electrical switching in organic molecules, an external electrical signal is applied to the sample sandwiched between metal electrodes. Many materials have been reported to show hysteretic impedance switching where a system in its high impedance state (OFF) is switched by a threshold voltage into a low impedance state (ON) and remains in the ON state even with the reversal of applied voltage. This phenomenon is also known as resistive switching. Switching mechanism depend on whether the contribution comes from thermal, electronic, or ionic effects (Waser et al., 2007). Resistive switching is generally dependent on number of mobile charges, their mobility and the average electric field. In the case where the current is highly localized within a small sample area, filamental conduction take place (Scott et al., 2007). This simply involves formation of metallic bridge connecting the two electrodes. Filamental conduction accounts for negative differential resistance (NDR) model which is the basis of many molecular switching processes (Ren et al., 2010). Although this model was originally applied to inorganic materials, it can also explain resistive switching in organic samples (Tseng et al., 2006). The model assumes a trap- controlled channel where tunnelling take place in between chains of metallic islands. Similarly a decrease in electron transport channels and weak coupling between electrodes and the contact molecule causes NDR switching behaviour. Electric field induced switching mechanism is common in literature (Waser et al., 2007). In general terms, built up of space charges trapped within the sample due to presence of defects create a field which is large enough to cause flow of mobile charge carriers. This phenomenon is sometimes called coulomb blockade (Tang et al., 2005). At high electric fields, charges are injected by Fowler-Nordheim tunnelling and subsequently trapped. As a result, electrostatic barrier character of the structure is modified and so is its resistance.

The most insightful switching mechanism in biomolecules is the redox process and the formation of charge transfer complex through donor-acceptor coupling. Aviram et al. (1988) suggested that electron- proton motion within hemiquinones molecules that comprised of catechol and o-quinone, molecules between two contacts switch the molecules to low impedance (ON) state due to the formation of semiquinones fee radicals. When an electron is injected into the molecules from the metal contact, it is gained by an electron acceptor molecule. An electron donor molecule then transfers the electron to the opposite contact thus allowing flow of charge.

The present chapter discusses structural characteristics (by use of Fourier Transform Infra-Red spectroscopy and Atomic Force Microscopy). Electrical conduction in cuticular membranes of Nandi flame (*Spathodea campanulata*, P. Beav) seeds hereafter referred just as cuticles. Fig. 1 shows the cuticle also presented still attached to the seed. The cuticles are thin (about 2 μm), translucent and very light. They are adapted to wind dispersion of the seeds.

4. Structural characterization

4.1 Fourier transform infra Red (FTIR) spectroscopy

Fig.2 shows the Fourier Transform Infra Red (FTIR) spectroscopy of the pristine cuticle. The samples were first annealed at 350K for 12 hrs before measurement. The wide band at 3348 cm^{-1} which has been observed in many other cuticular membranes (Bykov, 2008) is assigned to O-H stretching vibration. It is caused by presence of alcoholic and phenolic hydroxyl groups involved in hydrogen bonds. Methylene is the most repeated structural unit in the cutin biopolyester (Jose, et al., 2004) and these shows up in the spectra band around 2300 cm^{-1}. The band at 2916 cm is assigned to C-H asymmetric and symmetric stretching vibrations of methoxyl groups. Absorption around 1604 cm^{-1} and 1427 cm^{-1} are assigned to the stretching of C=C bonds and the stretching of benzenoid rings. Absorption bands in the range 1300-1150 cm^{-1} are related to asymmetric vibration of C-O-C linkages in ester to esters or phenolic groups. Fig. 3 show the infra red (IR) spectra of cuticle compared with the spectra of other biopolymers.

Fig. 1. Thin and translucent cuticle attached to the Nandi flame seed

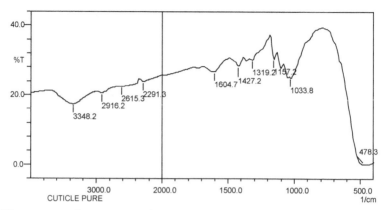

Fig. 2. FT-IR spectrum of pristine cuticle

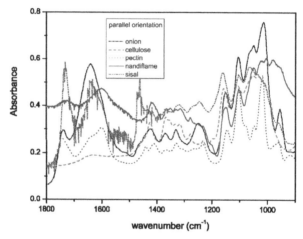

Fig. 3. Comparison of IR spectrum of Nandi flame cuticle with those of other biopolymers

4.2 Atomic force microscopy (AFM)

The AFM scans (Fig. 4) shows that NFSC has a highly oriented surface topography. The AFM permits measurement of the distance variations in the surface of the sample by use of two pairs of cursors. The cursors can be made to scale the surface in nanometer (nm) or micrometer (μm).

The interstitial region between the ridges represented in the dark area in Fig. 4 represents cavities in the membrane with a diameter of about 0.5nm. It is important to note that molecular basal spacing (cavities) in the range of 0.4-0.5nm has also been observed from molecular dynamic simulations of a cutin oligomer (Domınguez et al., 2011) and is attributed to the distance between the methylene groups of oligomeric chain. The presence of such cavities is a typical property of amorphous cross-linked polymers and may be important in explaining the interaction between cutin and endogenic low-molecular weight compounds such as phenolics and flavonoids.

Fig. 4. AFM topographic scans showing surface structure of the cuticle; (a) is a map from the scan on a larger area of about 3×10⁵ square pixels (b) is a scan on a single pixel of about 2.5×10⁶ nm². Doted lines on (a) represents the region shown by (b). Scale bars ~ 0.5nm.Taken from (Kipnusu et al., 2009).

5. Electrical characterization and charge transport mechanism

This section discusses electrical characteristics of the cuticle samples. Current-voltage (I-V) data measured as a function of annealing temperature, irradiation and pooling temperature was used in analyzing the electrical characteristics .Samples were separately annealed and irradiated before electrode coating was done. When annealing, cuticles were placed inside a temperature-controlled furnace, which was fitted inside an electrical shielded cage of a Lindberg/Blue Tube Furnace of model TF55035C. Samples were annealed at various temperatures of 320K, 350K and 400K for a constant period of 12hrs each. Irradiation of the sample was done with He-Ne laser beam of wavelength 632.8nm in a dark room each for a different period of 10minutes, 30minutes, and 60minutes. Electrode coating on the film of pristine, pre-annealed and pre-irradiated samples was done by using quick drying and highly conducting Flash-Dry silver paint obtained from SPI Supplies (USA). A mask of a circular aperture of 0.56 cm diameter was used while coating to ensure uniformity in size of coated surface. Circular aluminum foil of the same diameter was placed on freshly coated surface such that the sample was sandwiched between two aluminum electrodes. These metal-sample-metal sandwiches were left to dry at room temperature for a period of 24hrs to ensure that there was good ohmic contacts between aluminum electrode and the sample. The same Flash-Dry Silver paint was used to connect thin wires onto the aluminum electrodes. When measuring I-V at different temperatures, a sample sandwiched between aluminium electrodes was placed inside the Lindbarg/Blue Tube Furnace and temperature varied in steps of 5K between 350K and 500K at constant electric fields of 0.75V/cm, 1.50V/cm 2.25V/cm, 3.00V/cm, and 3.75V/cm.

Fig. 5(a) shows the I-V characteristics of pristine and annealed samples. These indicate clearly that there was electrical switching and memory effect in the cuticle samples. At certain threshold voltage, V_{th} current rises rapidly by an order of 2. There are two distinct regions for the increasing voltage. At low voltages the log I versus log V plots are approximately linear with a slope of 1; while at higher voltages, above a well-defined threshold voltage V_{th}, the plots are again approximately linear with a slope of 2.04 ± 0.07. These plots therefore show that at low voltages, OFF-state, current follows ohms law but after switching to ON-state at higher voltages, current follows a power law dependence given by $I\alpha V^n$ where $n = 2.04 \pm 0.07$ obtained from linear regression fitting parameters where the standard deviation was shown as 0.03 and coefficient of correlation as 0.0001. This shows that the ON-state region is governed by Space charge limited current (SCLC) controlled by single trapping level, the injecting carrier concentration dominating the thermally generated carriers. During the switching process the current increases appreciably leading to a local increase in temperature (Collines et al., 1993). The current does not follow the same path on decreasing applied electric field hence indicating that the samples exhibit memory switching that is not erased by annealing. The threshold voltage V_{th} for pristine samples is 5.0+0.5 volts. The width of V_{th} or transition voltage during switching from OFF to ON states is about 1.0 V. Inset of Fig 5 (a) shows non-uniform increase of V_{th} with the increase in annealing temperature and tends to attain a plateau at higher annealing temperature. Decrease in magnitude of the negative dielectric anisotropy during annealing is a major reason for the increase in V_{th} for the annealed samples (Katana& Muysoki, 2007). Annealing polymeric films at different temperatures causes structural changes which affects electrical conductivity. Annealing temperature increases grain size in the polymer films causing many changes in the electrical and other properties (Leszek et al., 2002). Threshold voltage V_{th} for pristine cuticles is higher than V_{th} reported for some synthetic polymers; PMMA (1.6V), PS (4.5V), Phthalocyanine (0.3V), 2,6-(2,2-bicyanovinyl) pyride (5.01V), Langmuir-Blodgett (1.0V) (Katana& Muysoki, 2007; Otternbacher et al., 1991; Xue et al., 1996; Sakai et al., 1988).

Fig. 5(b) shows I-V curves for cuticle samples that were pre-irradiated with laser light of wavelength 632.8nm for different duration of time. Just as noted for the annealed samples, I-V curves for irradiated samples shows electrical switching and memory effect with V_{th} that increases with the increase in irradiation time (see inset Fig. 5b). Increasing time of irradiation increases electrical conductivity. The increase in conductivity for irradiated samples can be attributed to dissociation of primary valence bonds into radicals. Dissociation of C-C and C-H bonds leads to degradation and cross linking which improves electrical conductivity (Ashour et al., 2006). Exposure of polymers to ionizing radiation produces charge carriers in terms of electron and holes which may be trapped in the polymer matrix at low temperatures (Feinheils et al., 1971). If the original conductivity is small, then the presence of these carriers produces an observable increase in conductivity of the polymer. Irradiation of polymers results in excitations of its molecules and creation of free electrons and ions that migrate through the polymer network till they are trapped. The electronic and ionic configurations created, cause changes in the electric conductivity. In the study of effect of gamma irradiation on the bovine Achilles tendon (BAT) collagen, Leszek et al. (2002) reported changes in electrical conductivity that is dose dependent. Higher concentration of free radicals generated by irradiation of collagen created charge carriers that increased electrical conductivity.

Fig.5(c) shows I-V curves of the cuticles obtained at different poling/measurement. These curves show that electrical current increases as measurement temperature increase. This is due to thermal excitation of the trapped charges across the potential barrier. The curves also show that forward bias characteristics have two regions which are typical examples of ohmic conduction for voltages below V_{th} (OFF-state) and a space charge limited current (SCLC) for voltages above V_{th} (ON-state). Increase in temperature facilitates diffusion of ions in the space charge polarization. Thermal energy may also aid in overcoming the activation barrier for orientation of polar molecules in the direction of the field. Charge carrier generation and transport in mitochondrial lipoprotein system has been investigated by electrical conductivity and the results show that increase in temperature causes a transition in conductivity where steady state conduction is correlated with chain segmental reorientations of phospholipid moiety below the transition and with an interfacial polarization process above it (Eley et al., 1977).

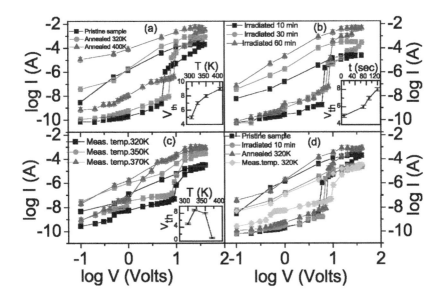

Fig. 5. I-V curves of the pristine cuticles and cuticles treated at different conditions. (a) pristine & samples pre-annealed at 320K and 400K for 12hrs each. (b) pristine & samples pre- irradiated with laser light of wavelength 632.8nm for the duration shown in the legends.(c) pristine samples measured at different temperatures. The insets in (a), (b) and (c) show variation of threshold voltage as a function of the annealing temperature, irradiation time, and measurement temperature respectively. (d) shows combined curves with the conditions shown in the legends.

V_{th} rapidly decreases with the increase in measurement temperature and that switching and memory effect almost disappears at higher temperatures (370K). This is due to the fact that the V_{th} decreases and that the gap between current in the forward bias and reverse bias in the ON-state region almost closed up such that the forward bias current nearly folllows the same path as the reverse bias current which indicates a loss of memory.

It is difficult to draw unambiguous conclusion from Fig.5 (d) because all the three imposed conditions affect conductivity in unique ways and also depends on duration of annealing, irradiation and the measurement temperature. However it is worthy mentioning that at measurement temperature of 320K conductivity is higher than conductivity of presitne samples at low electric field (OFF state). This observation is however reversed at higher electric fields where conductivity of pristine samples is higher.

Switching and memory behavior can be attributed to the fact that external electric field triggers embedded molecules with ridox centers hence creating some traps. Switching mechanism in these systems is by quantum interference of different propagation parts within the molecules which involve permutation of Lowest Unoccupied Molecular Orbitals (LUMO$_{+1}$) and Highest Occupied Molecular Orbitals (HOMO$_{-1}$)-the frontier orbitals. To get electric field induced switching effect, the relative energies of HOMO (localized on the donor group) and LUMO (localized on the acceptor group) must be permuted (Aviram et tal., 1988). This switching model supposes that both the permuting orbitals are initially doubly degenerate resulting to what is referred to as frontier orbitals. In the absence of electric field HOMO$_{-1}$ is localized on the acceptor group and LUMO$_{+1}$ on the donor group. When the external field is applied, electron orbitals are "pulled" towards the acceptor group reducing the HOMO-LUMO gap of frontier orbitals and the switching and hybridization between HOMO$_{-1}$ and LUMO$_{+1}$ takes place. While the strength of the field increases, the HOMO-LUMO gap in the molecular spectrum becomes smaller and the HOMO, HOMO$_{-1}$ and LUMO orbital split into the HOMO-LUMO gap and hence become delocalized. Electrical switching can also in part be explained by formation of quinoid and semiquinone structures from phenolic compounds accompanied by redox reactions. The quinoid form is planar, and is highly conjugated compared to phenyl groups. Changes in bond length and rotation of benzene ring during formation of quinoid structure results in activation barriers which are considered to be the origin of the temperature dependence conductance. Details of this explanation is found in Kipnusu et al. (2009b)

Fowler-Nordheim emission current is given in equation (1). To check for this current mechanism, experimental I-V data for annealed samples of NFSC were analyzed by potting $\ln(J/E^2)$ versus $1/E$. Four plots were made to represent different regime with different levels of measured current (Fig.6). Fowler-Nordheim tunneling mechanism is confirmed by straight lines with negative slopes given by; $4\left(2m^*\right)^{1/2}/3q\hbar(q\varphi_B)^{3/2}$ where m^* is the effective mass of the tunneling charge, q is the electron charge, $\hbar$ is a reduced planck constant, and ϕ_B is the barrier height expressed in eV units. Fig.6 shows that Fowler-Nodheim curves for low and high fields forward bias and low filed reverse bias were quite non-linear or had positive slopes therefore ruling out possibility of Fowler-Nordheim mechanism in these regimes. However in Fig.6 (d) the curves are relatively linear with negative slopes. This is the high fields' regime of the reverse bias where the current increased with decreasing voltage (see Fig.5). Making an assumption that m^* equals the electron rest mass (0.511MeV), and using the slope obtained from linear fits of Fig. 6 (d),the potential barrier height at the Al/cuticle junction is found to be 11.28 eV and 1.13 eV for pure samples and samples annealed at 400K respectively. Vestweber et al. (1994) noted that if the barrier height exceeds 0.3 eV tunnel process prevail with the consequence that high anodic fields are required in order to attain high current densities. It can therefore be concluded that Fowler-Nordheim quantum mechanical tunneling was responsible for the surge of current with decreasing field since the model assumes that once the carriers have tunneled into the insulator they are free to move within the material.

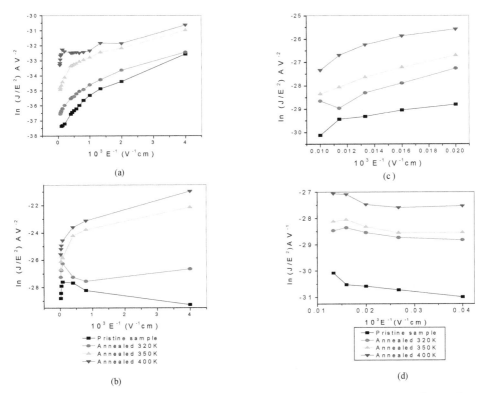

(a)   (c)

(b)   (d)

Fig. 6. Fowler-Nordheim curves: (a) Forward bias at low electric field (b) Reverse bias at low electric field (c) forward bias at high electric field and (d) reverse bias at high electric field.

Fig. 7. shows curves of current density versus square root of electric ($ln\ J\ versus\ E^{1/2}$) in the low field for forward bias regime. These curves neither support conduction mechanism by Poole-Frenkel nor Schottky emissions which predict linear graphs of $ln\ J\ versus\ E^{1/2}$ with positive slopes. Fig.8 shows linear fittings of $ln\ J\ versus\ E^{1/2}$ for forward and reverse bias at high fields (10^4 -10^5 V/cm). The current levels in the reverse bias are higher than forward bias. This behaviour may be interpreted either in terms of Poole-Frenkel effect which is due to thermal excitation of trapped charges via field assisted lowering of trap depth or by Schottky effect which is a field lowering of interfacial barrier at the blocking electrode (Deshmukh et al. 2007). The expressions for these processes are given in Eq.(2) and (3) respectively. Schottky coefficient (β_S) and Poole-Frenkel coefficient (β_{PF}) are related as shown in Eq. (4). Using the value of static dielectric permitivity (ε) of 3.0, (determined from dielectric spectroscopy) theoretical values of β_S and β_{PF} obtained from Eq. (4) are 3.51×10^{-24} J V$^{1/2}$ m$^{1/2}$ and 7.01×10^{-24} J V$^{1/2}$ m$^{1/2}$ respectively. Experimental values of β obtained from the slopes of plots of $lnJ\ versus\ E^{1/2}$ (Fig.8) at different temperatures are listed in Table 1. The standard deviation and coefficient of linear correlation were obtained as 0.34 and 0.005

respectively. The large discrepancy in experimental values of β listed in Table 1 and theoretical values of β_S and β_{PF} leads to a conclusion that current transport mechanism in our samples governing the high field at a temperature range of 320K-370K cannot be explained in terms of Shottky or Poole-Frenkel emission.

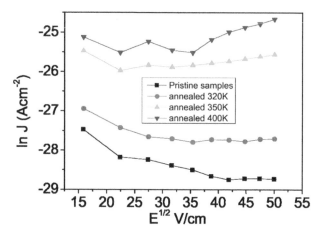

Fig. 7. Semi logarithmic plots of $\ln J$ versus $E^{\frac{1}{2}}$ for the low field of 225-2500 V/cm at a temperature range of 320K-400K

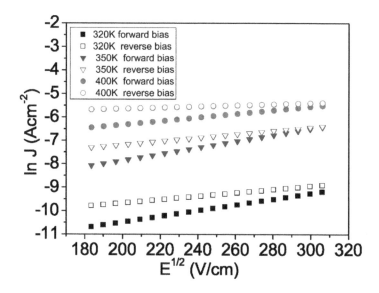

Fig. 8. Semi logarithmic plots of $\ln J$ versus $E^{\frac{1}{2}}$ for the high field of 3.2×10^4- 9.8×10^5 V/cm in forward and reverse biases at temperature range of 320K-400K

To analyze the effect of temperature on conductivity of the samples, Arrhenious curves ($\log \sigma$ vs $10^3/T$) for different polarizing fields (0.75kV/cm, 1.50kV/cm, 2.25kV/cm, 3.00kV/cm, and 3.75kV/cm) were plotted (Fig. 9) using the Arrhenius Eq. (11);

$$\sigma = \sigma_0 \exp\left(\frac{-E_a}{kT}\right) \tag{11}$$

where σ is conductivity, σ_0 the pre-exponential factor, E_a the activation energy, k is Boltzmann's constant, and T is temperature in Kelvin. Conductivity was obtained from the Eq. (12);

$$\sigma = \frac{I}{V}\frac{d}{A} \tag{12}$$

where I is measured current, V is measured Voltage, d is the thickness of the samples ($\approx 4.0 \times 10^{-4}$ cm), and A is the electrode active area (circular electrode of diameter 0.56cm was used). Initial increase in conductivity at low temperature is due to the injection of charge carriers directly from the electrodes. The increase in conductivity at selectively low field is due to the increase in the magnitude of the mean free path of the photon (Sangawar et al., 2006). At high temperatures, the increase in conductivity may be attributed to softening of the polymer which causes the injected charge carrier to move more easily into the volume of the polymer giving rise to a large current. Increased conductivity at higher temperatures could also be due to thermionic emission across the barrier potential.

Temperature (K)	Experimental $\beta \times 10^{-23}$ ($Jm^{1/2}$ $V^{1/2}$) values	
	Forward bias	Reverse bias
320	5.65 ± 0.12	3.17± 0.12
350	6.31 ± 0.14	3.56± 0.12
370	3.44± 0.04	0.93± 0.02

Table 1. Values of β obtained from experimental data

Variable range hopping mechanism predicts linear dependence of $\ln(\sigma T^{1/2})$ versus $T^{-1/4}$ with negative slope (Fig.9, inset b). The Mott parameters- T_d, γ and $N(E_F)$- are determined from equations 6 to 10 using the slope and intercept values of the plots in inset (b) of Fig. 9 and assuming a phonon frequency ((υ_{ph}) of 10^{13} s^{-1}. Other Mott parameters, the hopping distance R and average hopping energy W are determined from Eq. (9) and Eq. (10) respectively. Mott parameters from this calculation are listed in Table 2. Table 3 shows the variations of the Mott parameters with temperature in our samples. It is evident from Table 3 that $\gamma R > 1$ and $W > kT$, which agrees with Mott's condition for variable range hopping. It can therefore be concluded that the main conduction mechanism in NFSC is the variable range hopping.

Mott parameters	Value
T0 (K)	4.58×10^{10}
N(EF) (eV^{-1}cm^{-3})	9.04×10^{19}
α (cm^{-1})	3.0×10^{8}
R (cm)	1.44×10^{-7}
W (eV)	0.89

Table 2. Mott parameters at temperature range of (320-440K)

T (K)	R (cm^{-1})	W (eV)	kT (eV)	γR
300	1.54×10^{-7}	0.72	0.026	41.6
350	1.49×10^{-7}	0.81	0.030	40.1
400	1.44×10^{-7}	0.89	0.034	38.7
450	1.39×10^{-7}	0.98	0.039	37.6

Table 3. Variation of Mott parameters at temperature range of 300-450K

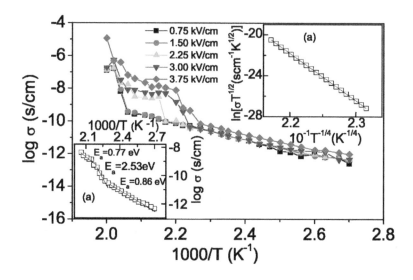

Fig. 9. Arrhenius plots showing variation of σ vs $1/T$ at different electric fields. Inset (a) average Arrhenius plot showing activation energy at low and intermediate temperature ranges. Inset (b) plot of ln ($\sigma T^{1/2}$) versus ($T^{-1/4}$) within a temperature of 350K-440K and average electric field of 2.25Kv/cm.

Degree of localization of the carriers in the trap states is indicated by $\gamma R > 1$ which shows that the charges are highly localized. Table 3 also shows that when the temperature decreases, the average hopping energy W decreases and the average hopping distance R increases, supporting the fact that when the phonon energy is insufficient (low temperature), carriers will tend to hop larger distances in order to locate in sites which are energetically closer than their nearest neighbours.

6. Conclusion

Current-Voltage characteristics of the cuticles, as a function of irradiation, annealing, and temperature, show electrical switching with memory effect. The threshold voltage increases with irradiation time and annealing temperature but it decreases with increase in measurement temperature. The threshold voltage of the annealed and irradiated samples ranges between 6-8 volts. Electrrical conduction in the OFF state follows Ohms' law but changes to space charge limited current after switching to ON state. A combination of Fowler-Nordheim field emission process and redox processes are responsible for electrical switching of the samples. Conduction at low temperatures takes place by variable range hopping mechanism. Since this biomaterial is biodegradable and is also considered to be biocompatible and immunologically inert, it has high potential in biomedical applications. It can be used in making contact eye lenses, scaffolds in tissue engineering, and in controlled release of drugs. Most notably, due to its switching properties, its use in the design of biosensors utilizing ion channels is very feasible.

7. References

Amit, P.;Watson, R; Lund, P; Xing, Y.; Burke, K.; Yufan H,; Borguet, E.; Achim, C; & Waldeck D. (2008). Charge Transfer through Single-Stranded Peptide Nucleic Acid Composed of Thymine Nucleotides. *J.Phys. Chem.*, Vol. 112, pp. 7233-7240

Armitage N., Briman M., & Gruner M. (2004). charge transfer and charge transport on the double helix. *Phys.stat. sol.*(b) Vol.241(1), pp. 69-75

Ashour, H., Saad, M., & Ibrahim, M. (2006). Electrical Conductivity for Irradiated, Grafted Polyethylene and Grafted Polyethylene with Metal Complex. *Egypt. J. Solids,* Vol. 29(2), pp. 351-362.

Ashutosh, T., & Singh, S. (2008). Synthesis and characterization of biopolymer-based electrical conducting graft copolymers. *J. Appl. Polym. Sci,* Vol.108(2), pp.1169-1177.

Aviram A., Joachim C. & Pomerantz M. (1988). Evidence of Switching and Rectification by a Single Molecule Effected with a Scanning Tunneling Microscope. *Chem. Phys. Lett.,* Vol.146(6), pp. 490-495

Boutelje, J. (1980). *Encyclopedia of world timbers, names and technical literature.* (Enc wTimber).p234

Bykov, I. (2008). *Characterization of Natural and Technical Lignins using FTIR Spectroscopy.* Master thesis, Lulea University of Technology, Lulea

Chun L.; Parvez, Y. Iqbal,Y.; Marzena A.; Lashkor, M.;Preece, J. & Mendes,P.(2010). Tuning Specific Biomolecular Interactions Using Electro-Switchable Oligopeptide Surfaces.*Adv. Funct. Mater.,* Vol.20, pp.2657–2663

Collins, R. & Abass, A. (1993). Electrical switching in Au-PbPc-Au thin film sandwich devices. *Thin Solid Films*, Vol.235, pp.22-24

Cordes, M. & Giese B. (2009). Electron transfer in peptide and proteins. Chemical society review, Vol. 38, pp. 892-901

Deshmukh, S., Burghate, D., Akhare, V., Deogaonkar, V., & Deshmukh, P. (2007). Electrical conductivity of polyaniline doped PPV-PMMA polymer blends. *Bull.Matter. Sci.* Vol. 30(1), pp.51-56

DiBenedetto, S.; Facchetti, A.; Rather, M.; & Marks, T. (2009). Molecular self-assembly monolayers and multilayers for organic and unconventional inorganic thin films transistor applications. Advanced materials, Vol.21, pp. 1407-1433

Dom´ınguez, E.; Alejandro, J.; Guerrero, H. & Heredia, A. (2011). The biophysical design of plant cuticles: an overview. *New Phytologist,* vol. 189, pp. 938–949

Eduardo, R., Margarita, D., & Pilar, A. (2005). Functional biopolymer nanocomposites based on layered solids. *J. Mater. Chem.* 15: 3650–3662 DOI: 10.1039/ b505640n

Eley, D., Lockhart, N., & Richardson, C. (1977). Electrical properties and structural transitions in the mitochondrion. *Journal of Bioenergetics and Biomembranes*, Vol.9(5), pp.289-301

Feinleib J., DeNeufville J., Moss S., and Ovshinsky S. (1971). Rapid reversible light-induced crystallization of amorphous semiconductors. *Applied Physics Letters*, Vol.18, pp. 254–257

Finkenstadt, V. & Willett, J. (2007). Preparation and characterization of electroactive biopolymers. *Macromolecular symposia* ISSN 0258-0322.

Galoppini, E, and Fox M. (1996). Role of dipoles in peptide electron transfer. *J.Am. Chem. Soc.* Vol.118, pp. 2299-2300.

Ghosh, M., & Meikap, A., Chattopadhyay ,S. and Chatterjee, S. (2001) Low temperature transport properties of Cl-doped conducting polyaniline. *J. Phys. Chem. Solids,* Vol.62, pp. 475–84

Gmati, F., Arbi, F., Nadra, B., Wadia, D. and Abdellatif, B. (2007). Comparative studies of the structure, morphology and electrical conductivity of polyaniline weakly doped with chlorocarboxylic acids. *J. Phys.: Condens. Matter,* Vol. 19, pp.326203

Hagen, J.; Li, W.; & Steckl, J. (2006). Enhanced emission efficiency in organic light-emitting diodes using deoxyribonucleic acid complex as an electron blocking layer. *Applied Physics Letters,* Vol. 88, pp. 171109(1-3)

Heredia,A. (2003). Biophysical and biochemical characteristics of cutin, a plant barrier biopolymer, *Biochimica et Biophysica Acta* Vol. 1620 pp. 1 – 7

Jose J. B., Matas, A. J., & Heredia, A. (2004). Molecular characterization of the plant biopolyester cutin by AFM and spectroscopic techniques. *Journal of Structural Biology.* Vol. 147, pp.179–184

Katana, G. & Musyoki, A. (2007). Fabrication and performance testing of gas sensors based on organic thin films. *J. Polym. Mater.* Vol.24 (4), pp.387-394

Khare, P., Pandey, R., and Jain P . (2000) .Electrical transport in ethyl cellulose –chloranil system. *Bull.Mater. Sci.* Vol.23 (4), pp.325-330

Kipnusu, W.; Katana, G.;Migwi, M.; Rathore, I.; & Sangoro J.(2009). Charge Transport Mechanism in Thin Cuticles Holding Nandi Flame Seeds, *International Journal of Biomaterials*, doi:10.1155/2009/548406

Kipnusu, W.; Katana, G.;Migwi, M.; Rathore, I.; & Sangoro J.(2009). Electrical Switching in Thin Films of Nandi Flame Seed Cuticles. *International Journal of Polymer Science*, doi:10.1155/2009/830270

Leszek, K., Ewa, M., & Feliks, J. (2002). Changes in the Electrical Conductivity of the - irradiated BAT collagen. *Polish J Med Phys & Eng*. Vol.8(3), pp.157-164

Lewis,T.&Bowen, P. (2007). Electronic Processes in Biopolymer Systems. Electrical Insulation, IEEE Transactions, El- Vol.19 (3),pp. 254-256

Lilly, G.; Whalley,A.; Grunder, S.; Valente,C.; Frederick, M.; Stoddart, J & Weiss, E. (2011). Switchable photoconductivity of quantum dot films using cross-linking ligands with light-sensitive structures. Journal of Materials Chemistry, in press match 2011 DOI: 10.1039/c0jm04397d

Long, Y., Chen, Z., Wang, N., Zhang, Z., & Wan, M. (2003). Resistivity study of polyaniline doped with protonic acids, *Physica* B Vol.325, pp.208–213

Luthra, V., Singh, R., Gupta, S., & Mansingh, A. (2003). Mechanism of dc conduction in polyaniline doped with sulfuric acid *Curr. Appl. Phys*. Vol.3, 219–222.

Mallick, H., & Sakar, A. (2000). An experimental Investigation of electrical conductivity in biopolymers.*Bull.mater.sci*.Vol.23.4, pp. 319-324

Mei Li, Z.; Turyanska, L.; Makarovsky, O.;Amalia Patan, A.; Wenjian Wu, W.; & S Mann, S. (2010). Self-Assembly of Electrically Conducting Biopolymer Thin Films by Cellulose Regeneration in Gold Nanoparticle Aqueous Dispersions. *Chem. Mater.* ,Vol. 22, pp. 2675–2680

Meisel, D., Pasveer, W., Cottaar, I., Tanase, I., Coehoorn, R., Bobbert, P., Blomp,W., De Leeuw D., & Michels, M. (2006). Charge-carrier mobilities in disordered semiconducting polymers: effects of carrier density and electric field. *phys. stat. sol.* (c) Vol.3, (2), pp. 267– 270.

Mihajlovića, G.; Brunet, N.; Trbović, J.; Xiong, P.; Molnár, S.; & Chase,P. (2004).All-electrical switching and control mechanism for actomyosin-powered nanoactuators, *Appl. Phys. Lett.*, Vol. 85, (6), pp. 1060-1062

Mott, N. & Davis, E. (1979). *Electronic Processes in Non-Crystalline Materials* (Oxford: Clarendon) p 157–60

Ottenbacher, D.,Schierbaum, K., & Gopel, W. (1991). Switching effect in metal/ Phthalocyanine/metal sandwich structures. *Journal of molecular electronics* Vol.7(7),pp.9-84

Pandey, P.C. (2006). Bacteriorhodopsin—Novel biomolecule for nano devices. *Analytica Chimica Acta*,Vol. 568, pp. 47–56

Priel, A.; Ramos, A.; Tuszynski, J.; & Horacio F.; Cantiello, H. (2006). A Biopolymer Transistor: Electrical Amplification by Microtubules. *Biophysical Journal* Vol. 90 , pp. 4639–4643

Pollard, M.; Beisson, F.;Li,and , Y.; & Ohlrogge,J. (2008). Building lipid barriers: biosynthesis of cutin and suberin. *Trends in Plant Science*, vol. 13, (5), pp. 236–246

Qun, L.;, Stuart, J.; Birge, R.; Xub, J.; Andrew Stickrath, A.; & Bhattacharya, P. (2004).Photoelectric response of polarization sensitive bacteriorhodopsin films. *Biosensors and Bioelectronics* ,Vol.19, pp.869–874

Radha, B., & Rossen, D. (2003). Nonlinear elastodynamics and energy transport in biopolymers. ArXiv:nlin.PS/ 0304060 vI

Ren,Y.; Chen,.;K.; He, J.; Tang, L.; Pan, A.; Zou, B.; & Zhang, Y. (2010). Mechanically and electrically controlled molecular switch behavior in a compound moleucular device. *Applied Physics Letters*, Vol 97, pp. 103506 (1-3)

Ronald, P., Peter, R., and Albert S. (1981). Water structure-dependence charge transport in proteins. *Proc.Natl. Acad. Sci USA*. Vol.78 (1), pp.261-265

Roy, S.; Prasad, M.; Topolancik, J.; & Vollmer, F . (2010). All-optical switching with bacteriorhodopsin protein coated microcavities and its application to low power computing circuits. *Journal of Applied Physics*,Vol. 107, pp. 053115(1-9)

Sakai, K., Matsuda H., Kawada H., Eguchi K., & Nakagiri T. (1998). Switching and memory phenomena in Langmuir-Blodgett films. *Appl. Phys. Lett.* Vol.53(14), pp.1274-1276

Sangawar, V., Chikhalikar P., Dhokne R., Ubale A., & Meshram S. (2006). Thermally stimulated discharge conductivity in polymer composite thin films. *Bull.Mater.Sci.* 29: 413-416

Singh,M.; Graeme, B.; Allmang, M.; (2010). Polaron hopping in nano-scale poly(dA)-poly(dT) DNA. *Nanoscale Res.Lett.* Vol. 5, pp. 501-504

Singh, R., Kaur A., Yadav K., and Bhattacharya D. (2003). Mechanism of dc conduction in ferric chloride doped poly(3-methyl thiophene) *Curr. Appl. Phys.* 3: 235–238

Shinwari, M.; Deen,M.; Starikov,E and Gianaurelio, C. (2010). Electrical conductance in biological molecules. Advance functional materials, Vol. 20. Pp. 1865-1883

Skourtis, S.; Waldeek, D. & Beratan, D. (2010). Fluctuation in biological and bioinspired electron transfer reaction. *Annual Review Phys. Chem.* March 2010, Vol. 61, pp. 461-485

Tang, W., Shi, H., Xu, G., Ong, B., Popovic, Z., Deng, J., Zhao, J. & Rao, G. (2005), Memory Effect and Negative Differential Resistance by Electrode- Induced Two-Dimensional Single- Electron Tunneling in Molecular and Organic Electronic Devices. *Advanced Materials*, 17: 2307–2311

Tao, L., Erfan, A., & Heinz-Bernhard, K. (2005). Peptide electron transfer. *Chem..Eur.J.*, Vol.11, pp.5186-5194.

Treadway , C.; Hill, M.G . & Barton, J.K . (2002). Charge transport through a molecular π-stack: double helical DNA.*Chemical Physics*, Vol. 281, pp.409–428

Tseng,R.; Ouyang, J.; Chu, C.; Huang, J.; & Yanga, Y.(2006). Nanoparticle-induced negative differential resistance and memory effect in polymer bistable light-emitting device. *Applied Physics Letters*, Vol. 88, pp. 123506 (1-3)

Waser, R. & Aono, M. (2007). Nanoionics-based resistive switching memories, *Nature materials*, Vol. 6, pp. 833-840

Weiss E.; Kriebel,J.; Maria-Anita Rampi, M.; & Whitesides, G. (2007). The study of charge transport through organic thin films: mechanism, tools and applications. *Philosophical Transaction of the Royal Society A*. Vol, 365 pp. 1509-1537

Xue, Z., Ouyang, M.,Wang, K., Zhang, H., and Huang, C. (1996). Electrical switching and memory phenomena in the Ag-BDCP thin film. *Thin Solid Films*, Vol.288, pp. 296-299

Zimmermann, R.;Osaki, T.;ller, T.; Gauglitz, G.; Dukhin, S. & Werner, C. (2006).Electrostatic Switching of Biopolymer Layers. Insights from Combined Electrokinetics and Reflectometric Interference. *Anal. Chem.*, Vol. 78, pp. 5851-5857

5

In Silico Study of Hydroxyapatite and Bioglass®: How Computational Science Sheds Light on Biomaterials

Marta Corno, Fabio Chiatti, Alfonso Pedone and Piero Ugliengo
Dipartimento di Chimica I.F.M. and NIS, Università di Torino, Torino
Dipartimento di Chimica, Università di Modena & Reggio Emilia, Modena
Italy

1. Introduction

Hydroxyapatite and Bioglass® are two well-known biomaterials, belonging to the vast class of ceramic supplies, both highly biocompatible and widely applied in the biomedical field. In spite of a huge research regarding engineering applications of both inorganic materials, still many aspects of their tissue integration mechanism have not been completely cleared at a molecular level. Thus, *in silico* studies play a fundamental role in the prediction and analysis of the main interactions occurring at the surface of these biomaterials in contact with the biological fluid when incorporated in the living tissue (prevalently bones or teeth).

Hydroxyapatite [HA, $Ca_{10}(PO_4)_6(OH)_2$] owes its relevance and use as a biomaterial since it constitutes the majority of the mineral phase of bones and tooth enamel in mammalians (Young & Brown, 1982). For sake of completeness, we mention that hydroxyapatite is also studied as an environmental adsorbent of metals and a catalyst (Matsumura & Moffat, 1996; Toulhat et al., 1996). One of the first applications of HA in biomedicine dates back to 1969, when Levitt *et al.* hot-pressed it in powders for biological experimentations (Levitt et al., 1969). From then on, several commercial forms of HA have appeared on the market. The material has also been utilized for preparing apatitic bioceramic, due to its bioresorbability which can be modulated changing the degree of cristallinity. There are so many examples of applications, from Mg^{2+}-substituted hydroxyapatite (Roveri & Palazzo, 2006) to the synthesis of porous hydroxyapatite materials by colloidal processing (Tadic et al., 2004), starch consolidation (Rodriguez-Lorenzo et al., 2002), gel casting (Padilla et al., 2002) and more. Furthermore, recent applications follow a biologically inspired criterion to combine HA to a collagen matrix aiming at the improvement of mechanical properties and bioactivity (Wahl et al., 2007). However, a complete review of all the practical as well as hypothetical uses of HA in the biomaterial area is outside the scope of this Chapter.

Inside the bone, a highly hierarchical collagen-mineral composite, hydroxyapatite is in the form of nano-sized mineral platelets (Currey, 1998; Fratzl et al., 2004; Weiner & Wagner, 1998) and contains carbonate ions for the 4-8 weight % (Roveri & Palazzo, 2006). In section 2.1 of this Chapter, two aspects of defects which can be encountered in a synthetic or natural HA sample will be presented. The first aspect deals with non-stoichiometric surfaces and their adsorptive behavior towards simple molecules (water and carbon monoxide). The second concerns the inclusion of carbonate ions in the pure HA bulk structure to simulate the apatite bone tissue. These examples of applying sophisticated computational techniques to the investigation of defects in HA represent a very recent progress achieved in our laboratory inside this biomedical research area, which has being carried out since 2003. For the interested reader, a summary of the last years work on simulation of HA in our research group has been recently published (Corno et al., 2010).

As for bioactive glasses, the first synthesis was performed in 1969-71 by Larry Hench in Florida (Hench et al., 1971). He had synthesised a silicate-based material containing calcium and phosphate and had implanted this composition in rats' femurs (Hench, 2006). The result was a complete integration of the inorganic material with the damaged bone. This very first composition was called Bioglass® 45S5 ($45SiO_2$ - $24.5Na_2O$ - $24.5CaO$ - $6P_2O_5$ in wt. % or 46.1 SiO_2, 24.4 Na_2O, 26.9 CaO and 2.6 P_2O_5 in mol %) and has been introduced in clinical use since 1985. The interest has been then to investigate the steps of the bioactivity mechanism leading to the formation of a strong bond between the material and the biological tissue. The most renowned hypothesis is the so-called Hench mechanism and its crucial step resides in the growth of a thin amorphous layer of hydroxy-carbonated apatite (HCA) (Hench, 1998; Hench & Andersson, 1993; Hench et al., 1971). Indeed, on that layer biological growth factors are adsorbed and desorbed to promote the process of stem cells differentiation. Moreover, before the growth of HCA, several other chemical reactions occur, dealing particularly with the exchange of sodium and calcium ions present in the Bioglass® with protons derived from the biological fluid. The influence of the chemical components of the inorganic material on its bioactivity has recently been object of scientific research and discussion. For instance, additives such as fluorine (Christie et al., 2011; Lusvardi et al., 2008a), boron, magnesium (A. Pedone et al., 2008) and zinc (Aina et al., 2011) were considered in a number of systematic studies. In section 2.2 of this Chapter, the role of phosphate concentration inside models with the 45S5 composition will be highlighted, since these changes in content can affect the crucial mechanism of gene activation and modify the local environment of the silicon framework and of Na and Ca sites, as well as the dissolution rate of silica (O'Donnell et al., 2009).

The joint use of experimental and theoretical techniques nowadays has reached a very large diffusion due to the completeness of the derived information. Particularly, in the biological or biochemical field, computational methods are essential to the investigation of interfacial mechanisms at a molecular level. Moreover, very often the interplay between experimental and calculated data allows researchers to improve both methodologies. A huge amount of examples could be reported, but for sake of brevity here we limit to our own experience of collaboration with a number of experimentalists. Indeed, in our research papers, dealing either with HA or with bioactive glasses, there is always a detailed comparison with measured data, for instance by means of NMR (Pedone et al., 2010), IR and Raman spectroscopy and of adsorption microcalorimetry (Corno et al., 2009; Corno et al., 2008). In our computational studies, we refer to quantum-mechanical techniques, which are very accurate but also quite heavy as for the need of resources. Usually, high parallel computing systems are required to run the simulations and we have successfully used the supercomputers of several HPC centers, such as the Barcelona Supercomputing Centre (Spain) or the CINECA Supercomputing Center (Italy).

The most used theoretical framework in the last decade's literature is the Density Functional Theory, which grants a good compromise in terms of accuracy of the representation and computational time. Our calculations are performed either with the pure GGA PBE functional (Perdew et al., 1996) or with the hybrid B3LYP (Becke, 1993), both well-known functionals. Two *ab initio* approaches are possible within DFT and they differ for the type of basis set functions. Indeed, a localized Gaussian basis set can be considered, as in the present case, or a plane waves one, also extremely diffuse. An *excursus* of advantages and disadvantages of these approaches is not useful in this context and will be omitted, by focusing exclusively on Gaussian type functions.

2. Hydroxyapatite and Bioglass® as computational case study

All the calculations mentioned in this Chapter have been performed using the CRYSTAL code in its latest release (Dovesi et al., 2005a; Dovesi et al., 2005b; Dovesi et al., 2009). This periodic quantum-mechanical software has been developed by the Theoretical Chemistry group of the University of Turin (Italy) together with the Daresbury Laboratory (UK) since 1988. CRYSTAL is capable of computing systems with every dimensionality, from molecules to real infinite crystals and it supports massive parallel calculations. This code uses local Gaussian basis sets and can deal with many electronic structure methods, from Hartree-Fock to Kohn-Sham Hamiltonians. Structural, electrostatic and vibrational properties of the studied materials have been characterized with the program. Another crucial aspect in modeling is the graphical visualization and representation of structures. For all the images displayed in this Chapter, MOLDRAW (Ugliengo et al., 1993), J-ICE (Canepa et al., 2011b) and VMD (Humphrey et al., 1996) programs were used. Further more precise computational details can be read in a number of our recent papers on both HA (Corno et al., 2009; Corno et al., 2006; Corno et al., 2007; Corno et al., 2010) and bioactive glasses (Corno & Pedone, 2009; Corno et al., 2008).

2.1 Defects in hydroxyapatite bulk and surfaces

Hydroxyapatite (HA) is a mineral which occurs in nature in two polymorphs, a monoclinic form, thermodynamically stable at low temperatures, and an hexagonal form, which can be easily stabilized by substitution of the OH⁻ ions (Suda et al., 1995). These ions are aligned along the c axis (the [001] direction), as highlighted in Fig. 1. The single crystal structure of the hexagonal form of HA is characterized by the $P6_3/m$ space group. The mirror plane, perpendicular to the [001] direction, is compatible with the column of OH⁻ ions because of an intrinsic static disorder of these ions, which can point, with no preference, in one of the two opposite directions ([001] or [00-1]). The result is a fractional occupation of the sites in the solved crystallographic structure (50% probability for each direction). As *ab initio* simulation cannot take into account the structural disorder, we reduced the symmetry to $P6_3$, removing the mirror plane and fixing the directions of the OH⁻ ions. In the most stable configuration found, both the OH⁻ ions point in the same direction, as reported in Fig. 1. The oxygen atom of the OH⁻ ion is close to three Ca ions, which form an equilateral triangle in the *ab* plane. Moreover, there are six phosphate ions inside the crystallographic cell, all symmetry equivalent.

The bulk structure of crystalline HA, fully characterized in the literature (Corno et al., 2006), has been considered as a starting point to model the surfaces which are experimentally found to be the most important: (001) in terms of reactivity, and (010) in terms of exposure in the crystal habit (Wierzbicki & Cheung, 2000). Those surfaces have already been fully characterized at an *ab initio* level, and all the structural, geometrical and electronic properties have been predicted and compared with the experimental values (Corno et al., 2007). More recently, these modeled surfaces were also employed to study the adsorption of different typologies of molecules that span from water (Corno et al., 2009) to organic acids (Canepa et al., 2011a), amino-acids (Rimola et al., 2008) and small peptides (Corno et al., 2010; Rimola et al., 2009), providing results comparable with the experimental heats of adsorption, when available.

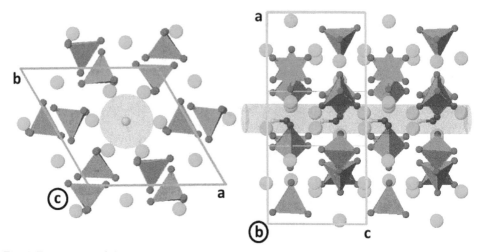

Fig. 1. Two views of the unit cell of the HA bulk structure: calcium in light green, oxygen red, hydrogen light grey and phosphate ions as orange tetrahedra. In the blue cylinders, the column of OH⁻ ions can be observed. Unit cell borders are drawn in light orange.

Nonetheless, the study of the possible terminations along the [010] direction is not yet complete, because two new kinds of (010) surfaces have been discovered experimentally. These surfaces are non-stoichiometric, because their composition is different from the bulk.

2.1.1 The (010) non-stoichiometric surfaces
In 2002, a HRTEM study highlighted three possible terminations for the (010) surface. As a matter of fact, HA shows an alternation of two layers, $Ca_3(PO_4)_2$ (A-type) and $Ca_4(PO_4)_2(OH)_2$ (B-type), which can be interrupted in three different ways (Sato et al., 2002). As the sequence is ...-A-A-B-A-A-B-A-A-B-..., the periodicity can be truncated by exposing as last layers ...-A-B-A or ...-A-A-B or ...-B-A-A, as shown in Fig. 2. As already done in previous work with other surfaces, we investigated these three possible structures with a slab approach. From the bulk, we extracted a piece of matter along the desired direction, in this case the [010], generating two faces which are exposed to vacuum. These faces ought to be the same to remove any possible dipole moment across the slab, due to geometrical dissimilarities. This necessity may bring the loss of stoichiometricity, in terms of the possibility of obtaining the bulk by replicating the slab along the non-periodic direction: neither B-A-A-B-A-A-B nor A-A-B-A-A-B-A-A can be replicated to regenerate the bulk. Only the stoichiometric surface, characterized by a Ca/P ratio of 1.67, maintains the bulk sequence A-B-A-A-B-A. The slab structure terminating in –A-A-B is called *Ca-rich* (010)

surface as its Ca/P ratio increases to 1.71. The last one is called *P-rich* (010) surface, with a Ca/P ratio of 1.62.

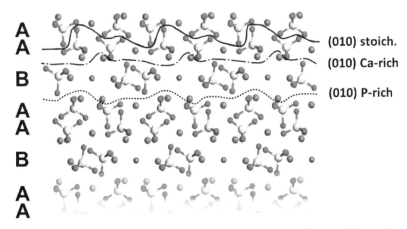

Fig. 2. The three possible terminations along the [010] direction of HA. Colour coding: calcium cyan, oxygen red, hydrogen grey and phosphorous yellow.

Clearly, due to the non-stoichiometric nature of these new surfaces, E_{surf} cannot be defined using the classical formula:

$$E_{surf} = (E_{slab} - n \cdot E_{bulk}) / (2 \cdot A) \tag{1}$$

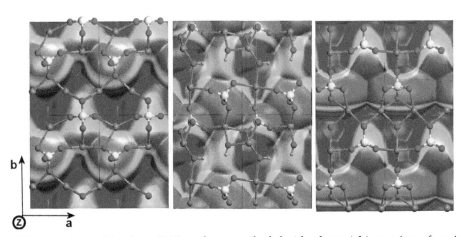

Fig. 3. Upper views of the three (010) surfaces: on the left side, the stoichiometric surface, in the middle, the Ca-rich surface and, on the right side, the P-rich surface. Each structure is superimposed on an isodensity surface colorcoded with the electrostatic potential (blue is most positive, red is most negative). The maps have been generated with the VMD program (Humphrey, Dalke & Schulten, et al., 1996). Calcium represented in cyan, oxygen in red, hydrogen in grey and phosphorous in yellow.

where A is the surface area (doubled, because two faces are generated), E_{bulk} and E_{slab} are the calculated energies of the related models and n is an integer number, required to match the chemical potentials of the two systems. Astala and Stott adopted a clever and rather involved scheme to evaluate the phase existence conditions for the two non-stoichiometric surfaces, showing that the region of their stability is outside the stability window defined by bulk HA, $Ca(OH)_2$ and β-$Ca_3(PO_4)_2$ (Astala & Stott, 2008).

The typology of the exposed layers differentiates the (010) stoichiometric surface from the non-stoichiometric ones. Indeed, the (010) Ca-rich surface exposes the OH⁻ ions belonging to the B layer, while both the stoichiometric and the P-rich surfaces contain the OH ions in an inner layer (see the structures displayed in Fig. 3 for details).

	Stoichiometric	Ca-rich	P-rich
a (Å)	6.98	6.93	7.00
b (Å)	9.28	9.27	9.33
γ (°)	89.87	90.01	90.02
Area (Å²)	64.83	64.24	65.24
* Thickness (Å)	13.28	20.05	20.23
<Ca-O> (Å)	2.41	2.39	2.40
<P-O> (Å)	1.55	1.55	1.55
<O-H> (Å)	0.97	0.97	0.97
<OPO> (°)	109.5	109.4	109.4
Band Gap (eV)	6.21	6.80	6.65
Dipole (Debye)	$3.5 \cdot 10^{-3}$	$4.0 \cdot 10^{-3}$	$1.4 \cdot 10^{-2}$

* The slab thickness is the perpendicular distance between the most exposed Ca ions on the upper and lower faces of the slab.

Table 1. Most important geometrical parameters of the HA (010) surfaces. The values in <...> correspond to the arithmetical averages of the considered feature.

In Fig. 3, the most exposed layers of the three terminations are reported, superimposed on the isodensity surface colorcoded with the electrostatic potential. The isodensity value is fixed to 10^{-6} electrons and the electrostatic potential spans from -0.02 a.u. (red) to +0.02 a.u. (blue). Positive values of the potential are visible in correspondence of the Ca ions while negative potential zones are mostly located upon superficial phosphate ions.

In Table 1, a geometrical analysis of the three surfaces is reported. The cell parameters are slightly different between the three cases, but the rectangular shape is mostly maintained. The slab thickness is not comparable, because of a different number of layers for each surface model. The interatomic distances and angles are, however, very similar.

Two important intensive parameters can classify the stability of a slab structure, the electronic band gap and the total dipole moment across the slab: each surface has a dipole moment close to zero and a band gap typical of electrical insulators. The three surfaces are, then, stable and can be adopted as a substrate to study the adsorption of molecules.

2.1.2 Adsorption of H_2O and CO upon the most exposed Ca ions of the (010) surfaces

The chemical reactivity of the most exposed cations of a surface is experimentally studied with the IR technique by monitoring the perturbation of the vibrational frequency of the probe molecule, which is compared to the value for the free molecule. Of the possible probe molecules, those commonly used are carbon monoxide and water, because of their selectivity towards cations, the ease of interpretation of their spectra, and the high sensitivity of IR instrumentation in the region where their vibrational frequencies fall (medium IR). The (010) stoichiometric surface has already been fully characterized by Corno *et al.* in relation to the adsorption of these molecules (Corno et al., 2009), whereas the non-stoichiometric surfaces are the subject of this work. Calculations provide binding energies and vibrational frequencies of the probe molecules which can be used as a future reference to be compared with experimental measurements. In Fig. 4, the optimized structures of the adducts are reported for water and CO.

The adsorptions of water upon the most exposed cations of the non-stoichiometric surfaces are characterized by BE values of 131 and 125 kJ/mol (BSSE ≈ 35 %), for the Ca-rich and P-rich surfaces, respectively. These values also take into account the formation of two hydrogen bonds between the water molecule and exposed anions, either phosphate only, or also hydroxyl anions in the case of the (010) Ca-rich surface. If the thermal and vibrational contributions are taken into account, these values decrease to 117 and 110 kJ/mol, respectively, allowing a consistent comparison with the experimental differential heat of adsorption of water of 110 kJ/mol (Corno et al., 2009). The latter value has been obtained from experimental micro-calorimetric studies of water adsorption on microcrystalline HA, for a loading of water comparable with our models.

The adsorption of CO upon the most exposed Ca ions gives results highly representative of the strength of the cationic site as no hydrogen bond can be formed: the BE values upon the Ca-rich and P-rich surfaces are, respectively, 38 and 40 kJ/mol (BSSE ≈ 20 %), showing an almost equivalence for the two Ca ions for the two surfaces.

When the vibrational features are considered, a crucial point is the comparison between the frequencies of the free and the adsorbate molecule. While the stretching mode of CO is easily identified in both cases, the modes of the adsorbed water molecule are no longer easily referable to those of the free molecule. The free water molecule is characterized by two stretching modes (the symmetric and, at higher frequencies, the anti-symmetric) and one bending mode. When the molecule interacts with the surface, these two kinds of stretching modes are no longer classifiable on a symmetry ground. The hydrogen bonds which are formed with the most exposed anions of the surface cause the loss of the C_{2v} symmetry of the molecule: each OH bond oscillates independently. As in previous work, we decided to compare the lower OH stretching frequency of the adsorbed molecule to the symmetric mode and *vice versa* (Corno et al., 2009).

With these remarks, the calculated symmetric stretching shifts are -761 and -329 cm⁻¹ for the (010) Ca-rich and (010) P-rich surfaces, respectively, while the anti-symmetric shifts are -310 and -328 cm⁻¹. The shifts of the bending frequencies are 113 and 88 cm⁻¹. When a hydrogen bond pattern occurs, the stretching mode shifts are negative due to the electronic density transfer and the resulting decrease of the bond strength. Instead, the bending frequencies increase because of the restraint caused by the hydrogen bond itself. The larger shift of the symmetric stretching of the H_2O adsorbed on the (010) Ca-rich surface, in combination with a larger bending shift, indicates the formation of a very strong hydrogen bond with an exposed anion of the (010) Ca-rich surface. Experimentally, the shift of the stretching is -400 cm⁻¹ while the bending shift is 40 cm⁻¹. The calculated values are, then, in agreement with the experimental ones as long as the signs and the trends are considered (Bertinetti et al., 2007).

(010) Ca-rich	(010) P-rich

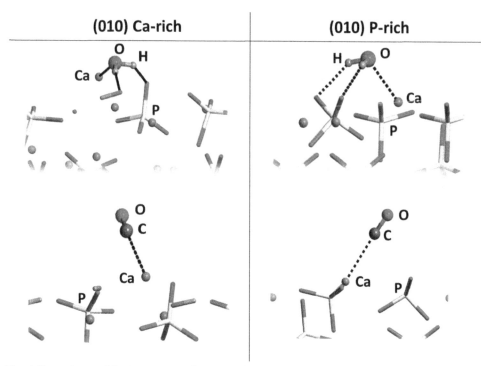

Fig. 4. Best views of the interactions between the most exposed ions of the (010) Ca-rich and (010) P-rich surfaces and the two probe molecules, water and carbon monoxide. The interactions with Ca ions and the hydrogen bonds are represented with dotted lines.

The stretching frequency shifts of the CO molecule are 40 cm^{-1} for the (010) Ca-rich surface and 41 cm^{-1} for the (010) P-rich surface. These BE and frequency shifts values indicate that the Ca ion can reasonably be considered as the major cause of the activity of HA towards polar molecules free from H-bond interactions.

A comparison with the experimental IR spectrum of CO adsorbed on HA reveals two main components, characterized by an upward shift of the stretching frequency of the adsorbed molecule of about +27 and +41 cm^{-1}, respectively. The highest shift is attributed to a tiny fraction of the most exposed calcium ions, which are those modeled in our studies, so proving a good agreement. On the other hand, the majority of calcium sites contribute to the lowest shift (Bertinetti et al., 2007; Sakhno et al., 2010).

2.1.3 Hydroxyapatite and carbonate ion defects

HA is the main component of the inorganic phase of bones and teeth, but, as many studies have demonstrated, the mineral present in those tissues is neither crystalline nor without defects (Fleet & Liu, 2003; Fleet & Liu, 2004; Fleet & Liu, 2007; Rabone & de Leeuw, 2005; Astala & Stott, 2005; Astala et al., 2006; Rabone & de Leeuw, 2007; de Leeuw et al., 2007). Indeed, it incorporates many other elements, ions or compounds, such as Mg or other alkaline earth metals instead of Ca, and, above all, the carbonate anion. Many studies, both theoretical and experimental, have already been conducted on the role of the carbonate ion on the hydroxyapatite structure, in order to comprehend where and how this ion is located. The results of these studies assert that the CO_3^{2-} can substitute an OH^- ion of the structure, a *type A* defect, or a PO_4^{3-}, a *type B* defect. This hypothesis has been confirmed by IR spectra in which the vibrational mode frequencies of the carbonate are different for each type of substitution.

As the net charge of the carbonate is different from those of the substitutional anions, charge compensation is required.

In the case of the type A defect, the charge excess can be compensated by creation of an ionic vacancy removing another OH ion, also because of the larger steric encumbrance of the carbonate ion with respect to the hydroxyl ion (Peroos et al., 2006).

In the case of the type B defect, the electro-neutrality can be maintained in five different ways (Astala & Stott, 2005):

1. Removal of one OH ion and creation of one Ca vacancy (B1 complex);
2. Removal of one Ca ion for every two carbonate ions (B2 complex);
3. Substitution of one Ca with one hydrogen (B3 complex);
4. Substitution of one Ca with one alkaline ion;
5. OH ion incorporation close to the carbonate ion.

When considering the type A defect, it is first necessary to notice that, in the hydroxyapatite unit cell, only two OH ions are present (Fig. 1). If they are both removed at once but only one carbonate ion is included, the resulting structure is no longer a hydroxyapatite, as it had lost all the hydroxyl ions, and has to be considered a carbonated apatite. In Fig. 5, this stable structure is reported. The similarity between carbonate apatite and calcite structures is clear: the averaged distance <Ca-OC> is 2.35 Å in the carbonate apatite and 2.34 Å in the calcite. From the *ab initio* calculation, it is also possible to predict the enthalpy of formation of the carbonate apatite from the calcium phosphate and the calcite structures, considering the reaction (2):

$$3Ca_3(PO_4)_2 + CaCO_3 \rightarrow Ca_{10}(PO_4)_6(CO_3) \qquad (2)$$

The B3LYP enthalpy of formation is -35 kJ/mol, directly comparable with the value of -32 kJ/mol obtained with the VASP code and the PW91 functional (Rabone & de Leeuw, 2007).

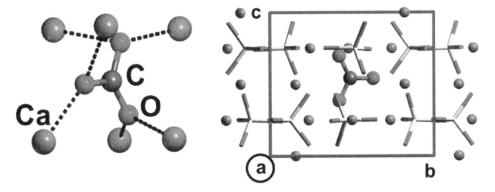

Fig. 5. Carbonated apatite. On the left side, a view of the carbonate ion is reported, highlighting its distances to the closest Ca ions. On the right side, the cell, where both OH are removed and one carbonate is substituted, is reported after full relaxation of the atoms.

As the main interest in the type A defect is the stability of the $OH^--CO_3^{2-}$ substitution as a function of CO_3^{2-} content, we modeled three cases using a 2x2 supercell:

$$(4-x)Ca_{10}(PO_4)_6(OH)_2 + xCa_{10}(PO_4)_6(CO_3) \rightarrow Ca_{40}(PO_4)_{24}(CO_3)_x(OH)_{8-2x} \qquad (3)$$

The most stable case occurs (see Fig. 6A) when three out of four pairs of OH ions (x=3) were substituted with carbonate ions (ΔE_r = -24 kJ/mol). It is notable that the percentage of carbonate related to this minimum energy structure (4.4%) is close to that found in dentine tissue (5.6%) and in tooth enamel (3.5%) (Dorozhkin, 2009).

Among the five typologies of B type defect, we investigated only the substitution of Ca with Na or H. As all the phosphate ions are equivalent by symmetry, the substitution with carbonate is easy. On the contrary, in the cationic substitution to restore the neutrality, there is a choice among ten Ca ions. These are not equivalent, as the symmetry is broken by the carbonate substitution. We selected only four Ca ions, those nearest to the carbonate, because from the experiment it is known that the two defective substituents are close to each other. Indeed, in the four selected cases, the most stable structures are those in which the distance between the substituents is minimum. The different stabilities were obtained by calculating the energy variation (ΔE_r) for the following reactions:

$$Ca_{10}(PO_4)_6(OH)_2 + NaHCO_3 \rightarrow Ca_9Na(PO_4)_5(CO_3)(OH)_2 + CaHPO_4 \qquad (4)$$

$$Ca_{10}(PO_4)_6(OH)_2 + H_2CO_3 \rightarrow Ca_9H(PO_4)_5(CO_3)(OH)_2 + CaHPO_4 \qquad (5)$$

The ΔE_r is 108 kJ/mol for reaction (4) and -94 kJ/mol for reaction (5), indicating that the inclusion of H is much more preferable than Na. The reason relies upon the formation of bulk water between the H and the OH ion of the column. This water molecule stabilizes the structure and its removal requires 136 kJ/mol, because of the occurrence of rather strong H-bonds. The two structures are reported in Fig. 6 (B/Na and B/H).

We also calculated some bulk structures in which both typologies (A and B) of substitutions were present at the same time, in order to better mimic the bone features. Among all the simulated structures, the most favorable situation is the one reported in Fig. 6 (A+B/H): a carbonate ion substitutes a pair of OH ions while another carbonate replaces a phosphate with formation of a water molecule. The hydrogen bonds formed by the water molecule are stronger than those of the B defect model, highlighting that the two defects interact and influence each other. The energy of formation of the mixed A+B/H structure is -752 kJ/mol for the reaction 6.

$$11.5\ Ca_3(PO_4)_2 + 1.5\ CaCO_3 + 3\ Ca(OH)_2 + 0.5\ H_2CO_3 \rightarrow Ca_{39}H(PO_4)_{23}(CO_3)_2(OH)_6 \qquad (6)$$

2.2 Bioglass: the effect of varying phosphorous content

As already described in the Introduction of this Chapter, bioactive glasses are extensively studied as prostheses for bone and tooth replacement and regeneration. In particular, the 45S5 composition has continuously been investigated, not only in its compact form, whose applications are limited to low load-bearing, but also as particulates and powders for bone filler use. Hence, the computational investigations of the variation in composition for the different glass components still represent a very interesting and stimulating task.

In the computational area, the most natural method to simulate glassy materials is usually classical molecular dynamics via the melt-and-quench procedure. Recently, much theoretical research work has been carried out on bioglasses using *ab initio* molecular dynamics, though the transferability of empirical potentials remains a challenging goal, typical of classical techniques (Tilocca, 2010).

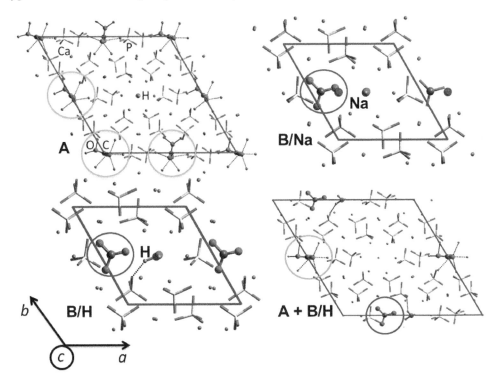

Fig. 6. Top views of the most stable carbonated hydroxyapatite structures. The carbonate can substitute a pair of hydroxyl ions (green circles) or a phosphate ion (blue circles). Top left, type A defect with three out of four pairs of OH ions substituted; top right, type B defect with Na as charge compensator; bottom left, type B with H as charge compensator; bottom right, type A and B, with H as charge compensator of the B type.

2.2.1 Multi-scale strategy for the modelling of 45S5 Bioglass®

The *ab initio* modelling of an amorphous material as the 45S5 Bioglass® has straightway caused a number of challenges. As the internal coordinates are not available from classical structural analysis, we have developed a multi-scale strategy to obtain a feasible amorphous bioglass model similar in composition to the 45S5. This approach has been largely illustrated in previous papers (Corno & Pedone, 2009; Corno et al., 2008; A. Pedone et al., 2008) so that here we only summarize the most important steps.

Firstly, we adopted classical molecular dynamics simulations to model a glassy bulk structure which could be close to the 45S5 composition keeping the size of the unit cell small enough for *ab initio* calculations, *i.e.* 78 atoms. In order to reproduce the correct ratio between SiO_2, P_2O_5, Na_2O and CaO components, the experimental density of 2.72 g/cm^3 has been maintained fixed in a cubic box of 10.10 Å per side. After randomly generating atomic positions, a melt-and-quench procedure was simulated: heating at 6000 K and then equilibrating for 100 ps. Next, continuously cooling from 6000 to 300 K in 1140 ps with a nominal cooling rate of 5 K/ps was performed. The temperature was decreased by 0.01 K every time step using Nose-Hoover thermostat with the time constant parameter for the frictional coefficient set to 0.1 ps (Hoover, 1985). Simulations were carried out in the constant volume NVT ensemble and 100 ps of equilibration at constant volume and 50 ps of data production were run at 300 K. On the derived structures, static energy minimizations were carried out at constant pressure and volume and the most representative model was chosen for *ab initio* calculations.

The final candidate structure was minimized both in terms of internal coordinates and lattice parameters, performing full relaxation runs. In Figure 7a the quantum-mechanical optimized structure of the selected Bioglass model is displayed. In the unit cell, which has become triclinic due to the lattice parameter deformation, two phosphate groups are present: one isolated (orthophosphate) and the other linked to one silicon atom. The structural analysis was followed by the simulation of the IR spectrum. Figure 7b reports the comparison between experimental (Lusvardi et al., 2008b) and computed spectra, which shows a very good agreement between the two spectra. The punctual assignment of each peak in the simulated case has been published in a previous paper (Corno et al., 2008), where the 45S5 model has been compared with an amorphous silica structure, to investigate the role of network modifier cations and phosphate groups in structural and vibrational properties of a pure SiO_2 framework. Hence, the reliability of the chosen multi-scale strategy has been proved.

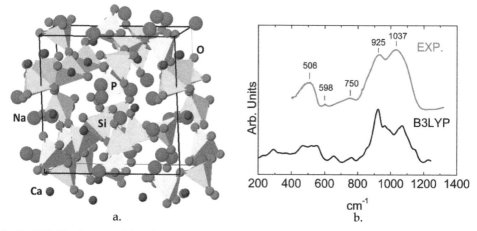

a.   b.

Fig. 7. 45S5 Bioglass model: a. best view of the optimized structure ($Na_{12}Ca_7P_2Si_{13}O_{44}$ composition), colour coding: silicon light blue, oxygen red, sodium pink, calcium dark blue and phosphorous yellow; cell parameters drawn in red for a, in green for b and in blue for c, while cell borders are in black; b. experimental (red line) and B3LYP (black line) IR spectra. (Corno et al., 2008; Lusvardi et al., 2008b)

2.2.2 Simulation of bioactive glasses with different P_2O_5 content

More recently, the influence of P_2O_5 content on the structure of bioactive glass compositions has been object of investigation (O'Donnell et al., 2009; O'Donnell et al., 2008a; O'Donnell et al., 2008b). Indeed, it is well known that structural and compositional features of bioactive glasses are strongly connected to their bioactivity (Clayden et al., 2005; Lin et al., 2005). In particular, it has been demonstrated (Tilocca, 2010) that in compositions less bioactive than the 45S5 the majority of phosphate groups are linked to one or two silicon atoms. Conversely, in bioactive glasses as the 45S5, almost all the phosphate groups are isolated and mobile (Tilocca & Cormack, 2007).

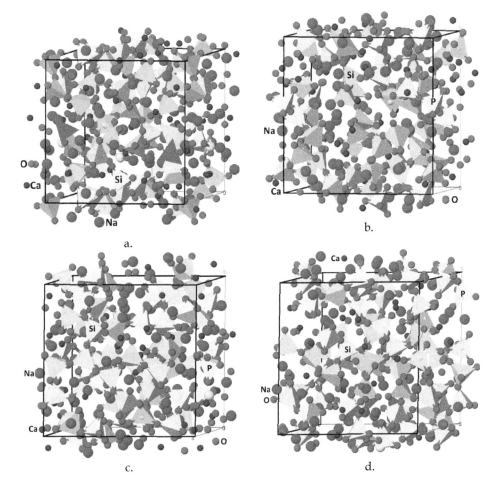

Fig. 8. Best views of the optimized structures of the four studied glass models: a. no phosphorous (P0); b. 2.5% phosphorous (P2.5); c. 5.5% phosphorous (P5.5) and d. 9.5% phosphorous (P9.5). Colour coding: silicon light blue, oxygen red, sodium pink, calcium dark blue and phosphorous yellow. Cell parameters drawn in red for *a*, in green for *b* and in blue for *c*, while cell borders are in black (their values listed in Table 5).

In our research work, we have aimed to correlate the change in phosphorous content with the change in structural and vibrational properties, these latter as a tool to detect the local coordination of the PO_4 group. To this extent, four models of phosphate soda-lime glasses were studied by applying the same melt-and-quench procedure used for the 45S5 Bioglass®. The unit cell size has also been increased from the former 78 atoms to new models containing an average of 250 atoms. The larger size has allowed us to derive models which could be more representative of the amorphous long-range disorder typical of glassy materials.

The main structural features of the four modelled structures, whose correspondent images are displayed in Figure 8., are listed in Table 2, together with their molar composition. The "P0" structure refers to a phosphorous-free soda-lime glass.

Model	SiO_2	CaO	Na_2O	P_2O_5	a	b	c	α	β	γ	Volume
P0	45	24	22	-	14.97	14.23	14.77	91.3	90.7	89.2	3144
P2.5	41	23	20.5	2.5	14.47	14.72	14.69	90.0	91.5	90.9	3128
P5.5	35	23	20.5	5.5	14.68	14.47	15.08	91.4	90.0	87.9	3199
P9.5	27	21	19.5	9.5	14.71	14.78	14.50	92.2	90.2	90.1	3150

Table 2. Molar per cent composition of the four studied models of glasses together with the unit cell parameter values of the optimized structures illustrated in Fig. 6. Lattice parameters expressed in Å, angles in degrees and volumes in Å^3.

A direct comparison of volume values for the four models is not reasonable, since there are a number of tiny differences in molar composition in order to maintain the desired ratios between components as well as the total electroneutrality. Indeed, no linear relationship exists between the increase in $\%P_2O_5$ and volume.

A comparison between the structural and vibrational features of the two models mostly similar in composition to the 45S5 Bioglass® has been carried out, i.e. the unit cell with 78 against that with 248 atoms (P2.5 of Figure 8b). In the smaller structure, as already described and illustrated in Figure 7, two phosphate groups are present: one isolated and the other connected to the silicon framework. In the larger model, five phosphate groups are located inside the unit cell, three of which are isolated, while the others linked to the siliceous network. In terms of Q^n species (a Q^n species is a network-forming ion, like Si or P, bonded to n bridging oxygens), the 60% of the total number of PO_4 groups is represented by Q^0 (orthophosphates), while the remaining 40% is equally divided among Q^1 and Q^2 (see Figure 9b, blue curve). If we analyse the total radial distribution function g(r) for the P2.5 model plotted in Figure 9., it clearly appears by the two peaks that the bond length of the P-NBO bond (NBO stands for non-bridging oxygen) is slightly shorter than that for the P-BO bonds (1.552 compared to 1.616 Å, respectively). Moreover, the P-BO bonds are numerically much less, as visible from the part b. of the same Figure 9.

Considering the Q^n distribution for the other two phosphorous-containing models, namely P5.5 and P9.5, it results: for P5.5 the 73% of the total 11 phosphate groups are isolated while the rest are Q_1 and for the total 19 phosphate groups of the P9.5 model, 37% are isolated, 58% are Q_1 and the 5% Q_2, in other words 7 Q_0, 11 Q_1 and a Q_2. The graph in Figure 9a. schematizes the different distribution for the larger models.

The P-O bond distances, both for bridging and non-bridging oxygens, vary according to the different Q^n species, as reported in Table 3.

As a general comment, P-NBO distances follow the trend: $Q^0 > Q^1 > Q^2$ while for P-BO values in case of Q^1 and Q^2 species there is no definite trend, probably due to the limited number of sites in the considered structures.

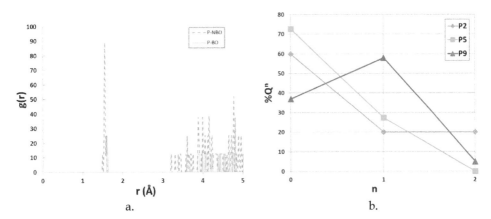

Fig. 9. a. Radial distribution function of the total P-O bonds, distinguishing between non-bridging oxygen (NBO, orange dotted line) and bridging oxygen (BO, blue line); b. percentage distribution of the phosphate groups in terms of Q^n species for the three large models P2.5 (blue), P5.5 (green) and P9.5 (red).

Model	<P-NBO> Q^0	<P-NBO> Q^1	<P-NBO> Q^2	<P-BO> Q^1	<P-BO> Q^2
P2.5	1.556	1.539	1.508	1.597	1.602
P5.5	1.559	1.523	-	1.660	-
P9.5	1.557	1.531	1.515	1.631	1.588

Table 3. Average P-O bond lengths for both bridging and non-bridging oxygen of the three P2.5, P5.5 and P9.5 models are reported in Å.

2.2.3 Effect of P_2O_5 content on the simulated IR spectra

A complete vibrational analysis was outside our computational facilities, due to the size of the simulated bioglass models (250 atoms inside the unit cell, no symmetry). An alternative approach, here adopted, is the so-called "fragment" calculation of frequency. It consists on the selection of the interesting atoms – in this case phosphate groups – to be considered for the calculation of vibrational normal modes. Obviously this approach is an approximation and needs to be first tested. Our test case was the 45S5 structure of Figure 7a, for which the full IR spectrum was available. In particular for the phosphate groups containing Q^1 and Q^2 species, the question was to decide whether to include or not the linked silicon atom with or without its connected oxygen atoms.

Figure 10 reports three simulated IR spectra of the 45S5 model: the full spectrum (black line), the full spectrum including only modes involving phosphate groups (red line) and the partial spectrum where the fragment contains only phosphate groups and silicon atoms linked to the Q^1 species.

In order to dissect the contribution to the full IR spectrum (black spectrum, Fig. 10) of modes involving the displacements of P atoms we rely on the Potential Energy Distribution (PED). All modes involving the P atom in the PED of the full spectrum are included whereas the remaining ones are removed from the spectrum (red spectrum, Fig. 10). The comparison with the spectrum (blue spectrum, Fig. 10) computed by including as a fragment the PO_4 (for fully isolated groups) and $PO_4(SiO_4)_{1,2}$ (for the other cases) shows a good agreement with the black spectrum, so that this methodology has been adopted to compute the spectra for the larger structures with variable P content. In other words, the differences in peaks between the red and the blue spectra of Figure 10 are due to the presence (blue line) or absence (red line) of the extra SiO_4 group.

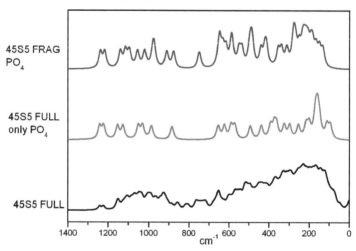

Fig. 10. Simulated IR spectra of the similar 45S5 Bioglass® model in the following sequence from bottom to top: full frequency calculation (black line), full frequency calculation including only PO_4-involved modes (red line) and fragment calculation considering in the fragment PO_4 and SiO_4 which is directly bonded to PO_4. No IR intensities are reported and the chosen band width is of 20 cm⁻¹.

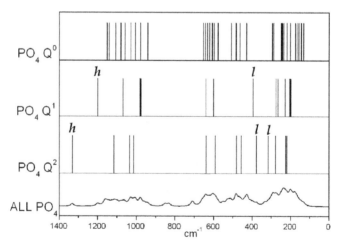

Fig. 11. IR peaks assignment for phosphate groups of the P2.5 model based on the different Q^n species. Label h and l refer to the peculiar bands at high and low frequencies, respectively, that allow us to distinguish the Q^0 species from the Q^1 and Q^2. In case of Q_2, see Fig. 12 for the schematic representation of the associated normal modes.

Figure 11 illustrates a specific example of how to detect the various phosphate groups in terms of Q^n (isolated or connected phosphates) by IR spectroscopy. We reported at the bottom of the graph the P2.5 IR spectrum of the phosphates computed applying the aforementioned procedure (fragment mode). The three bar charts refer each one to the Q^n species present inside the structure and show only the frequencies involving that specific class of phosphate groups (isolated or connected to the network). It is interesting to note the shift of the highest and lowest bands comparing Q^1 and Q^2 cases: the highest frequency – indicated with the label h – corresponds to the stretching of the P=O bond and is shifted to higher values in case of Q^2, with respect to Q^1 and Q^0. On the contrary, in the region of low frequencies, modes involving the Q^2 species are shifted to lower values compared to the Q^1 ones. The OPO bending region (600-700 cm^{-1}) remains almost unaffected.

Figure 12 illustrates the graphical representation of the normal modes displacement for the five stretching and bending modes of the Q^2 species present in the P2.5 glass.

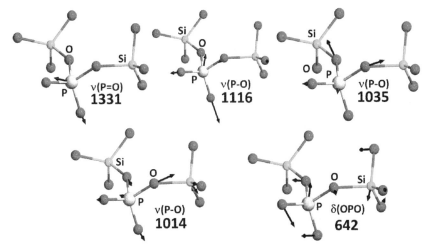

Fig. 12. Schematic representation of the normal mode displacement assigned to the Q^2 species for the P2.5 model (PO$_4$ structural unit connected to 2 SiO$_4$ groups). Colour coding: Si light blue, oxygen red, phosphorous yellow; frequencies expressed in cm^{-1}.

The simulated IR spectra for the phosphate groups of the three phosphorous-containing models have been then compared one to each other and to the phosphorous-free structure P0, as displayed in Figure 13.

The first evident difference between phosphorous-free P0 and the other models is the absence of bands in the spectral region at high frequencies (1200-1400 cm^{-1}). As we have discussed above, that is the typical region of the P=O stretching mode of phosphate groups. This mode is shifted to lower frequencies and the band is broadened when passing from P2.5 to P9.5. Another clear indication of the presence of phosphate groups is the band at about 600-700 cm^{-1}, which corresponds to the O-P-O bending region. The 1100-800 cm^{-1} spectral range, on the contrary, is not easily assigned to phosphate groups inside the bioglass since also Si-O stretching are located in that zone. However, the effect of increasing the P$_2$O$_5$ content inside the unit cell is reflected in a general broadening of the P-O stretching and O-P-O bending modes.

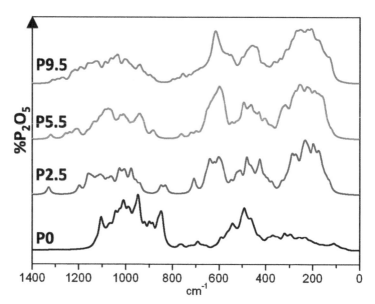

Fig. 13. Simulated IR spectra of the four models of glasses at increasing %P$_2$O$_5$ content.

2.2.4 Future perspectives: Surface modelling

The natural subsequent step in bioactive glass simulation deals with the modeling of surfaces. Indeed, each process of the Hench mechanism that leads to the implant integration typically occurs at the interface between the inorganic material and the biological fluid. Thus, the knowledge of surface properties, such as electrostatic potential and adsorptive behavior towards simple molecules as water, becomes essential in the investigation of bioglasses (Tilocca & Cormack, 2009).

Modeling surfaces is generally not a trivial task, particularly when the bulk material is amorphous. For an amorphous material the identification of a particular face by crystallographic indexes is rather arbitrary as the atomic density is statistically distributed in space in a rather uniform way. A second difficulty is the need of breaking both ionic and covalent bonds during the slab definition which may render the system non-neutral.

In Figure 14, the model of one of the many possible bioglass surfaces extracted from the P2.5 bulk of Figure 8b is presented. The surface was cut out from the bulk as a real 2D slab (infinite in the two dimensions), dangling bonds were saturated with hydrogen atoms and a full optimization run was performed. The resulting surface is very interesting per se, but much more considering its behaviour when hydrated, since water molecules are ubiquitously present in the biological fluids where the material is immersed. In particular, a key issue is to see whether H$_2$O will chemisorb by dissociating on the exposed Na$^+$ and Ca^{2+} cations, a step essential in the Hench mechanism.

In our laboratory a systematic study of the several possible surfaces of the structure with the 45S5 composition is on-going. The application of different methodologies, such as *ab initio* molecular dynamics, already used in the literature (Tilocca, 2010), will be considered to fully characterize the adsorption processes of water and even collagen occurring at the interface between bioactive material and the biological tissue.

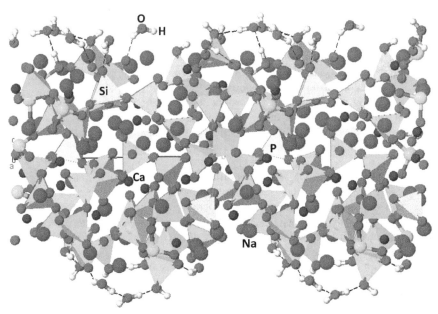

Fig. 14. Surface model of the P2.5 bioglass with adsorbed water molecules at both top and bottom faces. Colour coding: silicon light blue, oxygen red, sodium pink, calcium dark blue, phosphorous yellow, hydrogen bonds black dotted line.

3. Conclusion

In the present Chapter it has been explained how crucial the computational techniques are when applied together with experimentalist measurements in the understanding of biological complex systems and mechanisms dealing with biomaterials for a large number of reasons. Indeed, computational methods are extremely powerfully applied to predict structure formation and crystal growth as well as to describe at a molecular level the real interactions responsible of the attachment of the inorganic biomaterial to the organic tissue. In the investigation of phenomena related to a complex system such as the human body, many approximations are required, so a reductionist approach is employed also in the computational analysis.

In this Chapter, the approach has been explained for two typical biomaterials: hydroxyapatite and Bioglass® 45S5. In particular, for the first material, the aim was to describe the study of its (010) non-stoichiometric surfaces in interaction with water and carbon monoxide. For the latter, the adopted strategy has been analyzed and then a specific example has been reported, dealing with the spectroscopic characterization of computed vibrational features with the increasing amount of phosphorous in a sufficiently large unit cell starting from the well-know 45S5 Bioglass® composition.

The general knowledge gained in recent years through the use of computational techniques such as those described in this chapter is great, but not enough to fully understand the peculiar characteristics of the materials that make up the musculo-skeletal system and to provide appropriate care for important illnesses such as osteoporosis or degenerative and metabolic diseases, benign and malignant tumors and trauma.

To achieve this goal it is fundamental to understand the structure and properties of natural bone at a molecular level and to investigate the chemical-physical interaction between collagen and mineral phase comprising the bone composite.

This can only be achieved through the development and use of multiscale computational methods that combine quantum, classical and continuum approaches enabling to study chemical-physical-biological phenomena on large-scales both in space and time.

Regarding the study of the human bone, we believe that the key issues to be addressed by computational science researchers in the coming years will be the study of the structure and assembly of the collagen protein, the interaction at the molecular level of collagen with the mineral apatite, and finally the structure and mechanical properties of collagen-apatite composite.

As for the study of bioactive glasses, an important line of research that is developing in different research groups located in different nations involves the characterization of the chemical and physical properties and reactivity of the 45S5 Bioglass® surface. However, it will be wise not to neglect the study of the effect of composition on the structure and bioactivity of different systems and the study of the thermodynamics and crystallization kinetics of crystalline phases that are well-known to affect the bioactivity of the glass. Finally, the design of new bioactive glasses will also rely on a deep understanding of their fracture mechanism and the prediction of important properties such as brittleness and toughness, which determine the final use of glass.

4. Acknowledgment

The authors would like to thank the Distributed European Infrastructure for Supercomputing Applications (DEISA) for allowing of computational resources under the Extreme Computing Initiative (BIOGLASS Project). CINECA computing centre is also kindly acknowledged.

Colleagues V. Bolis (Dip. DiSCAFF, University of Eastern Piedmont) and G. Martra (Dip. Chimica IFM, University of Torino) are acknowledged for fruitful discussion and for providing the HA samples, synthesized and kindly supplied by ISTEC-CNR (Faenza, Italy).

R. Dovesi, B. Civalleri and the CRYSTAL team (Dip. Chimica IFM, University of Torino) are thanked for discussion and continuous support with the use of the code.

Part of the results on carbonated HA has been obtained by G. Ulian during his Master Degree Thesis, entitled "DFT study of carbonated defect in hydroxyapatite", 2010, University of Torino.

5. References

Aina, V., Bonino, F., Morterra, C., Miola, M., Bianchi, C. L., Malavasi, G., Marchetti, M. & Bolis, V. (2011) Influence of the Chemical Composition on Nature and Activity of the Surface Layer of Zn-Substituted Sol−Gel (Bioactive) Glasses. *The Journal of Physical Chemistry C*, Vol.115, No.5, pp.2196-2210, ISSN 1932-7447.

Astala, R. & Stott, M. J. (2005) First Principles Investigation of Mineral Component of Bone: CO_3 Substitutions in Hydroxyapatite. *Chemistry of Materials*, Vol.17, No.16, pp.4125-4133, ISSN 1520-5002.

Astala, R. & Stott, M. J. (2008) First-principles study of hydroxyapatite surfaces and water adsorption. *Physical Review B*, Vol.78, No.7, pp.075427, ISSN 1550-235X.

Astala, R., Stott, M. J. & Calderin, L. (2006) Ab Initio Simulation of Si-Doped Hydroxyapatite. *Chemistry of Materials*, Vol.18, No.2, pp.413-422, ISSN 1520-5002.

Becke, A. D. (1993) Density-functional thermochemistry. III. The role of exact exchange. *The Journal of Chemical Physics*, Vol.98, No.7, pp.5648-5652, ISSN 0021-9606.

Bertinetti, L., Tampieri, A., Landi, E., Ducati, C., Midgley, P. A., Coluccia, S. & Martra, G. (2007) Surface Structure, Hydration, and Cationic Sites of Nanohydroxyapatite: UHR-TEM, IR and Microgravimetric Studies. *The Journal of Physical Chemistry C*, Vol.111, No.10, pp.4027-4035, ISSN 1932-7447.

Canepa, P., Chiatti, F., Corno, M., Sakhno, Y., Martra, G. & Ugliengo, P. (2011a) Affinity of hydroxyapatite (001) and (010) surfaces to formic and alendronic acids: a quantum-mechanical and infrared study. *Physical Chemistry Chemical Physics*, Vol.13, No.3, pp.1099-1111, ISSN 1463-9084.

Canepa, P., Hanson, R. M., Ugliengo, P. & Alfredsson, M. (2011b) J-ICE: a new Jmol interface for handling and visualizing crystallographic and electronic properties. *Journal of Applied Crystallography*, Vol.44, No.1, pp.225-229, ISSN 0021-8898.

Christie, J. K., Pedone, A., Menziani, M. C. & Tilocca, A. (2011) Fluorine Environment in Bioactive Glasses: Ab Initio Molecular Dynamics Simulations. *The Journal of Physical Chemistry B*, Vol.115, No.9, pp.2038-2045, ISSN 1520-6106.

Clayden, N. J., Pernice, P. & Aronne, A. (2005) Multinuclear NMR study of phosphosilicate gels derived from $POCl_3$ and $Si(OC_2H_5)_4$. *Journal of Non-Crystalline Solids*, Vol.351, No.3, pp.195-202, ISSN 0022-3093.

Corno, M., Busco, C., Bolis, V., Tosoni, S. & Ugliengo, P. (2009) Water Adsorption on the Stoichiometric (001) and (010) Surfaces of Hydroxyapatite: A Periodic B3LYP Study. *Langmuir*, Vol.25, No.4, pp.2188-2198, ISSN 1520-5827.

Corno, M., Busco, C., Civalleri, B. & Ugliengo, P. (2006) Periodic *ab initio* study of structural and vibrational features of hexagonal hydroxyapatite $Ca_{10}(PO_4)_6(OH)_2$. *Physical Chemistry Chemical Physics*, Vol.8, No.21, pp.2464-2472, ISSN 1463-9084.

Corno, M., Orlando, R., Civalleri, B. & Ugliengo, P. (2007) Periodic B3LYP study of hydroxyapatite (001) surface modelled by thin layer slab. *European Journal of Mineralogy*, Vol.19, No.5, pp.757-767, ISSN 0935-1221.

Corno, M. & Pedone, A. (2009) Vibrational features of phospho-silicate glasses: Periodic B3LYP simulations. *Chemical Physics Letters*, Vol.476, No.4-6, pp.218-222, ISSN 0009-2614.

Corno, M., Pedone, A., Dovesi, R. & Ugliengo, P. (2008) B3LYP Simulation of the Full Vibrational Spectrum of 45S5 Bioactive Silicate Glass Compared to v-Silica. *Chemistry of Materials*, Vol.20, No.17, pp.5610-5621, ISSN 1520-5002.

Corno, M., Rimola, A., Bolis, V. & Ugliengo, P. (2010) Hydroxyapatite as a key biomaterial: quantum-mechanical simulation of its surfaces in interaction with biomolecules. *Physical Chemistry Chemical Physics*, Vol.12, No.24, pp.6309-6329, ISSN 1463-9084.

Currey, J. D. (1998) Mechanical properties of vertebrate hard tissues. *Proceedings of the Institution of Mechanical Engineers, Part H: Journal of Engineering in Medicine*, Vol.212, No.6, pp.399-411, ISSN 2041-3033

de Leeuw, N. H., Bowe, J. R. & Rabone, J. A. L. (2007) A computational investigation of stoichiometric and calcium-deficient oxy- and hydroxy-apatites. *Faraday Discussions*, Vol.134, 195-214, ISSN 1364-5498.

Dorozhkin, S. V. (2009) Calcium Orthophosphates in Nature, Biology and Medicine. *Materials*, Vol.2, No.2, pp.399-498, ISSN 1996-1944.

Dovesi, R., Civalleri, B., Orlando, R., Roetti, C. & Saunders, V. R. (2005a). Ab Initio Quantum Simulation in Solid State Chemistry. In *Review in Computational Chemistry, Vol. 21,*

K.B. Lipkowitz, R. Larter & T.R. Cundari (Ed.), pp. 1-127, John Wiley & Sons Inc., ISBN 9780471682394, New York

Dovesi, R., Orlando, R., Civalleri, B., Roetti, C., Saunders, V. R. & Zicovich-Wilson, C. M. (2005b) CRYSTAL: a computational tool for the ab initio study of the electronic properties of crystals. *Zeitschrift fur Kristallographie*, Vol.220, No.5-6, pp.571-573, ISSN 0044-2968.

Dovesi, R., Saunders, V. R., Roetti, C., Orlando, R., Zicovich-Wilson, C. M., Pascale, F., Civalleri, B., Doll, K., Harrison, N. M., Bush, I. J., D'Arco, P. & Llunell, M., CRYSTAL2009 User's Manual, Available from: <http://www.crystal.unito.it>

Fleet, M. E. & Liu, X. (2003) Carbonate apatite type A synthesized at high pressure: new space group (*P-3*) and orientation of channel carbonate ion. *Journal of Solid State Chemistry*, Vol.174, No.2, pp.412-417, ISSN 1095-726X.

Fleet, M. E. & Liu, X. (2004) Location of type B carbonate ion in type A-B carbonate apatite synthesized at high pressure. *Journal of Solid State Chemistry*, Vol.177, No.9, pp.3174-3182, ISSN 1095-726X.

Fleet, M. E. & Liu, X. (2007) Coupled substitution of type A and B carbonate in sodium-bearing apatite. *Biomaterials*, Vol.28, No.6, pp.916-926, ISSN 0142-9612.

Fratzl, P., Gupta, H. S., Paschalis, E. P. & Roschger, P. (2004) Structure and mechanical quality of the collagen-mineral nano-composite in bone. *Journal of Materials Chemistry*, Vol.14, No.14, pp.2115-2123, ISSN 0959-9428.

Hench, L. (2006) The story of Bioglass®. *Journal of Materials Science: Materials in Medicine*, Vol.17, No.11, pp.967-978, ISSN 0957-4530.

Hench, L. L. (1998) Biomaterials: a forecast for the future. *Biomaterials*, Vol.19, No.16, pp.1419-1423, ISSN 0142-9612.

Hench, L. L. & Andersson, O. H. (1993). Bioactive glasses. In *An introduction to bioceramics*. In *An introduction to bioceramics*, L.L. Hench & W. J. (Ed.), pp. 42-62, World Scientific, ISBN 978-981-02-1400-5 Singapore.

Hench, L. L., Splinter, R. J., Allen, W. C. & Greenlee, T. K. (1971) Bonding Mechanism at the Interface of Ceramic Prosthetic Materials. *Journal of Biomedical Materials Research A*, Vol.5, No.6, pp.117-141, ISSN 1552-4965.

Hoover, W. G. (1985) Canonical dynamics: Equilibrium phase-space distribution. *Physical Review A*, Vol.31, No.3, pp.1695-1697, ISSN 1094-1622.

Humphrey, W., Dalke, A. & Schulten, K. (1996) VMD: Visual Molecular Dynamics. *Journal of Molecular Graphics*, Vol.14, No.1, pp.33-38, ISSN 1093-3263.

Levitt, S. R., Crayton, P. H., Monroe, E. A. & Condrate, R. A. (1969) Forming methods for apatite prostheses. *Journal of Biomedical Materials Research A*, Vol.3, 683-684, ISSN 1097-4636.

Lin, C. C., Huang, L. C. & Shen, P. (2005) $Na_2CaSi_2O_6$-P_2O_5 based bioactive glasses. Part 1: Elasticity and structure. *Journal of Non-Crystalline Solids*, Vol.351, No.40-42, pp.3195-3203, ISSN 0022-3093.

Lusvardi, G., Malavasi, G., Cortada, M., Menabue, L., Menziani, M. C., Pedone, A. & Segre, U. (2008a) Elucidation of the Structural Role of Fluorine in Potentially Bioactive Glasses by Experimental and Computational Investigation. *The Journal of Physical Chemistry B*, Vol.112, No.40, pp.12730-12739, ISSN 1520-6106.

Lusvardi, G., Malavasi, G., Menabue, L., Menziani, M. C., Pedone, A., Segre, U., Aina, V., Perardi, A., Morterra, C., Boccafoschi, F., Gatti, S., Bosetti, M. & Cannas, M. (2008b) Properties of Zinc Releasing Surfaces for Clinical Applications. *Journal of Biomaterials Applications*, Vol.22, 505-526, ISSN 1530-8022.

Matsumura, Y. & Moffat, J. B. (1996) Methanol adsorption and dehydrogenation over stoichiometric and non-stoichiometric hydroxyapatite catalysts. *Journal of the Chemical Society, Faraday Transactions*, Vol.92, No.11, pp.1981-1984, ISSN 0965-5000.

O'Donnell, M. D., Watts, S. J., Law, R. V. & Hill, R. G. (2008a) Effect of P2O5 content in two series of soda lime phosphosilicate glasses on structure and properties - Part I: NMR. *Journal of Non-Crystalline Solids*, Vol.354, No.30, pp.3554-3560, ISSN 0022-3093.

O'Donnell, M. D., Watts, S. J., Law, R. V. & Hill, R. G. (2008b) Effect of P2O5 content in two series of soda lime phosphosilicate glasses on structure and properties - Part II: Physical properties. *Journal of Non-Crystalline Solids*, Vol.354, No.30, pp.3561-3566, ISSN 0022-3093.

O'Donnell, M., Watts, S., Hill, R. & Law, R. (2009) The effect of phosphate content on the bioactivity of soda-lime-phosphosilicate glasses. *Journal of Materials Science: Materials in Medicine*, Vol.20, No.8, pp.1611-1618, ISSN 1573-4838.

Padilla, S., Roman, J. & Vallet-Regi, M. (2002) Synthesis of porous hydroxyapatite by combination of gelcasting and foams burn out methods. *Journal of Materials Science: Materials in Medicine*, Vol.13, No.12, pp.1193-1197, ISSN 1573-4838.

Pedone, A., Charpentier, T., Malavasi, G. & Menziani, M. C. (2010) New Insights into the Atomic Structure of 45S5 Bioglass by Means of Solid-State NMR Spectroscopy and Accurate First-Principles Simulations. *Chemistry of Materials*, Vol.22, No.19, pp.5644-5652, ISSN 1520-5002.

Pedone, A., Malavasi, G., Cormack, A. N., Segre, U. & Menziani, M. C. (2008) Elastic and dynamical properties of alkali-silicate glasses from computer simulations techniques. *Theoretical Chemistry Accounts: Theory, Computation, and Modeling (Theoretica Chimica Acta)*, Vol.120, No.4-6, pp.557-564, ISSN 1432-2234.

Pedone, A., Malavasi, G., Menziani, M. C., Segre, U. & Cormack, A. N. (2008) Role of Magnesium in soda-lime glasses: insight into structure, transport and mechanical properties through computer simulations. *The Journal of Physical Chemistry C*, Vol.112, No.29, pp.11034-11041, ISSN 1932-7447.

Perdew, J. P., Burke, B. & Ernzerhof, M. (1996) Generalized Gradient Approxymation Made Simple. *Physical Review Letters*, Vol.77, No.18, pp.3865-3868, ISSN 1079-7114.

Peroos, S., Du, Z. M. & de Leeuw, N. H. (2006) A computer modelling study of the uptake, structure and distribution of carbonate defects in hydroxy-apatite. *Biomaterials*, Vol.27, No.9, pp.2150-2161, ISSN 0142-9612.

Rabone, J. A. L. & de Leeuw, N. H. (2005) Interatomic Potential Models for Natural Apatite Crystals: Incorporating Strontium and the Lanthanides. *Journal of Computational Chemistry*, Vol.27, No.2, pp.253-266, ISSN 1096-987X.

Rabone, J. A. L. & de Leeuw, N. H. (2007) Potential routes to carbon inclusion in apatite minerals: a DFT study. *Physics and Chemistry of Minerals*, Vol.34, No.7, pp.495-506, ISSN 1432-2021.

Rimola, A., Corno, M., Zicovich-Wilson, C. M. & Ugliengo, P. (2008) *Ab Initio* Modeling of Protein/Biomaterial Interactions: Glycine Adsorption at Hydroxyapatite Surfaces. *Journal of the American Chemical Society*, Vol.130, No.48, pp.16181-16183, ISSN 0002-7863.

Rimola, A., Corno, M., Zicovich-Wilson, C. M. & Ugliengo, P. (2009) Ab initio modeling of protein/biomaterial interactions: competitive adsorption between glycine and water onto hydroxyapatite surfaces. *Physical Chemistry Chemical Physics*, Vol.11, No.40, pp.9005 - 9007, ISSN 1463-9084.

Rodriguez-Lorenzo, L. M., Vallet-Regi, M., Ferreira, J. M. F., Ginebra, M. P., Aparicio, C. & Planell, J. A. (2002) Hydroxyapatite ceramic bodies with tailored mechanical properties for different applications. *Journal of Biomedical Materials Research A*, Vol.60, 159-166, ISSN 1097-4636.

Roveri, N. & Palazzo, B. (2006). Hydroxyapatite Nanocrystals as Bone Tissue Substitute. In *Tissue, Cell and Organ Engineering*, C.S.S.R. Kumar (Ed.), pp. 283-307, Wiley-VCH, ISBN 978-3-527-31389-1, Weinheim.

Sakhno, Y., Bertinetti, L., Iafisco, M., Tampieri, A., Roveri, N. & Martra, G. (2010) Surface Hydration and Cationic Sites of Nanohydroxyapatites with Amorphous or Crystalline Surfaces: A Comparative Study. *The Journal of Physical Chemistry C*, Vol.114, No.39, pp.16640-16648, ISSN 1932-7447.

Sato, K., Kogure, T., Iwai, H. & Tanaka, J. (2002) Atomic-Scale {101-0} Interfacial Structure in Hydroxyapatite Determined by High-Resolution Transmission Electron Microscopy. *Journal of the American Ceramic Society*, Vol.85, No.12, pp.3054-3058, ISSN 1551-2916.

Suda, H., Yashima, M., Kakihana, M. & Yoshimura, M. (1995) Monoclinic <--> Hexagonal Phase Transition in Hydroxyapatite Studied by X-ray Powder Diffraction and Differential Scanning Calorimeter Techniques. *The Journal of Physical Chemistry*, Vol.99, No.17, pp.6752-6754, ISSN 0022-3654.

Tadic, D., Beckmann, F., Schwarz, K. & Epple, M. (2004) A novel method to produce hydroxyapatite objects with interconnecting porosity that avoids sintering. *Biomaterials*, Vol.25, No.16, pp.3335-3340, ISSN 0142-9612.

Tilocca, A. (2010) Models of structure, dynamics and reactivity of bioglasses: a review. *Journal of Materials Chemistry*, Vol.20, No.33, pp.6848-6858, ISSN 0959-9428.

Tilocca, A. & Cormack, A. N. (2007) Structural effects of Phosphorus Inclusion in Bioactive SIlicate Glasses. *The Journal of Physical Chemistry B*, Vol.111, No.51, pp.14256-14264, ISSN 1520-6106.

Tilocca, A. & Cormack, A. N. (2009) Surface Signatures of Bioactivity: MD Simulations of 45S and 65S Silicate Glasses. *Langmuir*, Vol.26, No.1, pp.545-551, ISSN 0743-7463.

Toulhat, N., Potocek, V., Neskovic, M., Fedoroff, M., Jeanjean, J. & Vincent, V. (1996) Perspectives for the study of the diffusion of radionuclides into minerals using the nuclear microprobe techniques. *Radiochimica Acta*, Vol.74, No.1, pp.257-262, ISSN 0033-8230.

Ugliengo, P., Viterbo, D. & Chiari, G. (1993) MOLDRAW: Molecular Graphics on a Personal Computer. *Zeitschrift fur Kristallographie*, Vol.207, No.1, pp.9-23, ISSN 0044-2968.

Wahl, D. A., Sachlos, E., Liu, C. & Czernuszka, J. T. (2007) Controlling the processing of collagen-hydroxyapatite scaffolds for bone tissue engineering. *Journal of Materials Science: Materials in Medicine*, Vol.18, 201-209, ISSN 1573-4838.

Weiner, S. & Wagner, H. D. (1998) The material bone: structure-mechanical function relations. *Annual Review of Materials Science*, Vol.28, No.1, pp.271-298, ISSN 0084-6600.

Wierzbicki, A. & Cheung, H. S. (2000) Molecular modeling of inhibition of hydroxyapatite by phosphocitrate. *Journal of Molecular Structure: THEOCHEM*, Vol.529, No.1-3, pp.73-82, ISSN 0166-1280.

Young, R. A. & Brown, W. E. (1982). Structures of Biological Minerals. In *Biological Mineralization and Demineralization*, G.H. Nancollas (Ed.), pp. 101-141, Springer-Verlag, ISBN 978-0387115214, Berlin, Heidelberg, New York.

6

The Effects of Endurance Running Training on Young Adult Bone: Densitometry vs. Biomaterial Properties

Tsang-Hai Huang[1], Ming-Yao Chang[2], Kung-Tung Chen[3],
Sandy S. Hsieh[4] and Rong-Sen Yang[5]
[1]Institute of Physical Education, Health and Leisure Studies,
National Cheng Kung University, Tainan,
[2]Department of Biomedical Engineering, National Cheng Kung University, Tainan
[3]College of Humanities, Social and Natural Sciences,
Minghsin University of Science and Technology, Hsinchu
[4]Graduate Institute of Exercise and Sport Science,
National Taiwan Normal University, Taipei
[5]Department of Orthopaedics, National Taiwan University & Hospital, Taipei,
Taiwan

1. Introduction

Densitometric measurement of bone mineral parameters has been developed in recent decades. Since bone strength is associated with bone mineral density (BMD) and/or bone mineral content (BMC), densitometric measurement is widely accepted and used as one golden standard in clinical settings to determine bone health. Based on this concept, some human studies have suggested that endurance training, such as long distance running, provides no benefit and may even be harmful to bone health or bone mineral accretion during development, since long distance runners often have low BMD and/or BMC and may even exhibit conditions associated with bone loss or osteopenia.[1, 2] Conversely, serum bone marker assays in healthy distance runners show normal or positive bone metabolism status.[3, 4] Therefore, the definite role of endurance running training (ERT) on bone health remains a controversial issue. It would be valuable to further clarify whether ERT benefits bone health through a pathway other than absolutely increasing BMD or BMC.

Clinical observations of human subjects require further basic studies to investigate possible mechanisms. Animal studies can provide unique ways not feasible in studies using human subjects of assessing the effects of endurance running on bone. Generally, previous animal studies further verified benefits of ERT to bone health. However, the limitations of animal studies must be clarified before applying their findings to human beings.

The present article reviews the phenomena shown in bone of adolescent or young adult distance runners. Moreover, previous animal studies which adopted growing and young adult rats as subjects are reviewed, and the applicability of the findings to humans is also discussed.

2. The effects of endurance training on human bone: results and limitations

Conventionally, the extrinsic parameters of bone, such as BMD, BMC, and size-related measurements (*e.g.*, bone dimension, bone geometry), have been widely accepted as indicators of bone strength as well as predictors of fracture risk. Unfortunately, endurance running is usually considered an exercise mode that confers no benefit in terms of bone mineral accretion.[1, 2] Moreover, distance runners reportedly have low BMD and are often candidates for osteoporosis or stress fracture.[1, 2, 5-8] This section reviews studies on the effects of distance running training on both BMD and bone metabolism.

2.1 Results of human studies
2.1.1 BMD and BMC in distance runners

For references to human studies, the NIH website (http://www.ncbi.nlm.nih.gov /sites/entrez/) was searched by subject (adolescent, young adult runners) and research type (cross-sectional studies). Additionally, the major purpose of this article is to describe the long-term effects of ERT on bone in runners without concomitant health problems. Hence, reports describing energy deficiencies and/or serious menstrual cycle disorders in runners were excluded. The summary of previous cross-sectional studies in Table 1 indicates that distance runners usually reveal lower BMD and BMC values than those who engage in higher impact sports.[9] According to Frost's theory,[10] the slenderer body dimensions (Body Mass Index = 20 ~ 22) of runners who have a relatively lower body weight (BW) might partially contribute to a lower BMD and BMC. However, when compared to body-size matched control groups or another non-weight bearing exercise group, runners still do not seem to have much advantage on whole body, lumbar spine or regional cortical bone BMD.[11-18] Although oligomenorrhea or amenorrhea has been considered the cause of low BMD in female runners, even healthy female runners with normal menstrual cycles had lower BMD when compared to their size-matched control subjects.[12-14] Thus, ERT is usually concluded to be profitless for bone mineral accretion and bone health as well. However, if the analysis is limited to weight-bearing sites, runners do reveal higher site-specific regional BMD and/or BMC (*e.g.*, femoral neck, distal tibia, calcaneus) than do controls.[15, 19-21] Therefore, ERT is not entirely non-beneficial for bone mineral accretion when considering BMD and/or BMC as the major predictors of bone health.

Table 1 shows the findings of several studies indicating that distance runners have absolutely higher BMD values than do control groups.[3, 22-24] In the research publicized by Brahm and associates,[3] the runners showed only a slightly higher total-body BMD (3.6% higher, p=0.03), and no significant difference from the control group in total-body BMC was revealed. Interestingly, this study found that runners had distinctly higher BMD values in the legs and in the proximal femora. Regarding subject specificity, the training level of subjects or the normality of control subjects would be a major concern. Compared to elite distance runners, high school or club level runners may be trained at a more moderate intensity. Thus, these subjects did not really have typical body dimensions (e.g. slender body shape, low BMI) of elite distance runners.[22, 24] On the other hand, the BMI of 20.7 in the control group recruited in Kemmler *et al.* may have been too low for a normal control group.[23]

Authors	Subjects (gender, age, BMI, training status)	Results	Vs. control group
Grimston et al.[12]	♀, age (C:32.9, NR:32.2, LR:30.3), BMI (C: 20.6, NR:20.5, LR: 21.7), Training (NR:55.6km/wk for 7.7yr, LR: 53.0km/wk, 9.2yr)[a]	Most subjects reveal normal menstrual cycle (11-13 cycle/yr), LR group showed lower BMD in L2-L4 , femoral neck, and tibia than C group	↓
Robinson et al.[13]	♀, collegiate gymnasts (n = 21, age 16.2±1.7 years) runners (n = 20, 14.4±1.7 years), and nonathletic college women (n = 19, 13.0±1.2 years), No BMI data.	Lumbar spine BMD was lower in runners compared with both gymnasts and controls. Whole body BMD was lower in runners compared with gymnasts and controls.	↓
Taaffe et al.[14]	♀, 19.7±1.2yr, 19.5, 4-5d/wk, college level runners	Runners showed significantly lower BMD in femoral neck, lumbar spine but not in whole body BMD.	↓
Mudd et al. (2007)[9]	♀, 20.2±1.3yr, 21.0±1.6, college level athletes	Runners had lowest lumbar spine and pelvis BMD when compared to other athletes	↓
Emslander et al.[17]	♀, 20.3±0.6 yr, 21.9, 40mile/wk for 3yr	No significant difference was shown in total body, spine and proximal femoral BMD among runners, swimmers and control groups.	=
Duncan et al.[18]	♀, 17.8±1.4 yr, 21.3±1.6, 8.4±1.2 h/wk for 6.2±1.7 yr, high school level athletes	Areal BMD estimation was performed on mid-third femur. Runners had higher BMD only than cyclist. No difference was shown among groups of runners, swimmers, triathlete and controls.	=
Moen et al.[25]	♀, 15.1-18.8 yr, No BMI data, 58.1km/wk for >1.5yrs	There is no significant difference among amenorrheic runners, eumenorrheic runners, and controls in lumbar BMD.	=
Greene et al.[16]	♂, 16.8±0.6yr, 20.68±1.6, 6hr/wk for 2yr	After adjusting for lean tissue mass per kg of body weight, no difference in BMC was detected.	=
Jürimäe et al.[11]	♀, 22.6±4.3 yr, 20.6±1.6, 6h/wk for > 5yr	Endurance trained group showed no difference with normal-weight control group in BMD, but was lower than over-weight control group.	=
MacDougall et al.[15]	♂, 22-45yr, runners were divided into five groups per their training mile/wk[b]	Runners with running mileage 15-20mile/wk showed the highest BMD values in legs but not in total body and spine. Runners with higher	SS ↑

		training mileage did not show difference in BMD as compared to control group.	
Greene et al.[20]	♀, 16±1.7, 18.7±1.5, 6hr/wk for >2yr	Runner showed higher BMD and BMC in distal tibia, densitometric measurement performed only in distal tibia.	SS ↑
Egan et al.[19]	♀, 21.5±2.6 yr, 20.23, 8.4±3.4h/wk for 6.0±2.1yr	All sports groups had higher BMD values than did the controls. Runners showed a higher BMD only in legs and proximal femur, but lower than rugby athletes.	SS ↑
Fredericson et al.[21]	♂, 24.2±3.2 yr, 20.3±1.3, 70mile/wk for at least 1yr	Soccer player was higher in BMD of the skeleton at all sites measured. Runners only showed higher BMD in calcaneus than control group.	SS ↑
Brahm et al.[3]	♀&♂, 32yr, 22, 7h/wk for 12yr	Runners were significantly higher in total body, legs, femoral neck, trochanter wards triangle and calcaneus BMD than control group.	↑
Stewart & Hannan[24]	♂, 27.6±6.1yr, 21.9±1.3, 8.7±2.7h/week, club level runners	Runners showed higher total body and legs BMD.	↑
Duncan et al.[22]	♀, 17.6±1.4 yr, 21.3±1.6, 8.4±1.2 h/wk for 6.2±1.7 yr, high school level athlete	Runner were significantly higher in total body, lumbar spine, femoral neck and leg BMD as compared to BMI-matched control group	↑
Kemmler et al.[23]	♂, 26.6±5.5yr, 20.9, 9.25±2h/wk for 8.9 yr	Runners were higher in total body BMD, legs, pelvis, femoral neck, calcaneus BMD as compared to the BMI (20.7) matched control group	↑

Note: [a], subjects were divided into three groups (NR, runners with normal BMD; LR, runners with low BMD, C, control group); [b], runners were divided into five groups according to their training mileage per week, BMI value was not matched among groups that control group showed the highest value; ↓, runners were comprehensively lower than control group in total body and local bone; =, no significant difference was shown between runners and control group; SS ↑, runners showed site-specific increment in BMD; ↑, runners showed higher BMD in total body as well as in local bone.

Table 1. Summary of cross-sectional studies of BMD in adolescent or young adult distance runners

2.1.2 Results of human studies: Bone metabolism status in distance runners

As mentioned above, ERT conferred no clear benefits to bone health. An important inquiry is whether endurance running influences the physiology (e.g. exercise-induced acidosis) or causes related abnormalities in hormonal homeostasis (e.g. menstrual disorders in females or lower testosterone in males) that negatively affect the bone.

It's well known that patients with pathological acidosis suffer negative bone turn-over, which causes a net bone mineral loss.[26] Endurance exercise may induce acidosis, which negatively affects bone metabolism. However, transient acidosis caused by exercise is buffered by HCO_3^- and disappears within hours after exercise.[27] In addition, it has been suggested that acid buffer capacity is enhanced after a period of exercise training.[28] At the cellular level, a single bout of intense exercise induces transient increases in serum and urine calcium levels without showing cellular osteoclastical activities.[29,30]

With respect to the impact from abnormalities of hormone regulation, oligomenorrhea and amenorrhea related to bone loss are often reported in female runners undergoing intensive training. However, recent investigations suggest that endurance running does not directly cause menstrual disorders and the subsequent bone loss.[31-33] Menstrual disorders in endurance runners are more likely due to either energy or nutrition deficiencies. Therefore, dietary adjustment is usually more effective than hormone replacement therapy for restoring menstrual cycles and bone metabolism.[34] In males, ERT is known to lower testosterone, but lower testosterone does not necessarily correlate with lower BMD.[35] In addition, runners with different training mileage (from 5 to 75 mile/week) do not significantly differ in serum testosterone levels.[15]

Regarding studies of bone metabolism status, healthy distance runners at rest usually exhibit normal bone metabolism, and some studies even show a positive bone metabolism status, as revealed by serum bone markers.[3, 4] Brahm et al., in a study of bone metabolism in runners using various serum markers, found that runners had lower bone formation as well as lower bone resorption activities.[3] Moreover, triathletes reveal no difference in bone metabolism during the intense competitive season as compared to their non-training period.[4]

2.2 Limitation of human studies

To summarize the above, distance runners do not seem to acquire much benefit from their training when densitometric measurements are used to determine the bone health. However, as shown by serum bone marker assays, distance runners did not reveal an inadequate bone metabolism status. Actually, over the past decades, an increasing number of reports suggest that BMD does not accurately predict bone health or bone strength. Patients with fractures also show normal BMD values.[36] The BMD and BMC measures apparently correlate strongly with body mass and body size.[11, 37-39]

Today, "bone quality" is used as a new term to represent bone health, which is composed of various parameters, including tissue architecture, turnover, microfracture and mineralization, of a healthy bone.[40] Further, the organization of the bone matrix may also play an important role in bone strength.[41] Unfortunately, many of the bone quality measurements are too invasive to be feasible in human subjects. Thus, animal studies have been frequently used for further clarifying related issues.

3. Comprehensive results of animal studies: BMD and BMC

In biomedical science, animal experiments are performed to establish models that mimic human physiological phenomena. Either estimation methods or experimental designs, which may not be feasible in humans, can then be performed to further investigate possible mechanisms. Rodent models of treadmill activity are commonly used to investigate the effects and mechanisms of exercise on bone metabolism. This section reviews the findings of previous animal studies. Briefly, the results of animal studies using rodents as subjects showed gender differences, which might affect their further applicability in human subjects. As mentioned above, intensity-trained runners usually have a lower body mass and often have equal or even lower bone mass than non-athletes. Therefore, an animal model of ERT would be expected to reveal the same phenomena. Animal studies reviewed in the present article were selected according to training type (typically endurance treadmill training) and the age of animals (growing or young adult rats).

3.1 Rodents adapted to endurance exercise showed gender differences
3.1.1 Male rodent studies
Table 2 summarizes the outcomes of studies using male rats as subjects. The studies were reviewed and classified into two categories. The first category includes those using diet control or adjustment to achieve equivalent BW gains between exercise and control groups.[42-47] These studies demonstrated that trained animals have a higher BMD.[43-45] Tissue mechanical properties were not available in every study, and only one of them shows a higher load-withstanding capacity in the femoral diaphysis.[43] However, it must be mentioned that diet prohibition for the purpose of equalizing body weights among groups might cause an additional negative effect on tissue mechanical properties. Diet prohibition impairs the tissue levels (intrinsic) and mechanical properties of bone, suggesting that dietary manipulation of a control group might not be appropriate.[48, 49]

The second category of studies included animals fed *ad libitum*. In these studies, the exercise groups revealed significantly less BW gain after a programmed ERT.[50-55] With lower BW, exercise trained animals showed no difference or lower BMD values as compared to the sedentary control group. As in human subjects, male rats undergoing intense ERT exhibit lower BW gain and no benefits to bone health when considering BMD or BMC as a predictor. However, the higher load-withstanding and energy-absorption capacity of the bones in training rats introduced new research into how endurance exercise benefits bone quality (see section 4).

3.1.2 Female rodent studies
Compared with the treadmill training results for male rats, those for female rats are inconsistent with human subjects, and the data are somewhat controversial. Table 3 summarizes the results of ERT in growing or young adult female rats. Most studies indicate that female growing or young adult rats exhibit no change in BW after a period of ERT.[56-64] One study even reported increased BW in female rats after training.[58] Of the studies performing BMD analysis in female rats, many report positive effects from endurance running not only in site-specific increments but also in whole bone. Although densitometric measurements demonstrate this advantage, female rats rarely show improved biomechanical properties and may even reveal adverse effects after an intense training program (see Table 3). Therefore, female rats acclimate to ERT differently than do male rats. In human beings, distance runners are also expected to exhibit gender differences in physiological response to similar ERT. However, it seems inappropriate to use the gender difference found in rats to explain the one found in humans, since a period of programmed ERT would commonly reduce BW either in women or men. Thus, the phenomena observed in female rats may not be applicable to female humans. Based on the idea that animal models should mimic the phenomena shown in human subjects, studies using female rats may not be applicable to female humans, since female rats and women have been shown to respond differently to ERT. Possible reasons are discussed in the next section.

Author	strain, age	Protocol	BW control	BW or BW gain (g)	BMD or BMC	Biomechanical testing
Nordsletten et al.[53]	Wistar, 11 wk	27m/min, 60min/d, 10% inclination, 5d/wk, 4 wk	—	EXE<CO N*	N/A	Ultimate bending moment $(N \cdot m \times 10^{-2})$ EXE > CON
Brourrin et al.[50]	Wistar, 10wk	30m/min, 1.5h/d, 5d/wk, 5wk	—	EXE<CO N*	N/A	No data
Horcajada-Molteni et al.[51]	Wistar, 8wk	20m/min to 30m/min, 60min/d, 6d/wk, 90d	—	EXE<CO N*	Femoral BMD: NS	Femoral failure load (N): EXE > CON*
Huang et al.[54]	Wistar, 5wk	24m/min, 60min/d, 5 d/wk, 10wk	—	EXE<CO N*	Tibiae BMD (g/cm^2) :NS	No data
Huang et al.[52]	Wistar, 7 wk	22m/min, 60min/d, 5d/wk, 8wk	—	EXE < CON*	Tibia and femur BMD: NS	Three-point bending load (N), energy (mJ), stress (MPa), toughness (mJ/mm^3): EXE > CON*
Huang et al.[55]	Wistar, 7wk	Two groups: 22m/min, 60min/day vs. 30m/min, 5d/wk, 8wk	—	EXE < CON*	EXE < CON* in total femur (p=0.04)	Femoral midshaft bending energy and toughness (mJ & mJ/mm^3): EXE > CON
Joo et al.[43]	Wistar, 4 wk	30 m/min, 60 min/d, 5 d/wk, 10wk	+	NS	Femurs BMD (g/cm^2) EXE > CON	Femoral mid-diaphysis Bending stress (N/mm^2): NS Maximum load (N): EXE > CON
Kiuchi, et al.[44]	Wistar, 4 wk	35m/min, 5° inclination, 60min/d, 5d/wk, 10 wks	+	NS	BMC: EXE > CON*	No data
Notomi et al.[46]	S.D., 4 wk	24m/min, 60min/d, every other d, 4 wk	+	NS	NS	Vertebra and femoral maximal load (N): NS
Sakamoto & Grunewald[47]	Wistar, 4 wk	24m/min, 75min/d, 5d/wk, 8 wk	+	NS	N/A	Tibia breaking strength (kg): NS

| Newhall et al.[45] | S.D., 47 d | 10.2km/d, for 6 wks (voluntary running) | + | EXE > CON | Femur BMD & BMC: EXE > CON* | N/A |
| Ferreira et al.[42] | Wistar, 10 wk | 12m/min, 1h/d, 10 wks. | + | NS | NS | N/A |

Note: Protocol, training protocols were presented serially by the final training intensity (m or cm per minute), training time per day (minute or hour per day), training frequency (times per week), and training periods (day or week); d, day; wk, week; min, minute; NS, none significant difference; N/A, none available.

Table 2. Studies of endurance running training vs. growing or young adult male rats

Author	Strain and age	Protocol	BW control	BW or BW gain and tissue measurement	BMD or BMC	Biomechanical testing
Iwamoto et al.[56]	S.D., 4wk	24m/min, 60min/d, 5 d/wk, 8wk or 12wk	—	BW: NS Muscle weight & Femoral length (12wk): EXE > CON*	Femoral and L6 vertebral BMD (mg/µL): NS Femoral and L6 vertebral bone volume (µL), wet weight (mg): EXE > CON*	N/A
Iwamoto et al.[57]	Wistar, 4wk	24m/min, 60min/d, 5 d/wk, 7wk or 11wk	—	BW: NS Femoral length: EXE > CON*	Tibial BMC (g): EXE > CON* Tibial BMD (g/cm²): NS	N/A
Hagihara et al.[58]	Wistar rats, 8wk	A group: control group B-E group: 4~7d/wk, running at 15m/min, 30min/d, 8wk	—	BW gain: B > A*	Tibial trabecular BMD (mg/cm³): B, C, D, E > A group*. But, NS in tibial cortical BMD.	N/A
Wheeler et al.[59]	S.D., 120 d	55%, 65%, 75% $\dot{V}O_2$max, 30min/d, 60min/d, 90min/d, 4d/wk, 10wk	—	BW: NS	Tibial BMD (g/cm²): EXE > CON	Group trained at 75% $\dot{V}O_{2max}$ and 90min/d showed higher stiffness but lower energy to

						withstand torsion test.
Raab et al.[60]	Fischer 344, 2.5 month	36m/min, grade15%, 60min/day, 5days/wk for 10wk		BW: NS	N/A	Femur ultimate force (kg/mm): NS
Hou et al.[61]	S.D., 8 wk	49cm/s, 12% grade, 60min/d, 5d/wk, 10 wk (~75-80% of maximum oxygen capacity)	—	BW: NS	N/A	Femoral neck maximum load (N): NS Energy to maximum load (mJ): NS
Shimamura et al.[62]	Wistar, 6 wk	25m/min, 60min/d, 5d/wk, for 7 or 11 wk	—	BW: NS	BMC in Total tibia (mg): EXE > CON*	N/A
van der Wiel et al.[63]	Wistar, 5 month	20m/min,30 mim/d, 5° inclination, 5d/wk, 17wk	—	BW: NS	Total body BMC (g): NS CON : 9.3±1. 1 EXE : 9.9±1.0	Femoral neck maximal load (N): NS Femoral shaft maximal load (N): NS
Salem et al.[64]	S.D., 8 wk	45 cm/s, 5% grade, 60 min/d, 3d/wk, 10wk	—	BW: NS	N/A	Femoral neck maximal load (N):NS

Note: Protocol, training protocols were presented serially by the final training intensity (m or cm per minute), training time per day (minute or hour per day), training frequency (times per week), and training periods (day or week); d, day; wk, week; min, minute; NS, none significant difference; N/A, none available.3.1.3Gender differences revealed by animal studies

Table 3. Studies of endurance running training vs. growing or young adult female rats

As mentioned above, male and female rats adapt differently to endurance treadmill training, especially in densitometric measurements. The reasons for this gender difference in rodents have been comprehensively investigated elsewhere. According to the theory of Frost,[10] this difference may partially contribute to different adaptations in BW gain. Female rats usually exhibit a similar or sometimes higher body mass after training; they therefore may acquire a greater advantage from local mechanical loading than male rats with lower BW gain after forced endurance treadmill training or voluntary running.[65-68] The mechanisms of this gender difference in BW gain associated with ERT are unknown. A possible explanation is the involvement of gonadal hormones in BW regulation. Endurance exercise reportedly lowers plasma testosterone levels in male rats.[69] The down regulation of this anabolic hormone in growing male rats may account for the significantly lower protein mass gain and BW gain. In female rats, however, estrogens would suppress body mass, food consumption and fat deposition.[70, 71] Progesterone, on the other hand, has been verified to increase body fat and body mass.[72, 73] Moreover, a previous study suggests that regular treadmill training results in extended periods of progesterone secretion, which was associated with significant weight gain.[74] Women may respond to ERT with similar regulation in progesterone. However, the up-regulation of progesterone may be more pronounced in rats than in women, since female rats reveal no decrease in BW even under vigorous ERT.[60]

3.2 Studies of male rats mimic human practice
An analysis of gender differences observed in the above animal studies reveals that ERT increases BMD and BMC in female rats but not in male rats. However, the physiological response (*e.g.* BW gain) of female rats to ERT differs from that in female humans. Given that animal studies are intended to clarify the mechanisms of biological phenomena in humans, female rats may not be a suitable model for investigating the effects of ERT on developing or young adult bone. On the other hand, the changes in BW and densitometric parameters associated with ERT in male rats were similar to those in humans, indicating that male rats are a suitable model for investigating the effects of endurance running.

4. Effects of endurance training on bone biomaterial properties

Aside from BMD and BMC, biomaterial related analysis will provide more valuable information to predict the capacity that bone tissue can withstand extra mechanical loading generated by daily physical activity or accidents (e.g. fall), and thus, prevent bone from loading-induced damage.

Generally, biomaterial properties of bone tissue can be analyzed at a structural level and a tissue level. Structure biomaterial properties are size-dependent, that is, tissues bigger in size tend to be stronger than smaller ones. Conversely, tissue-level properties are analyzed under size-independent conditions using mathematic methods (e.g. normalized tissue size by cross-sectional moment of inertia) or mechanical methods (e.g. a specimen with consistent size is sectioned from a whole tissue).[75]

4.1 Effects on structural (whole bone) properties
Structural properties are calculated from original biomechanical testing raw data without any normalization. Related parameters are load (Nt), displacement (mm), energy (mini joule, mJ) and stiffness (Nt/mm) etc. As shown in previous studies, results of rodents' whole tissue biomechanical properties after a period of endurance running training were controversial. Some studies show that exercise groups were higher in load-withstanding capacity,[51-53] while others revealed a higher energy absorption capacity.[55] One possible explanation for these discrepancies may be differences in training protocol. Animals trained at a higher intensity tended to show higher bone strength (e.g. higher bending load or moment),[51, 53] suggesting that higher mechanical loading generated by intensively running may benefit to bone strength. Moreover, the specific testing conditions would also affect testing results; for instance, Nordsletten and colleagues measured bone strength *in vivo*, at which time bone strength may be affected by muscle strength gains achieved through training.[53]

However, in considering the applicability of exercise, a training program with moderate exercise intensity would be expected to show a higher compliance and therefore be more appropriate for the general population. As shown in Table 2, animals trained at a relatively moderate intensity (20-24m/min), which corresponds to 70% $\dot{V}O_2$ max,[76] also had lower body masses and slightly lower (~5% lower) total BMD (p = 0.04), but were not found to have enhanced structural bone strength.[55] The authors' previous study used body mass as a covariate to equalize raw data, which then revealed a comprehensively stronger bone tissue either in structural or tissue-level biomaterial properties.[52] With lower body mass, the data of the exercise group would be adjusted to a higher level, and the effects of ERT seemed to become "good" for animal bones. However, it would be a more relevant and natural study if animals were fed ad libitum and data were not adjusted

In aspects of biological efficiency, an athlete at her/his optimal physiological status will not necessarily be absolutely higher in every physiological parameter. Therefore, a smaller muscle mass or skeleton size seems to be a benefit, rather than a weakness, for a distance runner or an endurance athlete. With such a smaller bone size, moderate ERT rats did not show absolutely enhanced structural bending load values but, interestingly, they showed better energy absorption capacity in long bone tissue that ERT rats were found to have a four-fold increase in energy absorption after long bone tissue reached the yield-point (post-yield energy).[55] As mentioned in previous studies,[41] post-yield behaviors are highly correlated to tissue-level changes (e.g. collagen fiber orientation).

5. Effects on tissue-level (material) properties

In our previous studies, we used mathematical methods to estimate tissue-level biomaterial properties. Through calculating long bone's cross-sectional moment of inertia, we normalized load-displacement data to stress and strain. Under such conditions, ERT rats' worse structural material properties disappeared. Additionally, exercise and control groups showed no differences in yield stress, yield toughness or elastic modulus (Young's modulus),[55] suggesting that endurance training is not harmful for bone material properties.

ERT's benefits on the post-yield biomaterial behaviors seemed to be more size-independent and associated with tissue-level (e.g. bone matrix, collagen) changes. Because measuring post-yield mechanical properties using beam bending theory is only valid in the pre-yield regime,[75] reporting post-yield stress, strain or toughness is inappropriate. Therefore, we discussed this tissue-level adaptation base on the post-yield parameters measured from load-displacement data. As shown in our two recent studies, either moderate ERT or endurance swimming training benefits bone tissue more in terms of energy absorption capacity,[55,77] especially in post-yield energy. Similar results of enhanced post-yield behavior were provided by another ERT animal study, which showed a short-term treadmill running (21 days) enhanced tibia post-yield deformation in mice.[78] Moreover, such effects on post-yield behavior changes seem to apply not only to endurance training. After a short-term (5 days) freefall landing exercise, Wistar rats revealed an increased post-yield energy absorption in ulnae.[79] Such an enhanced absorption capability is more likely due to tissue-level (e.g. bone matrix, collagen orientation etc) changes rather than structural adaptation.

As mentioned in previous studies, tissue-level properties can be divided into the inorganic mineral phase (e.g., hydroxyapatite), which determines tissue stiffness and strength, [80, 81] and the organic bone matrix, which plays a key role in energy absorption, [82, 83] It has been suggested that the networks of collagen, one of the major components of bone matrix, could affect the energy dissipation between the yield point and fracture point in bone tissue.[84-86]

Collagen fiber orientation (CFO) has been measured by circularly polarized light microscopes as one parameter to represent the collagen network and to predict post-yield energy of bone tissue.[84] Hence, the post-yield behavior revealed by ERT rats' bone tissue could partially stem from a highly organized collagen fiber network. Though the information regarding collagen orientation in the present study is not available in our previous rodent studies,[55,77] it has been reported that dogs after one-year of intensive endurance running (40km/day) revealed a higher organization of collagen fibers in bone tissues.[87] Such highly organized collagen fibers seemed to be able to compensate for the 10% BMD decrease in running dogs. Thus, a highly organized CFO would be expected in rodent studies. As mentioned above, rats subjected to short-term freefall landing exercise (5 days, 10 or 30 times per day) from a height of 40cm also showed enhanced post-yield energy of ulnae.[79] In that study, authors tried to measure the CFO of cross-sectional ulnae. Unfortunately, no difference in CFO between exercise and control groups was found. One major reason for this lack of significant results could be species difference. That is, CFO analysis might not be sensitive enough to detect biomaterial differences in rodents. To date, CFO-related analysis in bone tissue specimens have all been obtained from big mammals,[88, 89] which have more mature Haversian's systems and visible osteons. However, in smaller mammals (e.g. young adult rodents), it is not possible to find Haversian's systems or complete osteon.[90, 91] Per our observation, the organization of collagen fiber tends to be relatively irregular in rats and, thus, CFO analysis might not be sensitive enough to predict post-yield material properties.

On the other hand, cross-links within collagen networks might be another contributor to changes in tissue post-yield behaviors.[92] In an exercise-related study, Kohn and colleagues verified that cortical toughness enhanced by a 21-day ERT could be correlated with the overall maturity of collagen cross-links.[93]

In addition to individually measure CFO or cross-links, the biomaterial properties (e.g. tissue strength or tissue post-yield behaviors) might benefit from better integration between collagen and its crosslinks. Related measurement methods are awaited and are worthy of further study.

Finally, microdamage is another factor influencing tissue's post-yield behavior. Accumulation of microdamages (or microcracks) would lead to a fragile bone with lower capability in post-yield energy dissipation.[84, 94] However, such accumulated microdamages seemed to be more related to aging. Also, as in CFO-related studies, microdamage studies have been more frequently done in big mammals. Whether microdamage measurement can be performed on exercise-related rodent study needs further verification.

6. Summary

Endurance running is a popular aerobic activity and typical training type. However, related human studies reveal no significant benefits to bone health based on densitometric measurement of bone mineral. On the other hand, animal studies apparently indicate that ERT enhances biomaterial of bone tissue in a size-independent way. The effects of endurance running on the organic bone matrix or other parameters, as well as their relationship to mechanical properties of bone tissues, are worthy of further study.

7. Acknowledgments

This study was supported by a grant from the National Science Council: NSC 99-2410-H-006 -114 -MY2. Miss Jae Cody is appreciated for her editorial assistance.

8. References

[1] Burrows M, Bird S. The physiology of the highly trained female endurance runner. Sports medicine (Auckland, NZ 2000;30:281-300.

[2] Burrows M, Nevill AM, Bird S, Simpson D. Physiological factors associated with low bone mineral density in female endurance runners. British journal of sports medicine 2003;37:67-71.

[3] Brahm H, Strom H, Piehl-Aulin K, Mallmin H, Ljunghall S. Bone metabolism in endurance trained athletes: a comparison to population-based controls based on DXA, SXA, quantitative ultrasound, and biochemical markers. Calcif Tissue Int 1997;61:448-54.

[4] Maimoun L, Galy O, Manetta J, Coste O, Peruchon E, Micallef JP, Mariano-Goulart D, Couret I, Sultan C, Rossi M. Competitive season of triathlon does not alter bone metabolism and bone mineral status in male triathletes. International journal of sports medicine 2004;25:230-4.

[5] Cooper LA, Joy EA. Osteoporosis in a female cross-country runner with femoral neck stress fracture. Current sports medicine reports 2005;4:321-2.

[6] Fredericson M, Kent K. Normalization of bone density in a previously amenorrheic runner with osteoporosis. Medicine and science in sports and exercise 2005;37:1481-6.

[7] Prather H, Hunt D. Issues unique to the female runner. Physical medicine and rehabilitation clinics of North America 2005;16:691-709.

[8] Wilson JH, Wolman RL. Osteoporosis and fracture complications in an amenorrhoeic athlete. British journal of rheumatology 1994;33:480-1.

[9] Mudd LM, Fornetti W, Pivarnik JM. Bone mineral density in collegiate female athletes: comparisons among sports. Journal of athletic training 2007;42:403-8.

[10] Frost HM. Bone "mass" and the "mechanostat": a proposal. The Anatomical record 1987;219:1-9.

[11] Jurimae T, Soot T, Jurimae J. Relationships of anthropometrical parameters and body composition with bone mineral content or density in young women with different levels of physical activity. Journal of physiological anthropology and applied human science 2005;24:579-87.

[12] Grimston SK, Tanguay KE, Gundberg CM, Hanley DA. The calciotropic hormone response to changes in serum calcium during exercise in female long distance runners. The Journal of clinical endocrinology and metabolism 1993;76:867-72.

[13] Robinson TL, Snow-Harter C, Taaffe DR, Gillis D, Shaw J, Marcus R. Gymnasts exhibit higher bone mass than runners despite similar prevalence of amenorrhea and oligomenorrhea. J Bone Miner Res 1995;10:26-35.

[14] Taaffe DR, Robinson TL, Snow CM, Marcus R. High-impact exercise promotes bone gain in well-trained female athletes. J Bone Miner Res 1997;12:255-60.

[15] MacDougall JD, Webber CE, Martin J, Ormerod S, Chesley A, Younglai EV, Gordon CL, Blimkie CJ. Relationship among running mileage, bone density, and serum testosterone in male runners. J Appl Physiol 1992;73:1165-70.

[16] Greene DA, Naughton GA, Briody JN, Kemp A, Woodhead H, Farpour-Lambert N. Musculoskeletal health in elite male adolescent middle-distance runners. J Sci Med Sport 2004;7:373-83.

[17] Emslander HC, Sinaki M, Muhs JM, Chao EY, Wahner HW, Bryant SC, Riggs BL, Eastell R. Bone mass and muscle strength in female college athletes (runners and swimmers). Mayo Clinic proceedings 1998;73:1151-60.

[18] Duncan CS, Blimkie CJ, Kemp A, Higgs W, Cowell CT, Woodhead H, Briody JN, Howman-Giles R. Mid-femur geometry and biomechanical properties in 15- to 18-yr-old female athletes. Medicine and science in sports and exercise 2002;34:673-81.

[19] Egan E, Reilly T, Giacomoni M, Redmond L, Turner C. Bone mineral density among female sports participants. Bone 2006;38:227-33.

[20] Greene DA, Naughton GA, Briody JN, Kemp A, Woodhead H, Corrigan L. Bone strength index in adolescent girls: does physical activity make a difference? British journal of sports medicine 2005;39:622-7; discussion 7.

[21] Fredericson M, Chew K, Ngo J, Cleek T, Kiratli J, Cobb K. Regional bone mineral density in male athletes: a comparison of soccer players, runners and controls. British journal of sports medicine 2007;41:664-8; discussion 8.

[22] Duncan CS, Blimkie CJ, Cowell CT, Burke ST, Briody JN, Howman-Giles R. Bone mineral density in adolescent female athletes: relationship to exercise type and muscle strength. Medicine and science in sports and exercise 2002;34:286-94.

[23] Kemmler W, Engelke K, Baumann H, Beeskow C, von Stengel S, Weineck J, Kalender WA. Bone status in elite male runners. Eur J Appl Physiol 2006;96:78-85.

[24] Stewart AD, Hannan J. Total and regional bone density in male runners, cyclists, and controls. Medicine and science in sports and exercise 2000;32:1373-7.

[25] Moen SM, Sanborn CF, DiMarco NM, Gench B, Bonnick SL, Keizer HA, Menheere PP. Lumbar bone mineral density in adolescent female runners. The Journal of sports medicine and physical fitness 1998;38:234-9.

[26] Kraut JA. The role of metabolic acidosis in the pathogenesis of renal osteodystrophy. Advances in renal replacement therapy 1995;2:40-51.

[27] Fahey TD, Baldwin KM, Brooks GA. Exercise Physiology: Human Bioenergetics and Its Applications: McGraw-Hill; 2005.

[28] Aoi W, Iwashita S, Fujie M, Suzuki M. Sustained swimming increases erythrocyte MCT1 during erythropoiesis and ability to regulate pH homeostasis in rat. International journal of sports medicine 2004;25:339-44.

[29] Ashizawa N, Fujimura R, Tokuyama K, Suzuki M. A bout of resistance exercise increases urinary calcium independently of osteoclastic activation in men. J Appl Physiol 1997;83:1159-63.

[30] Ashizawa N, Ouchi G, Fujimura R, Yoshida Y, Tokuyama K, Suzuki M. Effects of a single bout of resistance exercise on calcium and bone metabolism in untrained young males. Calcified tissue international 1998;62:104-8.

[31] Zanker CL, Swaine IL. Relation between bone turnover, oestradiol, and energy balance in women distance runners. British journal of sports medicine 1998;32:167-71.

[32] Zanker CL, Swaine IL. The relationship between serum oestradiol concentration and energy balance in young women distance runners. Int J Sports Med 1998;19:104-8.

[33] Zanker CL, Swaine IL. Bone turnover in amenorrhoeic and eumenorrhoeic women distance runners. Scandinavian journal of medicine & science in sports 1998;8:20-6.

[34] Zanker CL. Bone metabolism in exercise associated amenorrhoea: the importance of nutrition. British journal of sports medicine 1999;33:228-9.

[35] Smith R, Rutherford OM. Spine and total body bone mineral density and serum testosterone levels in male athletes. Eur J Appl Physiol Occup Physiol 1993;67:330-4.

[36] Cummings SR, Bates D, Black DM. Clinical use of bone densitometry - Scientific review. Jama-Journal of the American Medical Association 2002;288:1889-97.

[37] Rico H, Revilla M, Hernandez ER, Villa LF, Alvarez del Buergo M, Lopez Alonso A. Age- and weight-related changes in total body bone mineral in men. Mineral and electrolyte metabolism 1991;17:321-3.

[38] Siemon NJ, Moodie EW. Body weight as a criterion in judging bone mineral adequacy. Nature 1973;243:541-3.

[39] Soot T, Jurimae T, Jurimae J, Gapeyeva H, Paasuke M. Relationship between leg bone mineral values and muscle strength in women with different physical activity. J Bone Miner Metab 2005;23:401-6.

[40] Osteoporosis prevention, diagnosis, and therapy. Jama 2001;285:785-95.

[41] Viguet-Carrin S, Garnero P, Delmas PD. The role of collagen in bone strength. Osteoporos Int 2006;17:319-36.

[42] Ferreira LG, De Toledo Bergamaschi C, Lazaretti-Castro M, Heilberg IP. Effects of creatine supplementation on body composition and renal function in rats. Medicine and science in sports and exercise 2005;37:1525-9.

[43] Joo YI, Sone T, Fukunaga M, Lim SG, Onodera S. Effects of endurance exercise on three-dimensional trabecular bone microarchitecture in young growing rats. Bone 2003;33:485-93.

[44] Kiuchi A, Arai Y, Katsuta S. Detraining effects on bone mass in young male rats. International journal of sports medicine 1998;19:245-9.

[45] Newhall KM, Rodnick KJ, van der Meulen MC, Carter DR, Marcus R. Effects of voluntary exercise on bone mineral content in rats. J Bone Miner Res 1991;6:289-96.

[46] Notomi T, Okazaki Y, Okimoto N, Saitoh S, Nakamura T, Suzuki M. A comparison of resistance and aerobic training for mass, strength and turnover of bone in growing rats. Eur J Appl Physiol 2000;83:469-74.

[47] Sakamoto K, Grunewald KK. Beneficial effects of exercise on growth of rats during intermittent fasting. J Nutr 1987;117:390-5.

[48] Banu MJ, Orhii PB, Mejia W, McCarter RJ, Mosekilde L, Thomsen JS, Kalu DN. Analysis of the effects of growth hormone, voluntary exercise, and food restriction on diaphyseal bone in female F344 rats. Bone 1999;25:469-80.

[49] Mosekilde L, Thomsen JS, Orhii PB, McCarter RJ, Mejia W, Kalu DN. Additive effect of voluntary exercise and growth hormone treatment on bone strength assessed at four different skeletal sites in an aged rat model. Bone 1999;24:71-80.

[50] Bourrin S, Palle S, Pupier R, Vico L, Alexandre C. Effect of physical training on bone adaptation in three zones of the rat tibia. J Bone Miner Res 1995;10:1745-52.

[51] Horcajada-Molteni MN, Davicco MJ, Collignon H, Lebecque P, Coxam V, Barlet JP. Does endurance running before orchidectomy prevent osteopenia in rats? Eur J Appl Physiol Occup Physiol 1999;80:344-52.

[52] Huang TH, Lin SC, Chang FL, Hsieh SS, Liu SH, Yang RS. Effects of different exercise modes on mineralization, structure, and biomechanical properties of growing bone. J Appl Physiol 2003;95:300-7.

[53] Nordsletten L, Kaastad TS, Skjeldal S, Kirkeby OJ, Reikeras O, Ekeland A. Training increases the in vivo strength of the lower leg: an experimental study in the rat. J Bone Miner Res 1993;8:1089-95.

[54] Huang TH, Yang RS, Hsieh SS, Liu SH. Effects of caffeine and exercise on the development of bone: A densitometric and histomorphometric study in young Wistar rats. Bone 2002;30:293-9.

[55] Huang TH, Chang FL, Lin SC, Liu SH, Hsieh SS, Yang RS. Endurance treadmill running training benefits the biomaterial quality of bone in growing male Wistar rats. J Bone Miner Metab 2008;26:350-7.

[56] Iwamoto J, Yeh JK, Aloia JF. Differential effect of treadmill exercise on three cancellous bone sites in the young growing rat. Bone 1999;24:163-9.

[57] Iwamoto J, Shimamura C, Takeda T, Abe H, Ichimura S, Sato Y, Toyama Y. Effects of treadmill exercise on bone mass, bone metabolism, and calciotropic hormones in young growing rats. J Bone Miner Metab 2004;22:26-31.

[58] Hagihara Y, Fukuda S, Goto S, Iida H, Yamazaki M, Moriya H. How many days per week should rats undergo running exercise to increase BMD? J Bone Miner Metab 2005;23:289-94.

[59] Wheeler DL, Graves JE, Miller GJ, Vander Griend RE, Wronski TJ, Powers SK, Park HM. Effects of running on the torsional strength, morphometry, and bone mass of the rat skeleton. Medicine and science in sports and exercise 1995;27:520-9.

[60] Raab DM, Smith EL, Crenshaw TD, Thomas DP. Bone mechanical properties after exercise training in young and old rats. J Appl Physiol 1990;68:130-4.

[61] Hou JC, Salem GJ, Zernicke RF, Barnard RJ. Structural and mechanical adaptations of immature trabecular bone to strenuous exercise. J Appl Physiol 1990;69:1309-14.

[62] Shimamura C, Iwamoto J, Takeda T, Ichimura S, Abe H, Toyama Y. Effect of decreased physical activity on bone mass in exercise-trained young rats. J Orthop Sci 2002;7:358-63.

[63] van der Wiel HE, Lips P, Graafmans WC, Danielsen CC, Nauta J, van Lingen A, Mosekilde L. Additional weight-bearing during exercise is more important than duration of exercise for anabolic stimulus of bone: a study of running exercise in female rats. Bone 1995;16:73-80.

[64] Salem GJ, Zernicke RF, Martinez DA, Vailas AC. Adaptations of immature trabecular bone to moderate exercise: geometrical, biochemical, and biomechanical correlates. Bone 1993;14:647-54.

[65] Cortright RN, Chandler MP, Lemon PW, DiCarlo SE. Daily exercise reduces fat, protein and body mass in male but not female rats. Physiol Behav 1997;62:105-11.

[66] Lamb DR, van Huss WD, Carrow RE, Heusner WW, Weber JC, Kertzer R. Effects of prepubertal physical trainingon growth, voluntary exercise, cholesterol and basal metabolism in rats. Research quarterly 1969;40:123-33.

[67] Afonso VM, Eikelboom R. Relationship between wheel running, feeding, drinking, and body weight in male rats. Physiology & behavior 2003;80:19-26.

[68] Tokuyama K, Saito M, Okuda H. Effects of wheel running on food intake and weight gain of male and female rats. Physiology & behavior 1982;28:899-903.

[69] Dohm GL, Louis TM. Changes in androstenedione, testosterone and protein metabolism as a result of exercise. Proceedings of the Society for Experimental

Biology and Medicine Society for Experimental Biology and Medicine (New York, NY 1978;158:622-5.

[70] Dudley SD, Gentry RT, Silverman BS, Wade GN. Estradiol and insulin: independent effects on eating and body weight in rats. Physiology & behavior 1979;22:63-7.

[71] Gavin ML, Gray JM, Johnson PR. Estrogen-induced effects on food intake and body weight in ovariectomized, partially lipectomized rats. Physiology & behavior 1984;32:55-9.

[72] Guyard B, Fricker J, Brigant L, Betoulle D, Apfelbaum M. Effects of ovarian steroids on energy balance in rats fed a highly palatable diet. Metabolism: clinical and experimental 1991;40:529-33.

[73] Hervey E, Hervey GR. The effects of progesterone on body weight and composition in the rat. The Journal of endocrinology 1967;37:361-81.

[74] Chatterton RT, Jr., Hrycyk L, Hickson RC. Effect of endurance exercise on ovulation in the rat. Medicine and science in sports and exercise 1995;27:1509-15.

[75] Turner CH, Burr DB. Basic biomechanical measurements of bone: a tutorial. Bone 1993;14:595-608.

[76] Chen HI, Chiang IP. Chronic exercise decreases adrenergic agonist-induced vasoconstriction in spontaneously hypertensive rats. The American journal of physiology 1996;271:H977-83.

[77] Huang TH, Hsieh SS, Liu SH, Chang FL, Lin SC, Yang RS. Swimming training increases the post-yield energy of bone in young male rats. Calcif Tissue Int 2010;86:142-53.

[78] Wallace JM, Ron MS, Kohn DH. Short-term exercise in mice increases tibial post-yield mechanical properties while two weeks of latency following exercise increases tissue-level strength. Calcif Tissue Int 2009;84:297-304.

[79] Lin HS, Huang TH, Mao SW, Tai YS, Chiu HT, Cheng KYB, Yang RS. A short-term free-fall landing enhances bone formation and bone material properties. Journal of Mechanics in Medicine and Biology (In press) 2011.

[80] Currey JD. The effect of porosity and mineral content on the Young's modulus of elasticity of compact bone. J Biomech 1988;21:131-9.

[81] Currey JD. Changes in the impact energy absorption of bone with age. J Biomech 1979;12:459-69.

[82] Boskey AL, Wright TM, Blank RD. Collagen and bone strength. J Bone Miner Res 1999;14:330-5.

[83] Wang X, Bank RA, TeKoppele JM, Agrawal CM. The role of collagen in determining bone mechanical properties. J Orthop Res 2001;19:1021-6.

[84] Skedros JG, Dayton MR, Sybrowsky CL, Bloebaum RD, Bachus KN. The influence of collagen fiber orientation and other histocompositional characteristics on the mechanical properties of equine cortical bone. Journal of Experimental Biology 2006;209:3025-42.

[85] Burr DB. The contribution of the organic matrix to bone's material properties. Bone 2002;31:8-11.

[86] Zioupos P, Currey JD. Changes in the stiffness, strength, and toughness of human cortical bone with age. Bone 1998;22:57-66.

[87] Puustjarvi K, Nieminen J, Rasanen T, Hyttinen M, Helminen HJ, Kroger H, Huuskonen J, Alhava E, Kovanen V. Do more highly organized collagen fibrils increase bone

mechanical strength in loss of mineral density after one-year running training? Journal of Bone and Mineral Research 1999;14:321-9.

[88] Martin RB, Lau ST, Mathews PV, Gibson VA, Stover SM. Collagen fiber organization is related to mechanical properties and remodeling in equine bone. A comparison of two methods. J Biomech 1996;29:1515-21.

[89] Skedros JG, Mason MW, Nelson MC, Bloebaum RD. Evidence of structural and material adaptation to specific strain features in cortical bone. Anat Rec 1996;246:47-63.

[90] Martiniakova M, Grosskopf B, Omelka R, Vondrakova M, Bauerova M. Differences among species in compact bone tissue microstructure of mammalian skeleton: use of a discriminant function analysis for species identification. Journal of forensic sciences 2006;51:1235-9.

[91] Ruth EB. Bone studies. II. An experimental study of the Haversian-type vascular channels. The American journal of anatomy 1953;93:429-55.

[92] Garnero P, Borel O, Gineyts E, Duboeuf F, Solberg H, Bouxsein ML, Christiansen C, Delmas PD. Extracellular post-translational modifications of collagen are major determinants of biomechanical properties of fetal bovine cortical bone. Bone 2006;38:300-9.

[93] Kohn DH, Sahar ND, Wallace JM, Golcuk K, Morris MD. Exercise alters mineral and matrix composition in the absence of adding new bone. Cells Tissues Organs 2009;189:33-7.

[94] Augat P, Schorlemmer S. The role of cortical bone and its microstructure in bone strength. Age Ageing 2006;35 Suppl 2:ii27-ii31.

7

Effect of the Er, Cr: YSGG Laser Parameters on Shear Bond Strength and Microstructure on Human Dentin Surface

Eun Mi Rhim[1], Sungyoon Huh[2], Duck Su Kim[3], Sun-Young Kim[4],
Su-Jin Ahn[5], Kyung Lhi Kang[6] and Sang Hyuk Park[4,7]
[1]The Catholic University of Korea, St. Paul's Hospital,
Dept. of Conservative Dentistry, Seoul,
[2]Shingu University, Dept. of Dental Hygiene, Seongnam,
[3]Kyung Hee University, Dental Hospital, Dept. of Conservative Dentistry, Seoul
[4]Kyung Hee University, Dept. of Conservative Dentistry, Seoul,
[5]Kyung Hee University, Dental Hospital at Gandong,
Dept. of Biomaterials & Prosthodontics,
[6]Kyung Hee University, Dental Hospital at Gandong, Dept. of Periodontology,
[7]Kyung Hee University Dental Hospital at Gandong, Dept. of Conservative Dentistry
Korea

1. Introduction

The developments of laser technology have enabled their use in multiple dental procedures, such as soft tissue operations, composite restorations, tooth bleaching, root canal irrigation, caries removal and tooth preparations with minimal pain and discomfort (Turkmen et al., 2010) Recently, the Er,Cr:YSGG laser was recommended for minimally invasive purposes due to its precise ablation of the enamel and dentin without side-effects to the pulp and surrounding tissues. It has a 2780-nm wavelength and absorbed strongly by both water and hydroxyapatite. The sudden evaporation of bound water causes micro-explosions that blast away tiny particles of the tooth (Obeidi et al., 2010).

Previous studies reported that irregularities and the crater-shaped appearance of ablated dentin was comparable to the dentine surface after acid etching, which might promote micromechanical interlocking between dental restorative materials and the tooth surface (Visuri et al., 1996; Armengol et al., 1999; Martínez-Insua et al., 2000; Carrieri et al., 2007; Gurgan et al., 2008). Despite its efficiency, reports on the bond strengths of composite resin to a tooth substrate prepared by a laser are often confusing and contradictory. Some studies reported higher bond strengths to laser-prepared dentin than to acid-etched dentin (Visuri et al., 1996; Carrieri et al., 2007). Others have reported significantly lower bond strengths on laser preparation (Armengol et al., 1999; Martínez-Insua da Silva Dominguez et al. 2000; Gurgan Kiremitci et al. 2008) and others have reported no significant differences (Abdalla & Davidson, 1998).

Generally there are three adhesive systems. The first uses 30–40% phosphoric acid to remove the smear layer (etch-and-rinse (ER) technique). This bonding mechanism to dentin depends on the hybridization of the resin within the exposed collagen mesh as well as to the dentin tubules (Abdalla & Davidson, 1998), creating a micromechanical interlocking of the resin within the exposed collagen fibril scaffold. The second is the "self-etch" adhesives (SEA) which employs acidic monomers that simultaneously condition and prime dentin. The smear layer remains partially but is used to hybridize with the underlying dentin (Perdigao et al., 2000). The last is an all-in-one system, but the stability of the bond strength decreases with time by because it contains many hydrophilic monomers.

According to previous reports, a tiny flake surface formed by the laser irradiation of dentin can be removed with 30-40 % acid etching, but the primer of SEA cannot. This study hypothesized that the primer of SEA cannot improve the free surface energy of irradiated dentin. Therefore, it is unable to make a proper environment for sufficient bond strength. The null hypothesis was that the shear bond strength with "self-etch" adhesives (SEA) did not show a difference from that with the etch-and-rinse (ER) technique after a pretreatment with a laser in dentin.

This study examined the shear bond strength of a hybrid composite resin bonded with two different adhesive systems to the dentin surfaces prepared with Er,Cr:YSGG laser etching, and evaluated the morphologic structure of de-bonded dentin surface after Shear Bond Strength (SBS) Test by scanning electron microscopy.

2. Materials and methods

2.1 Tooth preparation

Tables 1 and 2 list the materials used in this study and the study design, respectively. Twenty four freshly extracted caries and restoration-free permanent human molars stored in distilled water were used. The teeth were embedded in improved stone with the occlusal surface of the crown exposed and parallel to the base of the stone, and the embedded teeth were sectioned at one third of the occlusal surfaces to expose the dentin surface. Each tooth was wet-ground with 320-grit silicon carbide paper and polished with 1200-grit to obtain a flat dentin surface. The specimens were stored in distilled water at 37 °C. The teeth were divided randomly into two groups, control and laser irradiated groups. The control groups without laser irradiation were divided randomly into two subgroups (SE bond and Single bond), and the laser irradiated groups were divided into four subgroups (SE bond with 1.4 W, 2.25 W and Single bond with 1.4W, 2.25W).

2.2 Laser irradiation

The Er,Cr:YSGG laser (Waterlase, BioLase Technology, Inc., San Clemente, CA) with a 2780 nm wavelength and 20 Hz of power with a sapphire tip was used. Laser irradiation was performed on the dentin surface with either 1.4 W or 2.25 W. The flattened dentin surfaces of the teeth were irradiated at 90° in non-contact mode with a fixed distance of 6 mm away from the laser tip in a sweeping motion to achieve an even surface coverage by overlapping the laser impact. The laser handpiece was attached to a modified surveyor to ensure a consistent energy density, spot size, distance and handpiece angle.

2.3 Bonding procedures

In all groups, including laser irradiated groups and control groups, the bonding procedures recommended by the manufacturer' instruction were followed strictly. In the single bond adhesive system, etching procedures were conducted using 37 % phosphoric acid (3M ESPE, St. Paul, MN, USA) for 15 seconds followed by rinsing with a water spray for 15 seconds, and blot dried, and bonded and light curing for 15 seconds. In the Clearfil SE bond system, a primer was applied to the dentin surface for 10 seconds followed by bonding and light curing, as shown table 1.

After light curing the bonding resin, a Teflon tube (GI tech, Seoul, Korea) with an inner diameter of 2 mm and a height of 2 mm was attached to each dentin surface and filled with A3 Body Shade of the hybrid composite Resin (Filtek Supreme Plus 3 M ESPE, MN, USA) followed by light curing for 40 s. After light curing, the teeth were stored in water at 37°C for 24h before the Shear Bond Strength (SBS) Test.

Brand name	Manufacturer	Material type	Composition	Application procedure
Clearfil SE bond	Kuraray, Tokyo, Japan	Self etching adhesive system	Primer: MDP, HEMA, N,N-Diethanol p-toluidine, water Bonding resin: Bis-GMA, CQ, HEMA, MDP, Micro filler	Apply primer and leave 20 seconds, air blow, apply bonding system
Single bond	3M ESPE, St. Paul MN, USA	Etch and rinse adhesive system	Bis-GMA, HEMA, Water, UDMA, Ethanol, Polyalkenoic acid copolymer	Acid-etch for 15 seconds, rinse, apply adhesive, gentle air blow, light curing for 10 seconds.

Table 1. Materials used in this study.

Laser intensity	Adhesive system	
	Single bond	SE bond
Control (no laser)	Group A	Group D
1.4W	Group B	Group E
2.25W	Group C	Group F

Table 2. Classification of the experimental groups.

2.4 Shear Bond Strength (SBS) test

Before loading, the tube mold was removed carefully with a sharp blade and the specimens were then placed in a custom-made fixture mounted on a Universal Testing Machine (INSTRON Model 5562, Norwood, MA). The specimens were loaded to failure under compression using a knife-edge loading head at a cross-head speed of 1 mm/min. The investigator, who was blinded to the group treatment, performed the load testing procedure. The maximum load to failure was recorded for each sample and the Shear bond strength (SBS) was expressed in megapascals (MPa), which was derived by dividing the imposed force (Newtons) by the bonded area (mm^2).

2.5 SEM evaluation

To evaluate the surface (treated with Etching, Laser, and Laser + Etching) before bonding and the fractured surface between the composite resin and dentin side after measuring shear bond strength, the representative dentin specimens were selected randomly and mounted on aluminum stubs. The samples were then sputter-coated with gold palladium and examined by scanning electron microscopy (SEM) (ISI SX-30 Cambridge, MA, USA) with 1000X magnification.

2.6 Statistical analysis

The obtained shear bond strength data were analyzed by two way ANOVA with a level of significance of $P < 0.05$.

3. Results

3.1 Shear Bond Strength (SBS) evaluation

The Mean SBS±SD (MPa) of each group followed by a Single bond was 7.35±1.65, 7.61±1.50 and 6.34±1.39, respectively, and the Mean SBS±SD (MPa) for each group followed by a SE bond was 7.73±1.17, 7.88±2.77, and 6.48±1.77, respectively (Table 3), but there were no significant differences between the groups. The maximum and minimum SBS was observed in groups E and C, respectively. The maximum and minimum SBS in the Single Bond group was observed in groups B and C, respectively. The maximum and minimum SBS in the SE Bond group was observed in group E and F, respectively (Figure 1). Some samples dislodged and de-bonded during preparation for the shear bond test due to the uneven samples distribution.

Adhesive system	Single bond			SE bond		
Laser intensity(W)	Control	1.4 W	2.25 W	Control	1.4 W	2.25 W
Sample	Group A	Group B	Group C	Group D	Group E	Group F
Mean (MPa)	7.35	7.61	6.34	7.73	7.88	6.48
SD	1.65	1.50	1.39	1.17	2.77	1.77

Table 3. Comparison of the mean shear bond strength of the control and laser irradiated groups (Single bond and SE bond). Mean value (MPa) and standard deviation (SD) of the Shear Bond Strength (SBS) Test for the specimens treated with the Er,Cr:YSGG laser are shown but there were no significant differences between the groups (P>0.05).

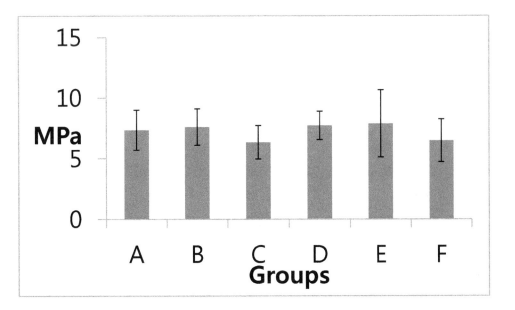

Fig. 1. Shear bond strength between the experimental groups.

3.2 SEM (Scanning Electronic Microscopy)

SEM revealed the fractured surface between the composite resin and dentin side, as shown in Figure 2. The representative SEM evaluation showed that thick layers of the adhesives were found on the fractured de-bonded surface of the composite resin (Figure 2-B, D, F). On the other hand, the fractured bonded surface of the dentin by the Clearfil SE bond adhesive treatment, as a result of the shear bond strength test followed by 37% phosphoric acid etching of some of the residues of the adhesives remaining on the surface of dentin side after the shear bond test (Figure 2-A, C, E).

Regarding the failure mode of the control group in the types of adhesive system, some residual chips were found (A) on the surface of dentin side fractured after 37% phosphoric acid etching and single bond adhesion but the fractured bonded surface of dentin by the Clearfil SE bond adhesive treatment as a result of the shear bond strength test followed by 37% phosphoric acid etching showed thick layers of composite resin attached to the dentin after the failure test (D).

In the irradiation groups, 1.4 W of irradiation followed by the Single bond procedure left the fractured dentin surface with small particles after the failure test (B). On the other hand, the fractured dentin surface irradiated with a power intensity of 2.25 W had exposed dentinal tubules after the test (C). On the other hand, 1.4 W of power intensity irradiation followed by the SE bond procedure revealed exposed dentinal tubules after the failure test (E), whereas 2.25 W of power intensity irradiation followed by the SE bond procedure resulted in a thick layer of composite resin fractured cohesively on the dentin side of the fractured surface (F).

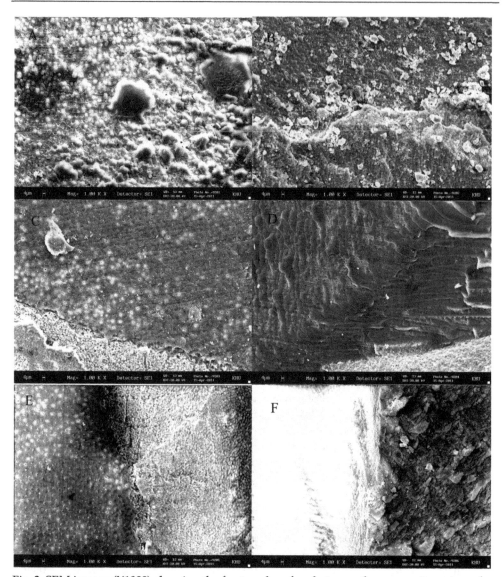

Fig. 2. SEM images (X1000) showing the fractured surface between the composite resin and dentin side. The fractured bonded surface of dentin by the single bond adhesive treatment as a result of the shear bond strength test followed by 37% phosphoric acid etching (A), and 1.4 W of power intensity irradiation followed by a Single bond procedure (B), and 2.25 W of power intensity irradiation followed by a Single bond procedure (C) were evaluated by SEM. The fractured bonded surface of the dentin by Clearfil SE bond adhesive treatment as a result of the shear bond strength test followed by 37% phosphoric acid etching (D), and 1.4 W of power intensity irradiation followed by the SE bond procedure (E), and 2.25 W of power intensity irradiation followed by the SE bond procedure (F) were evaluated by SEM.

4. Discussion

Recently, Er:YAG and Cr:YSGG lasers were introduced in dentistry. For physical and medical reasons, they are used for the treatment of hard tissue. The advantage of an Erbium wave is that it is well absorbed by water and dental hard tissue. Although dentin contains a high proportion of water, the depth is shallow for laser transmission. The strong absorption of water reduces the level of heat during tooth preparation. As water absorbs laser radiation better than dental hard tissue, it reduces the increasing temperature of the tissue during the preparation. Water reaches the boiling point and causes microexposure of the tooth. This action breaks up the surrounding tissue into small pieces and dissipates them at the same time. As this explosion occurs in water, it is so-called a preparation induced by water. Although most radiation is absorbed in water, a certain amount of heat transmission is unavoidable. Therefore, a water spray is used for cooling. The proper amount of water spray prevents pulpal damage. As heat transmission normally depends on the pulse shape and pulse maintenance time, the amount of water varies according to the laser system and treatment itself. A lack of enough water spray causes pulpal damage.

In contrast to initial studies, there was no immunological difference between the laser system and conventional etching method. Moreover, the positive effect of the laser to regenerate pulp was discussed. Essentially, cavity preparation should be performed with the proper instruments with the "method of minimal pulse energy" according to the manufacturer's protocol. This is because the preparation should be performed with the capability to minimize additional damage.

The bond strength of the adhesive system is one of the major factors for the success of restorations. Therefore, this restoration parameter can be measured accurately by the bonding test. The bonding of resin composite material to the tooth structure prepared with different types of lasers has been reported (Visuri et al., 1996; Armengol et al., 1999; Martínez-Insua et al., 2000; Carrieri et al., 2007; Ekworapoj et al., 2007; Lee et al., 2007; de Carvalho et al., 2008; Gurgan et al., 2008; Chou et al., 2009). These studies reported variable results comparing the bond strengths of composite resin to a laser-prepared and acid etched dentin surface, and suggested that the laser preparation is more effective than etching in bonding strength of the composite resin to the tooth structure (Obeidi et al., 2010; Turkmen et al., 2010; Navimipour et al., 2011). This study compared the in vitro SBS of two different adhesive systems to dentin treated with an Er,Cr:YSGG laser. These results showed that the type of adhesive system had no effect on SBS, whereas the laser intensity was a criterion to be considered.

Generally, the manufacturer of Er,Cr:YSGG lasers recommended 2.25 to 2.5 W for laser etching but other studies used 2 to 3 W referring from their results of a pilot study to obtain the proper effects (Çalışkan et al., 2010). The data from this study showed that the shear bond strength (SBS) of dentin treated with 1.4 W, was higher than that of SBS with 2.25 W, which was recommended by the manufacturer. These results were different from other study results showing that the shear bond strength of laser etching (1.25 W or 3 W) was higher than the diamond bur (Gurgan et al., 2008). The distance between the laser tip and laser irradiated dentin surface may not be strict in this kinds of study designs. On the other hand, in this experiment, the sweeping motion of the laser irradiation 6 mm away from the dentin surface was used before the application of adhesive procedures to employ consistent defocused irradiation with lower intensity. Therefore, the difference in the effect between the distance and irradiation time of the laser could not be performed and compared with other studies in this study (Chou et al., 2009; Obeidi et al., 2009; Obeidi et al., 2010).

Based on previous studies using Er,Cr: YSGG laser irradiation on the tooth structure, normal range of power intensity (2.5 W) was utilized on the surface of dentin while high power intensity of the laser (higher than 3 W) was more effective on the bonding strength of the enamel surface (Obeidi et al., 2010), (Dunn et al., 2005). On the other hand, Tagami *et al.* reported that 70mJ (1.75 W) of low laser intensity irradiation on the dentin surface followed by SE bond showed a higher micro shear bond strength than with 150mJ (3 W) (de Carvalho et al., 2008), which is similar to the present experiments. According to the manufacturer's recommendation for laser etching on dentin surface, it was decided to follow the laser protocol mode using two power intensities (2.5 W and 1.25 W). If a power intensity of 2.25 W and 1.4 W are converted to the mJ scale, they could be approximately 112.5 mJ and 70 mJ, respectively. In the present study, 1.4 W of laser intensity irradiation showed the highest shear bond strength between the composite resin and bonded dentin surface with the two types of adhesive systems (Single bond and Clearfil SE bond). Consequently, the SE primer of the SE bond probably could not remove the superficial layers of the irradiated dentin completely. On the other hand, phosphoric acid of three steps adhesive or two steps adhesive could remove most of the superficial layers of the irradiated dentin. In addition, one study reported the advantages of mechanical instruments or combining acid etching and mechanical instruments in removing the superficial layers (Obeidi et al., 2009) .

With the respect to the micromorphological changes after acid etching and laser etching, SEM revealed different characteristic features from those found in conventional acid etched surfaces. The dentin surfaces irradiated with the Er,Cr:YSGG laser had a scaly, irregular, and rugged appearance compared to the acid etched dentin surface (Chou et al., 2009). In addition, with the higher laser intensity or longer irradiation time, the condition was worse in terms of the irregularity of the dentin surface, which has a close relationship with the bonding strength of the composite resin to dentin.

The control groups showed a higher SBS than the 2.25 W of laser irradiated groups in all bonding systems, but, SEM evaluation revealed more adhesive chips remaining on the control group than on the 2.25 W laser irradiated group. On the other hand, thick, rough, and irregular collapsed composite resins were estimated on both the control and 2.25 W laser irradiated groups. SEM examinations of the two kinds of laser intensity irradiation (1.4 W and 2.25 W) revealed that the 1.4 W laser intensity group had more adhesive chips remaining on the dentin than the 2.25 W laser group in the Single bond adhesive system, which is opposite to that observed in the Clearfil SE bond system.

One possible explanation is that the thermomechanical effects of higher laser intensity probably have extended into the subsurface dentin and undermined the integrity of the resin–dentin interface resulting in a lower bond strength. The formation of fissures or cracks in the subsurface dentin might be a start point for the failure of resin-dentin adhesion. Obviously, all irradiated groups were affected by thermo-mechanical effect, but 1.4 W and 2.25W were all affected by thermomechanical effects. On the other hand, 1.4 W laser intensity might not be seriously affected. Laser irradiation with a high intensity may obstruct the dentinal tubules by melting and fail to produce a good hybrid layer. Whereas, laser irradiation with a low intensity may leave the dentinal tubule open and facilitate the infiltration of bonding agent. This may account for the lower SBS values with 2.25 W irradiated dentin than with 1.4 W laser intensity.

These results suggest that the acid etching of lased dentin can reinforce the hybrid layer and formation of resin tags, but the acidic monomer of SE bonds cannot function at its best due to the obstruction of dentinal tubules and denatured collagen fibrils network and the absence of smear layer. These results were not the same as expected. A self etching adhesive system and two steps adhesive system showed a similar SBS because the resin tags only contribute to the bond strength in small portions (15%) and the poorer quality of the hybrid layer appears to be the main reason for the lower bond strength.

5. Conclusion

An Er,Cr:YSGG laser was used to determine if the laser can increase the shear bond strength between the composite resin and surface treated dentine surfaces. On the other hand, the pretreatment of dentin with an Er,Cr:YSGG laser does not affect the shear bond strength of the two different adhesive systems under these experimental conditions. In addition, the shear bond strength of two different adhesive systems in dentin treated with a 1.4W laser intensity was higher than that treated with 2.25W but the difference was not significant.

6. Acknowledgment

This study was supported in part by a grant from Kyung Hee University in 2006 (KHU-20060930) and by the Basic Science Research Program through the National Research Foundation of Korea (NRF) funded by the Ministry of Education, Science and Technology (NRF-20100023448).

7. References

Abdalla AI, Davidson CL (1998) Bonding efficiency and interfacial morphology of one-bottle adhesives to contaminated dentin surfaces. *Am J Dent,* Vol. 11, No. 6, pp. 281-285, ISSN 0894-8275

Armengol V, Jean A, Rohanizadeh R, Hamel H (1999) Scanning electron microscopic analysis of diseased and healthy dental hard tissues after Er: YAG laser irradiation: in vitro study. *J Endod,* Vol. 25, No. 8, pp. 543-546, ISSN 0099-2399

Çalışkan MK, Parlar NK, Oruçoğlu H, Aydın B (2010) Apical microleakage of root-end cavities prepared by Er, Cr: YSGG laser. *Lasers in Medical Science,* Vol. 25, No. 1, pp. 145-150, ISSN 0268-8921

Carrieri TCD, Freitas PM, Navarro RS, P. Eduardo C, Mori M (2007) Adhesion of composite luting cement to Er:YAG-laser-treated dentin. *Lasers in Medical Science,* Vol. 22, No. 3, pp. 165-170, ISSN 0268-8921

Chou JC, Chen CC, Ding SJ (2009) Effect of Er, Cr: YSGG laser parameters on shear bond strength and microstructure of dentine. *Photomedicine and Laser Surgery,* Vol. 27, No. 3, pp. 481-486, ISSN 1549-5418

de Carvalho RCR, de Freitas PM, Otsuki M, de Eduardo CP, Tagami J (2008) Micro-shear bond strength of Er: YAG-laser-treated dentin. *Lasers in Medical Science,* Vol. 23, No. 2, pp. 117-124, ISSN 0268-8921

Dunn WJ, Davis JT, Bush AC (2005) Shear bond strength and SEM evaluation of composite bonded to Er:YAG laser-prepared dentin and enamel. *Dent Mater,* Vol. 21, No. 7, pp. 616-624, ISSN 0287-4547

Ekworapoj P, Sidhu SK, McCabe JF (2007) Effect of different power parameters of Er, Cr: YSGG laser on human dentine. *Lasers in Medical Science*, Vol. 22, No. 3, pp. 175-182, ISSN 0268-8921

Gurgan S, Kiremitci A, Cakir FY *et al.* (2008) Shear Bond Strength of Composite Bonded to Er,Cr:YSGG Laser-Prepared Dentin. *Photomedicine and Laser Surgery*, Vol. 26, No. 5, pp. 495-500, ISSN 1549-5418

Lee BS, Lin PY, Chen MH *et al.* (2007) Tensile bond strength of Er, Cr: YSGG laser-irradiated human dentin and analysis of dentin-resin interface. *Dental Materials*, Vol. 23, No. 5, pp. 570-578, ISSN 0109-5641

Martínez-Insua A, da Silva Dominguez L, Rivera FG, Santana-Penín UA (2000) Differences in bonding to acid-etched or Er: YAG-laser-treated enamel and dentin surfaces. *The Journal of prosthetic dentistry*, Vol. 84, No. 3, pp. 280-288, ISSN 1097-6841

Navimipour EJ, Oskoee SS, Oskoee PA, Bahari M, Rikhtegaran S, Ghojazadeh M (2011) Effect of acid and laser etching on shear bond strength of conventional and resin-modified glass-ionomer cements to composite resin. *Lasers in Medical Science*, pp. 1-7, ISSN 0268-8921

Obeidi A, Liu P-R, Ramp LC, Beck P, Gutknecht N (2010) Acid-etch interval and shear bond strength of Er,Cr:YSGG laser-prepared enamel and dentin. *Lasers in Medical Science*, Vol. 25, No. 3, pp. 363-369, ISSN 0268-8921

Obeidi A, McCracken MS, Liu PR, Litaker MS, Beck P, Rahemtulla F (2009) Enhancement of bonding to enamel and dentin prepared by Er,Cr:YSGG laser. *Lasers Surg Med*, Vol. 41, No. 6, pp. 454-462, ISSN 1096-9101

Perdigao J, May KN, Jr., Wilder AD, Jr., Lopes M (2000) The effect of depth of dentin demineralization on bond strengths and morphology of the hybrid layer. *Oper Dent*, Vol. 25, No. 3, pp. 186-194, ISSN 0361-7734

Turkmen C, Sazak-Ovecoglu H, Gunday M, Gungor G, Durkan M, Oksuz M (2010) Shear Bond Strength Of Composite Bonded With Three Adhesives To Er, Cr: YSGG Laser-Prepared Enamel. *Quintessence International*, Vol.41, pp. e119-e124, ISSN 0033-6572

Visuri SR, Gilbert JL, Wright DD, Wigdor HA, Walsh JT (1996) Shear Strength of Composite Bonded to Er:YAG Laser-prepared Dentin. *Journal of Dental Research*, Vol. 75, No. 1, pp. 599-605, ISSN 1544-0591

8

The Use of Vibration Principles to Characterize the Mechanical Properties of Biomaterials

Osvaldo H. Campanella[1], Hartono Sumali[2],
Behic Mert[3] and Bhavesh Patel[1]
*[1]Agricultural and Biological Engineering Department and
Whistler Carbohydrate Research Center, Purdue University, West Lafayette, IN
[2]Sandia National Laboratories, Albuquerque, NM
[3]Department of Food Engineering, Middle East Technical University, Ankara
[1,2]USA
[3]Turkey*

1. Introduction

Mechanical properties are a primary quality factor in many materials ranging from liquids to solids including foods, cosmetics, certain pharmaceuticals, paints, inks, polymer solutions, to name a few. The mechanical properties of these products are important because they could be related to either a quality attribute or a functional requirement. Thus, there is always a need for the development of testing methods capable to meet various material characterization requirements from both the industry and basic research.

There is a wide range of mechanical tests in the market with a wide price range. However, there is an increasing interest in finding new methods for mechanical characterization of materials specifically capable to be adapted to in-line instruments. Acoustic/vibration methods have gained considerable attention and several instruments designed and built in government labs (e.g. Pacific Northwest National Laboratory and Argon National Laboratory) have been made commercially available.

To measure mechanical properties of material a number of conventional techniques are available, which in some cases may alter or change the sample during testing (destructive testing). In other tests the strains/deformations applied are so small that the test can be considered non-destructive. Both types of test are based on the application of a controlled strain and the measurement of the resulting stress, or viceversa. Different types of deformations, e.g. compression, shear, torsion are used to test these materials.

Depending on the type of material, different conventional techniques utilized to measure its mechanical properties can be grouped as viscosity measurement tests (liquid properties), viscoelasticity measurement tests (semiliquid/semisolid properties), and elastic measurement tests (solid properties).

Acoustics based techniques can be used for all types of material and the following sections discuss in detail how these techniques have been adapted and used to measure materials whose properties range from liquids to solids. Some of the applications discussed in this chapter are based on the basic impedance tube technique. Applications of this technique for material characterization in air have been around for almost a century. The technique is described in basic acoustics text books such as Kinsler et al., (2000) and Temkin (1981). Impedance tube methods based on standing waves and the transfer function method have been accepted as standard methods by the American Society for Testing and Materials (ASTM 1990 and 1995), thus they will not be discussed in this chapter. Instead the chapter will focus on liquids, semiliquid and semisolid materials, many of them exhibiting viscoelastic properties, i.e. those properties that are more representative of biomaterials behavior.

2. Vibration fundamentals and analysis

The theoretical background that supports mechanical characterization of materials using vibration/acoustic based methods is mainly based on the characterization of acoustic waves propagating though the material. In that sense the analysis can be classified on the type of material being tested, i.e. liquid, viscoelastic semifluid, and viscoelastic semisolid.

2.1 Liquid materials

The analysis of liquid samples can be further classified based on the type of container used to confine the testing liquid. One of more important aspects to consider in this classification is the rigidity of the container walls. Two cases are considered: containers with rigid walls and containers with deformable/flexible walls.

2.1.1 Rigid wall containers

Since acoustic waves reveal useful information on the characteristic of the material through which they travel, measurement of acoustical properties such as velocity, attenuation, and phase changes resulting mainly from wave reflections in the transfer media are often used as a tool for mechanical characterization of materials. Specifically, ultrasound has found a wide range of applications in the measurement of the viscosity of liquids. Mason et al. (1949) first introduced an ultrasonic technique to measure the viscosity of liquids. They used the reflection of a shear wave in the interface between a quartz crystal and the sample liquid. Since then many other ultrasonics related techniques have been developed to measure viscosity of liquids [Roth and Rich (1953), Hertz et al. (1990), and Sheen et al., (1996)]. However, acoustical techniques that use sonic frequency have been rather limited. The main reason for this has been the lack of practical approaches that can employ frequencies in the sonic range to study the rheology of liquids. Tabakayashi and Raichel (1998) tried to use sonic frequency waves for rheological characterization of liquids. They affixed a hydrophone and a speaker to both ends of a cylindrical tube and analyzed the effect of the liquid non-Newtonian behavior on the propagation of the sound waves. The approach used by these authors is similar to the well known impedance tube method commonly applied to gases contained in cylindrical tubes, known as waveguides, but their analysis did not include the effect of the tube boundaries on sound propagation which can be of importance. In that sense, the application of the impedance tube techniques to test liquids has been limited because the loss of wall rigidity, i.e. the tube boundary, which normally it does not occur in tubes filled with air or air waveguides. Thus, when the tube is filled with a liquid, the tube may become an elastic waveguide and the rigid wall approximation loses its validity. The key assumption of having rigid walls is related to the shape of the acoustic wave moving through the liquid. In order to have manageable equations to estimate the viscosity of the liquid from acoustic measurements using these systems it is important to generate planar standing waves from which acoustic parameters can be readily obtained. Mert et al. (2004) described a model and the experimental conditions under which the assumption of standing planar waves stands.

If a tube with rigid walls is considered the propagation of unidirectional plane sound waves though the liquid contained in the rigid tube can be described by the following equation (Kinsler et al., 2000):

$$\frac{\partial^2 p}{\partial t^2} - C_1^2 \frac{\partial^2 p}{\partial x^2}$$

(1)

where p is the acoustic pressure, t is time, x distance and C_1 the speed of the sound in the testing liquid. The solution of Equation (1) can be expressed in terms of two harmonic waves travelling in opposite directions and whose composition gives place to the formation of standing waves (Kinsler et al., 2000):

$$p(x,t) = p^+ e^{\left[i\omega t + \hat{k}(L-x)\right]} + p^- e^{\left[i\omega t - \hat{k}(L-x)\right]}$$

(2)

p^+ is the amplitude of the wave traveling in the direction $+x$ whereas p^- is the amplitude of the wave traveling in the direction $-x$ and i is the imaginary number equal to $\sqrt{-1}$. The complex wave number $\hat{k} = \omega / C_1 - i\alpha$ includes the attenuation α due to the viscosity of the liquid, ω is the frequency of the wave. The corresponding wave velocity can be obtained from integration of the pressure derivative respect to the distance x, an equation that is known as the Euler equation (Temkin, 1981) and calculated as $v(x,t) = -\frac{1}{\rho_0}\int \frac{\partial p}{\partial x} dt$), which

yields:

$$v(x,t) = P^+ e^{\left[i\omega t + \hat{k}(L-x)\right]} - P^- e^{\left[i\omega t - \hat{k}(L-x)\right]}$$

(3)

By applying the boundary conditions such that at $x = 0$ the fluid velocity is equal to the velocity of the piston that creates the wave, which is $v(0,t) = v_0 e^{i\omega t}$. The other boundary condition is derived from the practical situation of using an air space on the other end of the tube, which mathematical provides a pressure release condition at $x = L$, written as $p(L, t) = 0$. With these boundary conditions the coefficients P^+ and P^- in Equation 3 can be calculated as:

$$P^+ = \frac{\rho_0 C_1 v_0}{2\cos \hat{k} L} \quad \text{and} \quad P^- = -\frac{\rho_0 C_1 v_0}{2\cos \hat{k} L}$$

(4)

where ρ_0 is the density of the liquid, in repose, and v_0 the amplitude of the imposed wave. An analysis made by Temkin (1981) showed that two absorption mechanisms produce the attenuation of the sound energy. One is due to the attenuation effects produced by the dilatational motion of the liquid during the acoustic wave passage. The other term arises from the grazing/friction motion of the liquid on the wall of the tube. It can be shown that the attenuation due to wall effects is significantly larger than the attenuation due to the dilatational motion of the liquid (Herzfeld and Litovitz, 1959). Thus, a simplified form can be used to obtain the complex wave number ($\hat{k}$) and ultimately the attenuation (α) from which the fluid viscosity can be extracted. The complex wave number is given by the following equation:

$$\hat{k} = k_1 - i\alpha \tag{5}$$

where the real number k_1 is calculated as $k_1 = \dfrac{\omega}{C_1}$ and α is the attenuation of the wave due to the viscosity of the testing liquid.

For measurement purposes it is convenient to estimate the acoustic impedance as p/v, which from Equations (2) and (3) after application of the Fourier transform to the pressure and velocity variables yields:

$$Z_{a0} = \frac{\tilde{p}}{\tilde{v}} = \frac{i\rho_0\omega}{\hat{k}}\tan\hat{k}L = \frac{i\rho_0\omega}{\hat{k}}\frac{\sin\hat{k}L}{\cos\hat{k}L} \tag{6}$$

Where $\tilde{p}$ and $\tilde{v}$ are the Fourier transformed pressure and velocity respectively. The acoustic impedance is a complex number, as well as its inverse, which is known as mobility. Thus, the magnitude or absolute value of the mobility $Abs(1/Z_{a0})$ can be calculated. From values of the acoustic impedance or mobility the complex wave number $\hat{k}$ can be obtained as well as its real and imaginary components, from which the viscosity of the liquid and the intrinsic sound velocity in the liquid of interest can be estimated from the following equation (Temkin, 1981):

$$\hat{k} = k_1 - \frac{i}{RC_1}\left\{\left(\frac{\omega\mu}{2\rho}\right)^{0.5} + \left(\frac{\omega^2}{2C_1^2}\right)\left[\left(\frac{\mu}{\rho}\right)^{0.5} + \left(\frac{\mu}{3\rho}\right)^{0.5}\right] + \frac{\omega^2}{4C_1^2}\left(\frac{\mu}{\rho}\right)^{0.5}\right\} \tag{7}$$

Viscosities of the liquids can be also estimated from the Kirchhoff's equation, which has been derived for waveguides containing a gas and are given by the following equation (Kinsler et al., 2000):

$$\alpha = \frac{1}{RC_1}\sqrt{\frac{\mu\omega}{2\rho_0}} \tag{8}$$

The parameter R in Equations 7 and 8 is obtained from to the magnitude of the mechanical impedance at the resonance frequency (see Figure 2).

From Equation (6) can be observed that plots of the acoustic impedance becomes a maximum when $\cos(\hat{k}L)$ is a minimum, i.e. when the frequencies are given by the relationship $f = \dfrac{C_1}{2L}(n + \dfrac{1}{2})$, where $n = 0,1,2,3,....$, which is in terms of the length of liquid L and the velocity of sound in the liquid C_1, can provide a location of those maxima. Those theoretical observations were experimentally validated by Mert et al. (2004) using an experimental setup as that described in Figure 1.

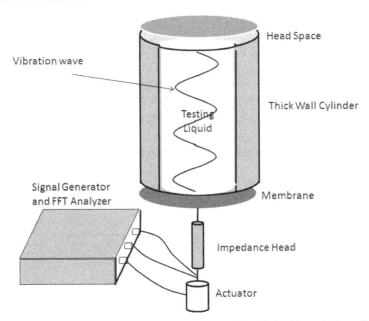

Fig. 1. Schematic of the apparatus to measure viscosity of liquids inside a thick wall rigid tube.

Further details of this experimental setting can be found elsewhere in Mert et al. (2004).
It has been hypothesized that plots of absolute mobility as a function of frequency should exhibit peaks and valleys corresponding to the resonance and anti-resonance frequencies of the standing waves from which the rheological properties of the liquid, notably its viscosity, can be obtained. To validate the theory described above, liquids with known physical properties including viscosity, density and sound velocity though them (the latter measured by a pulse-echo ultrasound method) were tested. Table 1 list viscosity of the liquids (liquids 1 to 4) used for this test along with their relevant physical properties.

Liquid	Viscosity $mPa.s$	Density kg/m^3	Intrinsic sound velocity m/s
1	17	800	1246
2	96	971	1010
3	990	969	967
4	4900	963	943
5	10,050	980	-

Table 1. Physical properties of the liquids used to validate theory; properties are at 25°C.

Figure 2 illustrates measured magnitude of the mechanical impedances Abs ($1/Z_{a0}$) liquids 2, 3, and 4 (see Table 1). It is clearly shown in the figure the effect of the liquid viscosity on the measured impedance, in general the higher is the viscosity of the liquid the lower is the value of the magnitude of the mechanical impedance at the peak/resonance frequency. The resonance frequency (frequency where peaks in impedance are obtained) is also affected by the viscosity of the liquid and tend to decrease when the viscosity of the liquid increases, which would be indicating a decrease in the sound velocity though the liquid. The latter in fully agreement with the ultrasonic pulse-echo measurements used to estimate the sound velocities in the different liquids, which are reported in Table 1.

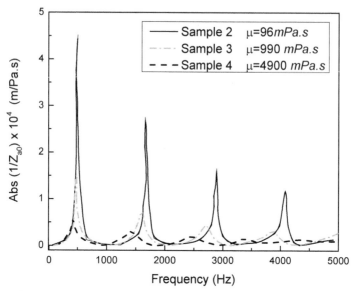

Fig. 2. Effect of viscosity on the magnitude of acoustical impedances measured in a setup as illustrated in Figure 1 at the driver position x = 0. Liquids 2, 3 and 4 were tested and properties and given in Table 1.

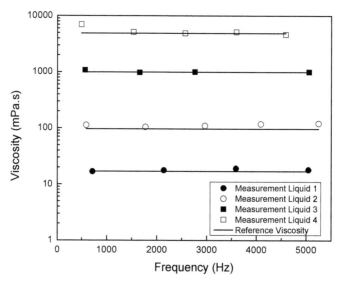

Fig. 3. Viscosities obtained from the different resonance frequencies for 4 different liquids obtained from the measured mechanical impedance and from Equation (7).

Extracted viscosities from impedance measurements along reported values of the liquid viscosities are illustrated in Figure 3.

Liquid viscosities were estimated from Equations 7 and 8 but better accuracy was obtained with Equation 7 that was derived for waveguides containing liquids (Mert et al., 2004). Results of the calculated viscosities are illustrated in Figure 3.

2.1.2 Flexible wall containers

When the walls of the waveguides are not rigid, during propagation of the waves the walls of the container expand and the overall liquid bulk modulus decreases leading to reduced sound velocity in the liquids. In addition, the fluid velocity cannot be assumed to be 0 at the can wall due to the expansion of the wall. These conditions make the governing equations used to estimate the wave attenuation and other relevant acoustic parameters very elusive, thus it is not possible estimate the liquid viscosity directly. One can overcome this problem by an empirical approach, which is by defining the quality factor Q, which is determined from the following equation:

$$Q = \frac{\omega_o}{\omega_2 - \omega_1} \tag{9}$$

ω_1 and ω_2 are the frequencies at which the amplitude of mechanical impedance response is equal to half the actual value at resonance and ω_0 is the resonance frequency (Kinsler et al., 2000). If the quality Q is known in a given container-liquid system it would be possible to have an estimation of the viscosity of the liquids contained in the container, which would serve for quality control purposes. To prove that concept Mert and Campanella (2007) performed a study where a shaker applied vibrations to a cylindrical can containing a liquid using a system, schematically shown in Figure 1. The vibration was able to move the can containing the liquid, and a wave was generated through the liquid, which reflected back in the interface between the liquid and the headspace to form standing waves resulting from composition of the forward and reflected wave. Properties of the standing waves were measured in the frequency domain, and in particular the resonance frequency and the amplitude of the wave at that resonant frequency were obtained and using a calibration curve approach related empirically to the rheological properties of the liquid. It is important to note that these measurements do not provide the true viscosity of the testing liquid because the walls of the can are not rigid and deform significantly due to the vibration. Despite of that the properties of the standing waves, measured by the quality Q, were highly correlated with the rheology of the liquid, which was tested offline in a rheometer (Mert and Campanella, 2007). Results shown in Figure 4 are a frequency spectra for the liquids reported in Table 1.

Quality factors Q for the testing liquids can be estimated from the two peaks observed in Figure 4 and potted as a function of the liquid viscosity as illustrated in Figure 5. However, as shown in Figure 4, the first resonances peaks do not seem to provide sufficient resolution and the second resonance peaks, at higher frequency, are used to find a relationship with the liquid viscosity.

Although interesting from an academic standpoint correlations like the one shown in the Figure 5 are not of practical applicability. However, if the can/cylinder contains a product

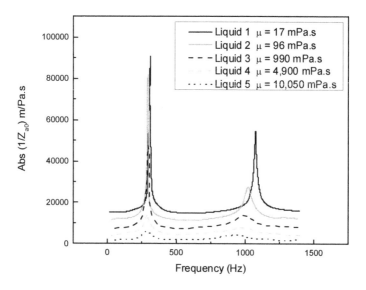

Fig. 4. Amplitudes at resonance frequencies for standard liquids given in Table 1. Curves were shifted up for clarity.

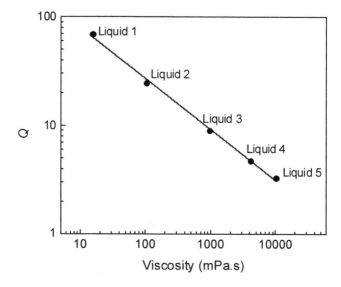

Fig. 5. Quality factor Q versus viscosity of liquid standards

whose viscosity is an important quality parameter, the present method would be of great applicability because the viscosity of the liquid could be assessed quickly without the necessity of opening the can (i.e. non-destructive method). One example of such application is testing cans of tomato products, which are widely produced and consumed in most part of the world and the viscosity of these products is a key parameter associated to their quality in terms of sensory evaluation and processing applications. Viscosity of tomato products is evaluated using a number of rheological techniques that range from empirical to more fundamental methods. One standard method is the use of the Brookfield[TR] viscometer. This viscometer is based on the rotation of a particular element (spindle) inside a can containing the product. The viscosity of the product is obtained basically from the resistance offered by the product to the rotation of the spindle provided shear rates and shear stresses can be accurately calculated. Since the geometries of the spindles often are not regular it is difficult to estimate the rheological parameters, i.e. shear stress and shear rate, which enable the calculation of the liquid viscosity. Tomato paste is a very viscous product that exhibits an important structure that can be destroyed during testing. One approach to overcome this problem is the use of a helical-path spindle, which through a helical movement continuously is touching a fresh sample. Given the complicated rheology, results of this test are considered empirical. The other approach is to use technically advanced rheological equipment, which though the use of geometries such as parallel plates the sample can be minimally disturbed. For these cases the true viscosity of these tomato purees can be obtained. These materials are known as non-Newtonian to indicate that their viscosities are a function of the shear rate. Most of the non-Newtonian liquids can be described by a rheological model known as the power-law model which can be expressed by the following equation:

$$\eta = k\dot{\gamma}^{n-1} \tag{10}$$

where η is the apparent viscosity of the material, which is a function of the applied shear rate $\dot{\gamma}$ and k and n are rheological parameter known as the consistency index and the flow behavior , respectively. The value of k is an indication of the product viscosity whereas n gives a relation between the dependence of the viscosity with the applied shear rates. For tomato products $n<1$, and the smaller is the value of n the largest is the effect of the shear rate on the viscosity of the liquid. To test the feasibility of vibration methods, cans of tomato puree and dilutions ranging from 23% to 3.5% solids were tested in an apparatus similar to the one shown in Figure 1. Results of the tests can be observed in Figure 6, where frequency spectra resulting from some the tests are shown. Figure 7 shows possible correlations between the quality Q measured from the frequency spectra data and the parameter k determined using a standard rheological technique and a parallel plate geometry.

As indicated in Figure 6 the frequency spectra peaks shift to lower frequencies when the solid content of the concentrates increases. In the range of frequency tested the two peaks are visible for concentrates with low solid content whereas the second peak at higher frequencies disappears for concentrates with higher higher solid contents (23 Brix). The amplitude of the Absolute value of the mobility is considerably decreased when the solid content and the viscosity of the concentrate increase (Figure 6). Given the existence of two peaks quality values can be extracted from the two peaks and relate them with the measured rheological properties, in this case the value of the consistency index (Figure 7). It can be seen in the figure that correlations of the liquid rheological properties with its acoustic properties measured by the quality factor Q are strong and that the quality evaluated at the first peak provides a better representation of the liquid viscosity. This is also enhanced due to the presence of two peaks for low and moderate solid contents (measured as Brix), which provides more experimental points to establish the correlation (see Figure 7 – and compare first peak and second peak data).

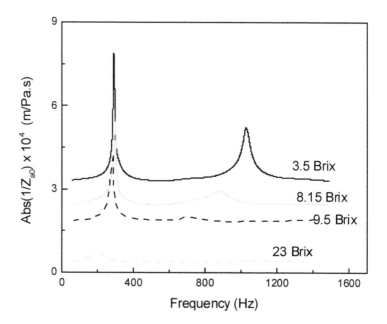

Fig. 6. Frequency response spectra of tomato concentrates at different soluble solids concentrations, measured as Brix. Curves were shifted up for clarity.

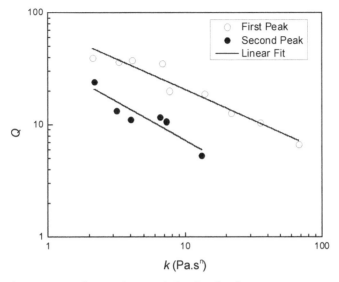

Fig. 7. Quality factor versus the consistency index k value for tomato puree and its dilutions

2.2 Viscoelastic materials
2.2.1 Semifluid materials

Squeezing flow is a well-known technique that has been applied to characterize the properties of various biomaterials ranging from liquids to semisolids. The traditional method involves measuring the force required to squeeze a sample between two cylindrical disks either at a constant velocity or by applying a constant force or stress (Campanella and Peleg 2002).

The oscillatory squeezing low method (OSF) uses the same geometry as the standard squeezing flow method but it involves the application of small amplitude oscillations at random frequencies up to 20 kHz (Mert and Campanella, 2008). The method allows one to calculate both the viscous and elastic components of the sample viscoelasticity by measuring the response of the material in terms of force and acceleration to those oscillations. Transformation of the measured force and acceleration to the frequency domain yields a frequency spectrum for the sample and, ultimately, its resonance frequency. From analysis of this frequency response, two important viscoelastic properties of the samples, the loss modulus G" (viscous) and the storage modulus G' (elastic), can be obtained.

The application of acoustic principles to the squeezing flow method is a novel technique, which convert the squeezing flow method into OSF methos can measure the rheological properties of materials that range from pure liquids to solids. Advantages of this method include it being non-invasive, little to no sample preparation, and its ability to monitor rapid changes during dynamic processes.

A schematic of the OSF testing apparatus is illustrated in Figure 8. The design uses a piezo-electric crystal stack attached to an impedance head. Upon the application of voltage, the upper plate oscillates, and the force and acceleration at the oscillating plate are measured through the impedance head and transformed into the frequency domain using a Fast Fourier Transformation (FFT) routine. This transformation is very useful because from the inspection of the frequency response of the measurement it is possible to identify a characteristic resonance frequency for the sample.

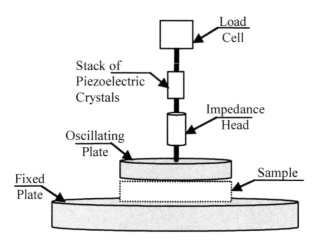

Fig. 8. Schematic of the OSF testing apparatus

Since the samples are considered viscoelastic they can be represented by a combination of elastic component, with a stiffness S, and viscous components with a damping R. A schematic of the elastic-viscous system used for analysis of the measurements is illustrated in Figure 9. For the testing, the sample is placed between the oscillating plate and a fixed plate. The oscillating plate is then brought down to touch the sample and the gap between

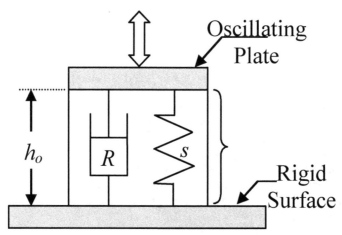

Fig. 9. Schematic of the spring-dashpot system, used for the analysis of the data. h_o, R, and S are the height, the viscous resistance (damping), and the stiffness of the sample respectively.

the plate (h_0) is noted. A load cell is attached above the oscillating plate to control the squeezing force applied to the sample prior the application of the oscillation. The squeezing force is simply applied to make sure that a good contact is established between the oscillating plate and the sample. It will be shown below, however, that the results are independent of that squeezing force.

The frequency response data is obtained using software that interfaces the results obtained with the signal generator device (see also details in Figure 1). The transformed mechanical impedance $\hat{Z} = \dfrac{\hat{F}}{\hat{u}}$ is then calculated from the force and velocity measured on the oscillating plate with the impedance head. $\hat{F}$ and $\hat{u}$ are the Fourier transformed variables and since they have been transformed into the frequency domain, they, as well as the mechanical impedance, are complex variables. The measured complex mechanical impedance at the driving point can be defined as:

$$\hat{Z}_{meas} = \hat{Z}_{instrument} + \hat{Z}_{sample} \tag{11}$$

where $\hat{Z}_{meas}$ is the measured impedance, $\hat{Z}_{instrument}$ is the instrument impedance that can be calculated simply as $\hat{Z}_{instrument} = im_{plate}\omega$ because the instrument does not have any spring mechanism or internal damping. From Equation (11) the sample impedance $\hat{Z}_{sample}$ can be obtained by subtracting the instrument impedance from the measured impedance.

The rheological behavior of the sample can be described in terms of the viscous component (R), which provides the damping of the oscillation and the elastic component (S), which provides the sample elasticity (Figure 9). The relationship between the mechanical impedance and the damping (viscous component) and stiffness (elastic component) of the sample can be described by Equation (12) below:

$$\hat{Z}_{sample} = R_{sample} + i \cdot (\omega \cdot m - \frac{S_{sample}}{\omega}) \tag{12}$$

where ω is the angular frequency of the oscillation, m is the mass of the system, and i is $\sqrt{-1}$.

The mobility of the sample, can be plotted as a function of frequency to provide the resonance spectrum of the sample. The resonance frequency, f_{res} of the sample, which is obtained as the frequency at which the mobility is a maximum is directly related to the stiffness and the mass of the system by Equation (13):

$$f_{res} = \sqrt{\frac{S_{sample}}{m}} \tag{13}$$

A typical plot of Mobility versus frequency for different concentrations of xanthan gum, a biopolymer that produces viscoelastic suspensions, is illustrated in Figure 10.

The higher is the concentration of xanthan gum the higher is its elasticity, which is clearly illustrated in the Figure 10 by a shifting to the right of the resonance frequency. That shifting of the frequency is a clear indication on increase in the stiffness of elasticity of the sample with concentration (see Equation 13).

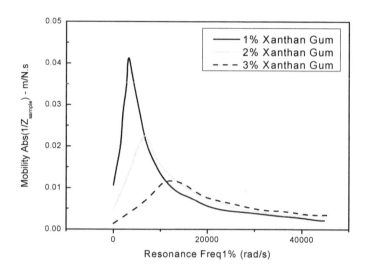

Fig. 10. Mobility plots obtained with the OSF method for xanthan gum dispersions of different concentrations

Although the mobility and the elastic and viscous components of a viscoelastic sample are often useful as quality and processing parameters (Gonzalez et al., 2010), for modeling purposes it is necessary to get quantitative and fundamental rheological information of a sample in terms of elastic and viscous modulus. Phan-Thien (1980) demonstrated that for viscoelastic materials subjected to squeezing flow under an oscillating plate, the squeezing flow force can be calculated as:

$$F_o \cos(\omega t) = \frac{3\pi a^4}{2h_o^3} \eta^* u_o \cos(\omega t) \tag{14}$$

where a is the radius of the top plate, h_o is the distance between plates and $\eta^* = \eta' - i\eta''$ is the complex dynamic viscosity of the sample. $\eta' = \dfrac{G''}{\omega}$ and $\eta'' = \dfrac{G'}{\omega}$ are the viscous and elastic components of the sample, respectively. Application of FFT to Equation (15) yields:

$$\hat{Z}_{sample} = \frac{3\pi a^4}{2h_o^3} \eta^* \tag{15}$$

Expressions can be rearranged and using the definition of the complex viscosity η^*, expressions for the viscoelastic moduli, i.e. the storage and loss modulus, G' and G'', can be obtained as:

$$G' = \left[Im\left(\hat{Z}_{sample}\right) + \frac{3 m_{effective}\, \omega a^2}{20 h_o^2} \right] \cdot \frac{2h_o^3 \omega}{3\pi a^4} \tag{16}$$

$$G'' = Re\left(\hat{Z}_{sample}\right) \cdot \frac{2h_o^3 \omega}{3\pi a^4} \qquad (17)$$

$Im\left(\hat{Z}_{sample}\right)$ and $Re\left(\hat{Z}_{sample}\right)$ are the imaginary and real part, respectively, of the sample complex mechanical impedance $\hat{Z}_{sample}$. It is important to note that for the calculation of the storage modulus G' by Equation (16) is required to know the mass of the system, which consist of an effective mass that includes the mass of the sample and the squeezing force imposed to achieve good contact between the oscillating plate and the sample. If Equation (12) is rewritten in terms of those masses we can obtain an equation for the measured complex mechanical impedance:

$$\hat{Z}_{meas} = R_{sample} + i(\omega m_{plate} + \omega m_{effective} - S_{sample} / \omega) \qquad (18)$$

The inertia produced by the plate instrument can be easily estimated as $\hat{Z}_{instrument} = i\omega m_{plate}$. However, to estimate the inertia produced by the effective mass $m_{effective}$ some additional calculations are required. If the imaginary part of the sample impedance is divided by the frequency, the following equation is obtained:

$$\frac{Im(\hat{Z}_{sample})}{\omega} = (m_{effective} - S_{sample} / \omega^2) \qquad (19)$$

The term S_{sample} / ω^2 in Equation (19) approaches to zero as the frequency is very high, which results in $Im(\hat{Z}_{sample}) / \omega \approx m_{effective}$. Results of this calculation show that the parameter $Im(\hat{Z}_{sample}) / \omega$ reaches an asymptotic value, which is independent of the squeezing force applied to the sample, S equals to $m_{effective}$. That value can then be used to calculate the storage modulus by Equation (16). Results for a xanthan gum suspension of concentration 2%, and whose mobility versus frequency data is shown in Figure 10, are shown in Figure 11 in terms of the viscoelastic storage and loss moduli defined by Equations (16) and (17). As shown in the figure the data compares well with those obtained with a conventional rheometer at comparable frequencies. It must be also noted that the range of frequencies of the OSF method is significantly higher than those applied by conventional rheological methods where inertia may play an important role.

2.2.2 Semisolid materials

Physical properties of solid viscoelastic foods and other biological products are very important in food production, storage, handling, and processing. The importance of the knowledge of the physical properties of biomaterials is demonstrated in the case of fruits. One of the most important quality parameter of fruits is its texture. Texture is the first judgment a purchaser makes about the quality of a fruit, before sweetness, sourness, or flavor. Since fruit texture is such an important attribute, one would expect the changes in texture during maturation and cool-storage to be well understood, however this is not the case. One of the main limitations to the study of fruit texture is the accurate and precise measurement of texture as perceived by a consumer.

Traditionally, the texture of fruits is measured by a Magness Taylor pressure tester (Magness and Taylor, 1925). This simple device measures the force required to insert a metal

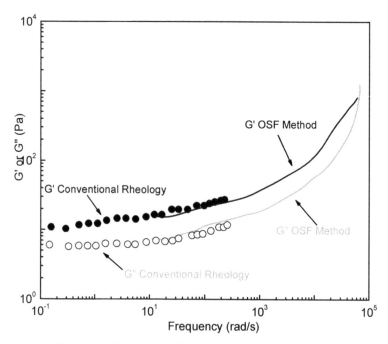

Fig. 11. Storage and loss modulus of a xanthan gum suspension of 2% concentration obtained with the OSF and conventional rheological methods

cylindrical plunger for a given distance into the fruit flesh. Although this device is cheap, quick, and easy to use, it has some disadvantages. Firstly, it is a destructive test, and it is not possible to have repeated measurements on a single fruit. Due to extremely high fruit-to-fruit variability, repeated measures on the same fruit would be highly desirable, especially from a research perspective. Secondly, there is high variability between operators, depending on the speed and force that the plunger is inserted into the fruit. However the greatest limitation of the Magness Taylor fruit pressure tester is that it measures the compression or crushing force that the cells of the fruit cortex are able to withstand. This force is very different from the force that is exerted on a fruit by the teeth of a consumer or the pressure that the consumer can exert with the fingers to assess the fruit texture. Biting an apple, for instance, involves cleaving cells apart, the exact opposite of the force measured by pressure testers. Universal Testing Machines (UTM), like for example the Instron™ instrument, are devices that can measure cleaving force, but they are slow, expensive, and again the method used with these machines is a destructive test.

Throughout the literature, there is abundant evidence that the ripeness or softness of intact fruits is related to their vibration properties. A device that uses vibrations to characterize intact fruits was patented in 1942 (Clark et al., 1942). Vibration properties of some fruits are correlated with firmness and ripeness. Finney (1970) explored methodologies for measuring and characterizing the vibration response of many fruits, and concluded that Young's modulus and shear modulus of apple flesh were correlated with the product between the resonance frequency squared and the mass of the whole fruit. It was also found that the stiffness coefficient measured by vibration was closely associated with sensory panel subjective evaluations of Red Delicious apples (Finney, 1970, 1971). Garrett (1970) argued that the stiffness factor should be proportional to the frequency squared times mass to the two-thirds power. Cooke (1972) confirmed Garrett's theory using elastic theory relating the stiffness factor to the shear modulus of the fruit flesh. Both Finney's and Garrett's stiffness factors have been shown to correlate well with the fruit modulus (Clark and Shackelford, 1973; Yamamoto et al., 1980; and De Bardemaeker (1989); Abbott, 1994; Abbott and Liljedahl, 1994). Differences in shear moduli between green and ripe fruits can be detected non-destructively, and relatively easily, by vibration testing.

Through a few decades of development, a vibration-based characterization method called Experimental Modal Analysis (EMA) has advanced to become a very efficient tool for obtaining the dynamic properties such as natural frequency, damping and mode shapes in aerospace and mechanical engineering. For fruits, natural frequency is related to the shear modulus of the tissue (Cooke, 1972), which in turn determines the firmness of the fruit. For instance, in apples, high damping prevents them from giving a nice them from giving a nice crispy ringing sound when tapped (De Baerdemaeker and Wouters, 1987). Mode shapes are an important indicator of the whole fruit's conditions such as ripeness, bruises, or defects (Cherng, 2000). It is obvious that there is abundant information on the texture and quality of fruits that could be associated with their acoustic parameters. This section of the chapter is meant to give an idea of the direction that researchers have taken in the last decade or so. To give specific examples, a few details are given on the measurement of Young's modulus, finite element modeling, and experimental modal analysis.

2.2.2.1 Measuring modulus of elasticity

An instrument that measures force as a function of displacement, such as a UTM, can be used to obtain force-displacement curves of the samples. For a melon, the flesh and the rind should be tested differently. Melons have a rind that is significantly stiffer than the flesh. The cylindrical samples of the flesh can be tested with a compression test. The rind samples can be tested using a three point jig, and treated as a simply supported beam for analysis.

For modeling vibrations, where only small displacements are of interest, the tests should be limited to small displacements and changes in the cross-section and in the length can be ignored in the data processing. Also, the mechanical properties of biological materials in general are not constant. Because of the structures at the micro scale, the Young's modulus is a function of strain. The following method can be used to facilitate the computation of the Young's modulus at zero deformation. The stress-strain curve can be fit to a cubic function

$$\sigma = a_3 \varepsilon^3 + a_2 \varepsilon^2 + a_1 \varepsilon + a_0 \qquad (20)$$

The coefficients a_0 through a_3 can be obtained by cubic regression analysis. The Young's modulus is the derivative of the stress with respect to strain

$$E = \frac{d\sigma}{d\varepsilon} = 3a_3 \varepsilon^2 + 2a_2 \varepsilon + a_1 \qquad (21)$$

For vibration analysis and modeling, the most important value of the Young's modulus is at zero strain, which is simply a_1. Therefore, the constitutive relation for vibration models is

$$\sigma = E\varepsilon \tag{22}$$

where $E = a_1$ from the test above.

In the bending test where the rind is treated as a simply supported beam, the equation for the young's modulus can be derived accordingly. If the force is applied in the middle of the span, then the deflection at that point is

$$x = \frac{FL^3}{48EI} \tag{23}$$

If the material is linear, the constant Young's modulus is

$$E = \frac{L^3}{48I} \frac{F}{x} \tag{24}$$

Where L is the span, and I is the moment of inertia of the rind sample, which is given by

$$I = \frac{base.height^3}{3} \tag{25}$$

For nonlinear materials, E can be obtained by taking the derivative of the force with respect to x. In particular, the Young's modulus at zero strain is

$$E(x) = \frac{L^3}{48I} \frac{dF}{dx}\bigg|_{x=0} \tag{26}$$

The modulus of elasticity of a melon was measured statically using an UTM machine, which can record the force as a function of deflection. For the flesh, cylindrical core samples were cut out of the melon; whereas for the rind, beam-like samples were cut out (Ehle, 2002). The force-displacement curve for the cylindrical core compression test was transformed into a stress-strain curve for small deflections by dividing the force by the cross-section area and dividing the deflection by the length. The small changes in cross-section area and the length were neglected because the displacement was small. Cubic regression of the stress-strain curve obtained the coefficients in Equation 20. Then Equation 21 was used to calculate the slope, which was the Young's modulus. The value at zero strain is the value used for modeling the vibration properties of the melon. Figure 12 shows that a cubic function fits the data very well, and that the slope at zero strain can be obtained accurately. In the above compression test of the flesh samples, the Young's modulus could also be obtained from the load-displacement curve without transforming into the stress-strain curve numerically. A little algebra gives:

$$E(x) = \frac{L}{A} \frac{df}{dx}\bigg|_{x=0} \tag{27}$$

For the bending test, the force-displacement data were processed according to Eq. 24 to obtain the Young's modulus of the rind. The result shows characteristics similar to those illustrated in Figure 12.

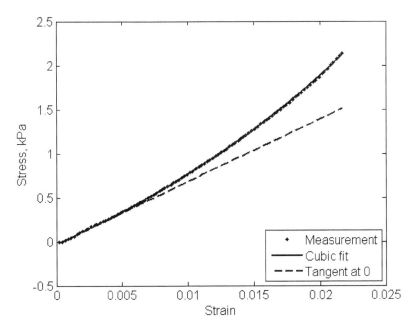

Fig. 12. Stress-strain curve of a core sample from compression test.

2.2.2.2 Numerical modal analysis with finite element modeling (FEM)

The application of stresses and strains in biological materials not only has a component associated to consumer evaluation but also it is related to quality control. Many of the stresses and strains are applied locally either during processing or consumption of the fruits. But they are distributed over the entire surface, thus it is necessary to use methods that can estimate the distribution of stresses over the product. Finite Element Modeling (FEM) is a well developed and proven mathematical and simulation tool to apply to the study of quality of fruits during consumption, storage and processing.

Cherng (2000) stated that the lowest natural frequency for hollow fruit like melons corresponded to the elongation along the major axis. Cooke (1972) concluded that the most important mode was the first twisting mode because it corresponds to the shear modulus of the flesh, which in turn corresponds to the fruit ripeness. Finite element models published in Cherng (2000) did not model the rind separately but assumed that the rind has the same properties as the flesh. Nourain et al (2004) compared a finite element model to experimental modal analysis results.

As an illustration, a finite element mesh used for a melon fruit is shown in Figure 13, with an octant removed to show the inside. The idealized melon-like geometry is a prolate ellipsoid with the major axis in the vertical direction. The major radius of the flesh is 77.5mm. The minor radius is 75mm. The hollow radii are 52.5mm (major) and 50mm (minor). The rind thickness is 2.5mm. For simulation purposes assumptions for the rind Young's modulus was 4.0 MPa whereas values used for the flesh Young's modulus were 1.0MPa, 1.5MPa, 2MPa, and 3MPa. The modal frequencies will be given in the order corresponding to those Young's moduli. The Young's moduli selected above are somewhat arbitrary and do not represent the values shown in Figure 12 from the test applied to a melon. The purpose of this model is to give insight into how the vibration modes change with the stiffness of the flesh.

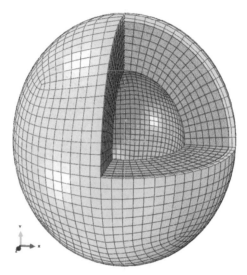

Fig. 13. A finite element mesh of a melon model

Figure 14 shows the elastic mode with the lowest natural frequency, which is elongation in the minor axis direction. Figure 15 shows the second mode, which is elongation in the major axis direction. Figure 16 shows the twisting mode. When the Young's modulus is 1.5MPa or lower, this mode changes shapes into the shape shown in Figure 17

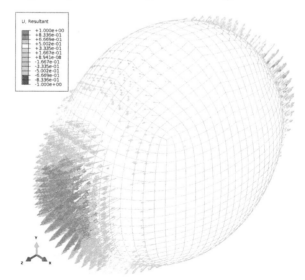

Fig. 14. Mode 1, "sideways elongation", 75.2Hz, 90.4Hz, 97.4Hz, 115.0Hz

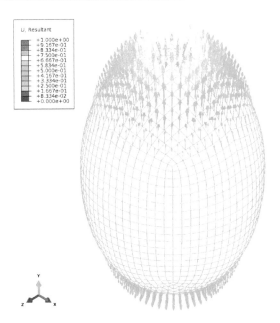

Fig. 15. Mode 2, elongation along the major axis, 79.4Hz, 91.5Hz, 102.0Hz, 120Hz

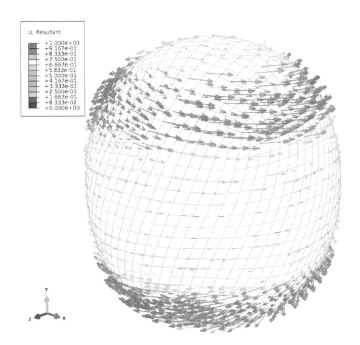

Fig. 16. Twisting about vertical axis, 101.4Hz, 125.1Hz, 135.0Hz, 152.3Hz

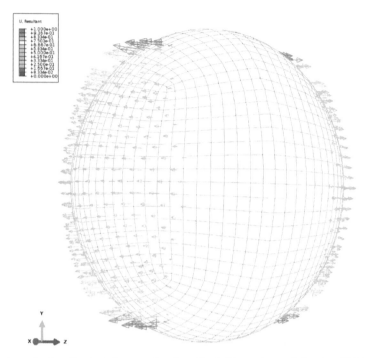

Fig. 17. Mode replacing the twisting mode when flesh Young's modulus is 1.5MPa

2.2.3 Experimental modal analysis of the fruit

Experimental modal analysis is a technique to obtain natural frequency, mode shapes and damping ratios of an elastic structure by application of a vibrational stimulus to the sample and sensing the resulting vibration at various locations on the sample surface. This section illustrates an experimental modal analysis on a melon fruit (Ehle, 2002). The excitation was a pulse force delivered by an impact hammer. A force transducer at the tip of the hammer measured the force. An accelerometer was used to measure the acceleration response at several points all over the surface of the melon. The measurement points were designed to follow a grid pattern where the distances between adjacent points were almost uniform throughout the surface of the fruit. The melon most closely resembled a sphere. Table 2 and Fig. 18 show the measurement points. A tri-axial accelerometer must be used to sense the important torsional mode (Fig. 8) as well as radial motions.

Layer	1	2	3	4	5	6	7	8	9
Points	1	2-7	8-19	20-37	38-55	56-73	74-85	86-91	92
Longitudinal Angle (deg)	90	67.5	45	22.5	0	-22.5	-45	-67.5	-90
Latitudinal Angle (deg)	N/A	60	30	15	15	15	30	60	N/A

Table 2. Locations of the points where acceleration was measured on the melon.

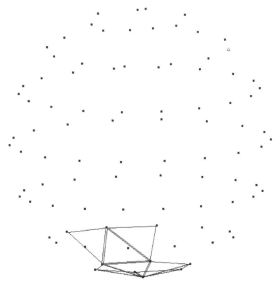

Fig. 18. Measurement points for the above described testing

Fig. 19. A long elastic cord and rubber bands supporting the melon.

Any method used to hold the melon is a boundary condition which will alter the vibration modes. However, the free-free boundary condition has been proven to not alter the vibration modes. This boundary condition can be approximated by suspending the melon from a rigid structure with elastic cords. Soft elastic cords are used in the suspension so that the rigid-body vibration of the melon and its suspension has very low natural frequency and does not alter the elastic modes. Figure 19 shows the suspension the setup used for the holding of the sample.

The porous surface of the melon made it difficult to affix the accelerometer to the surface of the melon using conventional methods. To make sure the accelerometer was affixed snugly to the surface of the melon, wax was applied to the bottom of the accelerometer. This allowed for the mating of the accelerometer to the surface of the melon, which would result in better vibration transmission. Also the accelerometer was held in place by a "cradle" made from rubber bands (Figure 20). The width of the rubber bands was approximately 6 mm, just wide enough to cover the top surface of the accelerometer. The rubber band was positioned in such a way as to apply sufficient pressure so that the surface of the accelerometer was always held normal to the surface of the melon. This configuration allowed for an effective and easy-to-move attachment method to install the accelerometer.

Fig. 20. Accelerometer attachment

To excite the fruit a modal hammer with a force transducer was used. A plastic tip was used as opposed to the metal tip. The plastic tip allowed for the force from the hammer to stay within the crucial 20dB range past the cut off frequency of 500 Hz. The accelerometer then measured the resulting accelerations of the surface of the melon. An FFT analyzer transformed the acceleration and force into the frequency domain, and obtained the ratio of acceleration to force, which was the Frequency Response Function (FRF) at each sense point.

To reduce random noise, the average of data from ten impacts was used to compute each FRF. The *coherence* is a function in the frequency domain that indicates the quality of the FRF. A perfect coherence is 1.0. A coherence less than that at a given frequency means that the vibration sensed by the accelerometer at that particular frequency is not a linear response to the excitation from the impact hammer. A low coherence in general indicates excessive noise, nonlinearity, or other causes of bad measurement.

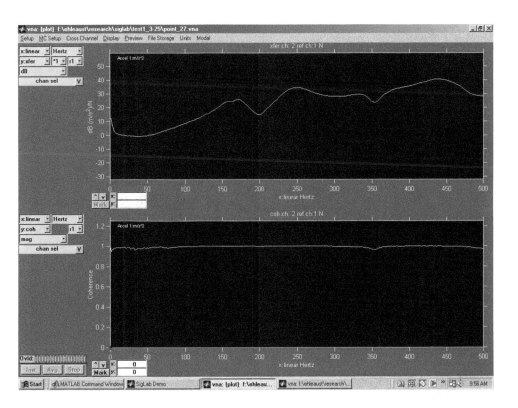

Fig. 21. Frequency response function obtained by hitting the melon rind directly

Figure 11 shows a typical FRF (top curve) and coherence (bottom curve) from the measurement. Each measurement point in Table 1 resulted in one FRF. A peak at a certain frequency means large vibration at that frequency, which indicates a mode at that frequency. If the FRF at a particular point shows a valley or low response at a modal frequency, it is indicating that the point is a node (point of no motion) of the corresponding mode shape. Using those rules, the mode shapes at any resonant frequency could be visually determined. A modal analysis program uses mathematical algorithms to compute the natural frequency, damping and mode shapes from the FRFs. It was used to analyze data from this test, but may not be necessary if the tester can figure out the modes by careful visual examination of the FRFs.

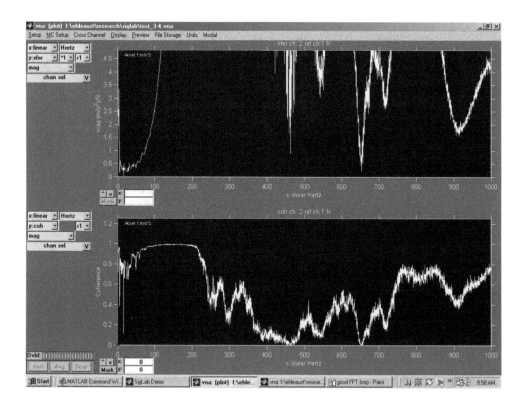

Fig. 22. An FRF (top) and coherence (bottom) obtained without a local stiffener on the surface of the melon.

Fig. 23. A stiffening metal disk on the melon rind.

It is important that the excitation force contain energy at all the frequencies of interest. That means that the force imparted by the hammer on the fruit must be a sharp enough pulse in the time domain. Experimental modal analysis is most commonly done on hard structures such as vehicles, buildings or metal structures. The surface of the fruit is much softer than most other structures for which the hammer tip was designed. Hitting the fruit with a hammer would result in a broad pulse in time domain, which translates into a narrow-band excitation in the frequency domain. As a result, the impulse spectrum of the excitation force would not have enough energy to excite vibration modes higher than 200 Hz without its intensity dropping more than 20 dB from its DC magnitude. When that condition is violated, the coherence of the data is very poor as seen in figure 22. To overcome this narrow-band excitation problem, a small metal disk (Figure 23) was attached to the fruit at the point of impact. This resulted in a significantly broader-band force excitation.

3. Concluding remarks

This chapter has presented a few examples of research that has been done to take advantage of the advancement in vibration analysis along with applications to characterize the rheological properties of biomaterials. The literature shows that the rheological properties of biomaterials are associated to quality indicators, specifically for foods to their texture and their sensory evaluation, thus many of the applications described in this chapter deal with food materials. In particular, this chapter has shown examples of application of basic vibration theories to measure the rheology of liquids as well as viscoelastic semi fluids and semi solid materials. The static measurement of modulus, finite element computation of the vibration natural frequencies and mode shapes, and an experimental modal analysis of a melon fruit are also described.

The authors believe that research on testing of biomaterials using vibration methods may help achieve:

- Higher quality foods due to better selection, testing, handling, grading and processing
- Lower prices due to quick nondestructive tests in harvesting, selection and grading of raw materials as well as testing finished products like for example during canning operations
- Increase in food safety due to, for example, the ability to detect dangerous salmonella in eggs from the change in physical properties detected by quick and accurate vibration based methods (Sinha et al., 1992) or infestation in fruits.
- Better design of food packaging, handling and transportation.
- Better tools to monitor the rheological properties of raw materials, e.g. during cooking and drying of cereal (Gonzalez et al., 2010)
- More interdisciplinary research combining engineering and food and biological sciences.

4. References

Abbott, J.A., (1994). Firmness measurement of freshly harvested delicious apples by sensory methods, sonic transmission, magness-taylor, and compression. Journal of American Society of Horticultural Science, 119(3), 510-515.

Abbott, J.A., Liljedahl, L.A., (1994). Relationship of sonic resonant frequency to compression test and magness-taylor firmness of apples during refrigerated storage. Transactions of the ASAE, 37(4), 1211-1215.

American Society for Testing and Materials, (1995). Standard test method for impedance and absorption of acoustical materials by the impedance tube method. ASTM Designation: C, 384-95.

American Society for Testing and Materials, (1990). Standard test method for impedance and absorption of acoustical materials using a tube, two microphones, and a digital frequency analysis system, ASTM Designation: E, 1050-90.

Campanella, O.H., Peleg, M., (2002). Squeezing flow viscometry for nonelastic semiliquid-foods theory and applications. Critical Reviews in Food Science and Nutrition, 42(3), 241–264.

Chen, H., De Baerdemaeker, J., (1993). Modal analysis of the dynamic behavior of pineapples and its relation to fruit firmness. Transactions of the ASAE, 36(5), 1439-1444.

Cherng, P.A., (2000). Vibration modes of ellipsoidal shape melons. Transactions of the ASAE, 43(5), 1185–1193.

Clark, H.L., Lake, B., Mikelson, W., (1942). Fruit ripeness tester, U.S. Patent 2277037.

Clark, R.L., Shackelford, P.S., (1973). Resonant and Optical Properties of Peaches as Related to Flesh Firmness. Transactions of the ASAE, 16(6), 1140-1142.

Cooke, J.R., (1972). An interpretation of the resonant behavior of intact fruits and vegetables. Transactions of the ASAE, 15(4), 1075-1080.

De Baerdemaeker, J., Wouters, A., (1987). Mechanical properties of apples: II. Dynamic measurement methods and their use in fruit quality evaluation, In: Jowitt R.

(Eds). Physical properties of foods 2, Elsevier Applied Science, Pp.417-428, London , U.K.,

Ehle, A., (2002). Measuring modulus of elasticity and vibration modes of a melon. Undergraduate Research Report, (unpublished), Purdue Univerisity, West Lafayette, IN, USA.

Finney, E., (1970). Mechanical resonance within red delicious apples and its relation to fruit texture. Transactions of the ASAE, 13(2), 177-180.

Finney, E., (1971). Dynamic elastic properties and sensory quality of apple fruit. Journal of Texture Studies, 2, 62-74.

Garrett, R.E., (1970). Velocity of Propagation of Mechanical Disturbance in Apples, PhD Thesis, Cornell University, Ithaca, New York.

D.C. Gonzalez, D.C, Khalef, N., Wright, K, Okos,M.R., Hamaker, B.R and Campanella, O.H. (2010). Physical aging of processed fragmented biopolymers. Journal of Food Engineering, 100(2), 187-193.

Hertz, T.G., Dymling, S.O., Lindstrom, K.,. Persson, H.W., (1990). Review of Scientific Instruments, 62, 457.

Herzfeld, F.K., Litovitz, T.A, (1959). Absorption and Dispersion of Ultrasonic Waves. Academic Press, New York.

Kinsler, L.E., Frey, A.R., Coppens, A.B. and Sanders, J.V., (2000). Fundamentals of Acoustics (4th edn). John Wiley and Sons, Inc, New York.

Magness, J.R., Taylor, G.F., (1925). An improved type of pressure tester for the determination of fruit maturity. USDA Agri. Circular , 350, Pp.8. Washington, D.C., USDA.

Mason, W.P., Baker, W.O., McSkimin, H.J., Heiss, J.H., (1949). Measurement of shear elasticity and viscosity of liquids at ultrasonic frequencies. Phys. Rev. 75, 936.

Mert, B., Campanella, O.H., (2007). Monitoring the rheological properties and solid content of selected food materials contained in cylindrical cans using audio frequency sound waves. Journal of Food Engineering, 79 (2), 546-552.

Mert, B., Campanella, O.H., (2008). The study of the mechanical impedance of foods and biomaterials to characterize their linear viscoelastic behavior at high frequencies. Rheologica Acta, 47, 727-737.

Mert, B., Sumali, H., Campanella, O.H., (2004). A new method to measure viscosity and intrinsic sound velocity of liquids using impedance tube principles at sonic frequencies. Review of Scientific Instruments, 75(8), 2613–2619.

Nourain, J., Ying, Y.B., Wang, J., Rao, X., (2004). Determination of acoustic vibration in watermelon by finite element modelling. Proceding of SPIE, 5587, 213, doi:10.1117/12.576953

Phan-Thien, N.,(1980). Small strain oscillatory squeeze film flow of simple fluids. Journal of the Australian Mathematical Society (Ser. B), 32, 22–27.

Roth,W., Rich, S.R., (1953). A new method for continuous viscosity measurement. General theory of the ultra-viscoson. Journal of Applied Physics, 24, 940-950

Sheen, S.H., Chein, H.T., Rapis, A.C., (1996). Measurement of shear impedances of viscoelastic fluids, IEEE Ultrasonic Symposium Proceedings, IEEE, New York.1, pp. 453.

Sinha, D.N. Johnston, R.G. Grace, W.K., Lemanski, C.L., (1992). Acoustic resonances in chicken eggs. Biotechnology Progress, 8, 240-243.

Takabayashi, K., Raichel, D.R., (1998). Discernment of non-Newtonian behavior in liquids by acoustic means, Rheol. Acta, 37, 593–600.

Temkin, S., (1981). Elements of Acoustics. John Wiley and Sons, New York.

Yamamoto, H.M., Iwamoto, M., Haginuma, H., (1980). Acoustic Impulse Response Method for Measuring Natural Frequency of Intact Fruits and Preliminary Applications to Internal Evaluation of Apples And Watermelons. Journal of Texture Studies, 11, 117-136.

9

Elaboration and Characterization of Calcium Phosphate Biomaterial for Biomedical Applications

Foued Ben Ayed

Laboratory of Industrial Chemistry, National School of Engineering, Box 1173, 3038 Sfax
Tunisia

1. Introduction

Calcium phosphates constitute an important family of biomaterials resembling the part of calcified tissues. This study is based on calcium phosphate such as hydroxyapatite ($Ca_{10}(PO_4)_6(OH)_2$, Hap), fluorapatite ($Ca_{10}(PO_4)_6F_2$, Fap) and tricalcium phosphate ($Ca_3(PO_4)_2$, TCP) phases because their chemical composition is similar to that of bone mineral (Hench, 1991; Legeros, 1993; Uwe et al.,1993; Elliott, 1994; Landi et al., 2000; Varma et al., 2001; Destainville et al., 2003; Wang et al., 2004; Ben Ayed et al., 2000, 2001a, b; 2006a, b; 2007; 2008a, b; Bouslama et al., 2009; Chaari et al., 2009). The most frequent is β-TCP because it is resorbable and osteoinductive (Gaasbeek et al., 2005; Steffen et al., 2001). β-TCP is resorbed in vivo by osteoclasts and replaced by new bone (Schilling et al., 2004). The tricalcium phosphate has been used clinically to repair bone defects for many years. However, mechanical properties of calcium phosphates are generally inadequate for many load-carrying applications. The tricalcium phosphate has a low density decreasing the mechanical properties. But, the efficiency of bi-phasic calcium phosphate (BCP) has been fully importantly efficiency its clinical efficacy to combat the chronic osteomyelitis in the long term. To our knowledge, if it is possible to vary the composition of bioceramic materials (composed of Hap and β-TCP) with its inherent porosity, it could be a solution owing to the faster resorption of this BCP together with the sustained release of the antibiotic.

In the literature Hap, β-TCP or the combination of both (Hap/TCP) was the most commonly used synthetic augments in high tibial osteotomy (Haell et al., 2005; Koshino et al., 2003; Gaasbeek et al., 2005; Van Hemert et al., 2004; Gutierres et al., 2007). The use of bone cement as a temporary spacer for bone defects has been described, but secondary biological reconstruction was performed after cement removal (DeSilva et al., 2007). However, permanent acrylic bone cement has been used as an interface in the postero-medial part of high tibial osteotomy to maintain the opening angle and good results have been achieved (Hernigou et al., 2001). However, due to the different biomechanical features between bone and bone cement and missing bony remodelling and incorporation, the use of bone cement as a permanent spacer was not recommended, if one aims to achieve biological regeneration. Recently, Jensen and colleagues described that rapid resorption of β-TCP might impair the regenerative ability of local bone, especially in the initial stage of bone healing (Jensen et al.,

2006). The microstructure of the used β-TCP has important influence on the osteogenic effects (Okuda et al., 2007). This has recently been confirmed by Fellah and colleagues (Fellah et al., 2008). They show that the Hap/TCP with different micropores was evaluated in a goat critical-defect model.

Several research studies dealt with the question where and how to perform the osteotomy and which fixation material is most beneficial (Brouwer et al., 2006; Agneskirchner et al., 2006). Aryee et al. demonstrate histologically and radiologically that the complete rebuilding of lamelliform bone in patients without synthetic augmentation, whilst bony in growth into the Hap/TCP wedge of augmented osteotomies just slowly progressed (Aryee et al., 2008). In contrast to diminished osteotomies, there was no advantage in using Hap/TCP wedges or the combination of Hap/TCP wedges and platelet rich plasma (PRP) as supporting material after 12 months. In cases where augmentation is performed, either autologous spongious iliac bone graft or an Hap/TCP wedge of appropriate size was inserted into the osteotomy opening and pushed laterally until it is firmly aligned to the tibial bone. The Hap/TCP wedge utilised by us consists of 60% micro–macroporous biphasic Hap and 40% β-TCP. The average total porosity is 65–75%, whilst two different sizes of porosity are found within this material. The microporous part consists of pores with a diameter less than 10μm. The macroporous part consists of pores with a diameter between 300μm and 600μm (same as autograft macropores).

As a result of limited autologous bone availability and to minimise the problem of donor-site morbidity, many efforts have been made to find adequate supporting material for augmentation after osteotomy (Bauer et al., 2000; De Long et al, 2007). In this context, we chose biomaterials on base of calcium phosphates as solution for the biomedical applications. Thus, β-TCP or Hap-TCP combination has been clinically used to repair bone defects for many years (Elliott, 1994). Whereas, β-TCP or Hap-TCP have poor mechanical properties (Elliott, 1994; Wang et al., 2004). The usage at high load bearing conditions was restricted due to its brittleness, poor fatigue resistance and strength. Hence, there was a need for improving the mechanical properties of these materials by suitable biomaterials for clinical applications. We offer the study of the mixtures of tricalcium phosphate (β-TCP) and synthetic Fap in order to obtain a bioceramic with better mechanical properties than Hap-TCP combination or β-TCP as separately used. In fact, Fap is an attractive material due to its similarity in structure and bone composition in addition to the benefit of fluorine release (Elliott, 1994; Ben Ayed et al., 2001a). In Vitro studies we have shown that Fap is biocompatible (Elliott, 1994). It also has better stability and provides fluorine release at a controlled rate to ensure the formation of a mechanically and functionally strong bone (Elliott, 1994; Ben Ayed et al., 2006b).

Most studies have been devoted to the knowledge of the mechanical properties and biomedical applications of TCP-Hap (Elliott, 1994; Landi et al., 2000; Gutierres et al., 2007). On the contrary little work has been devoted to the sintering, mechanical properties and clinical applications of TCP-Fap (Ben Ayed et al., 2007). So, the aim of this study is to prepare a biphasic calcium phosphates composites (tricalcium phosphate and fluorapatite) at various temperatures (between 1100°C and 1450°C) with different percentages of fluorine (0.5 wt %; 0.75 wt %; 1 wt %; 1.25 wt % and 1.5 wt % respectively, to the mass Fap percentage: 13.26 wt %; 19.9 wt %; 26.52 wt %; 33.16 wt % and 40 wt %). It also aims to characterize the resulting composites with density, mechanical resistance, infrared spectroscopy, X-ray diffraction, nuclear magnetic resonance (^{31}P) and scanning electron microscopy measurements.

2. Materials and methods

In this study the main used materials are commercial tricalcium phosphate (Fluka) and synthesized fluorapatite. The Fap powder was synthesized by the precipitation method (Ben Ayed et al., 2000). A calcium nitrate ($Ca(NO_3).4H_2O$, Merck) solution was slowly added to a boiling solution containing diammonium hydrogenophosphate (($NH_4)_2HPO_4$, Merck) and ammonium fluorine (NH_4F, Merck), with continuous magnetic stirring. During the reaction, pH was adjusted to the same level (pH 8-9) by adding ammonia. The obtained precipitate was filtered and washed with deionised water; it was then dried at 70°C for 12h.

Estimated quantities of each powder (β-TCP and Fap) were milled with absolute ethanol and treated by ultra-sound machine for 15 min. The milled powder was dried at 120°C in a steam room to remove the ethanol and produce a finely divided powder. Powder mixtures were moulded in a metal mould and uniaxially pressed at 150 MPa to form cylindrical compacts with a diameter of 20 mm and a thickness of about 6 mm. The green bodies were sintered without any applied pressure or air at various temperatures (between 1100°C and 1450°C). The heating rate was 10°C min^{-1}. The green compacts were sintered in a vertical resistance furnace (Pyrox 2408). The relative densities of the sintered bodies were calculated by the dimensions and weight. The relative error of densification value was about 1%.

The received powder was analyzed by using X-ray diffraction (XRD). The X-rays have used the Seifert XRD 3000 TT diffractometer. The X radiance was produced by using CuK$_\alpha$ radiation (λ = 1.54056 Å). The crystalline phases were identified from powder diffraction files (PDF) of the International Center for Diffraction Data (ICDD).

The samples were also submitted to infrared spectrometric analysis (Perkin-Elmer 783) using KBr.

Linear shrinkage was determined by dilatometry (Setaram TMA 92 dilatometer). The heating and cooling rates were 10°C min^{-1} and 20°C min^{-1}, respectively.

Differential thermal analysis (DTA) was carried out using about 30 mg of powder (DTA-TG, Setaram Model). The heating rate was 10°C min^{-1}.

The ^{31}P magic angle spinning nuclear magnetic resonance (^{31}P MAS NMR) spectra were run on a Brucker 300WB spectrometer. The ^{31}P observational frequency was 121.49 MHz with 3.0 µs pulse duration, spin speed 8000 Hz and delay 5 s with 2048 scans. ^{31}P shift is given in parts per million (ppm) referenced to 85 wt% H_3PO_4.

The microstructure of the sintered compacts was investigated by scanning electron microscope (Philips XL 30) on fractured sample surfaces. Because calcium phosphates are insulating biomaterial, the sample was coated with gold for more electronic conduction.

The particle size dimension of the powder was measured by means of Micromeritics Sedigraph 5000. The specific surface area (SSA) was measured by the BET method using azotes (N_2) as an adsorption gas (ASAP 2010) (Brunauer et al., 1938). The main particle size (D_{BET}) was calculated by assuming that the primary particles are spherical (Ben Ayed et al., 2001b):

$$D_{BET} = \frac{6}{s \cdot \rho} \qquad (1)$$

Where ρ is the theoretical density of β-TCP (3.07 g.cm^{-3}) or Fap (3.19 g.cm^{-3}) and s is the SSA. The Brazilian test was officially considered by the International Society for Rock Mechanics (ISRM) as a method for determining the tensile strength of rock materials (ISRM, 1978). The Brazilian test was also standardised by the American Society for testing materials (ASTM) to obtain the tensile strength of concrete materials (ASTM, 1984). The diametrical compression test also called the Brazilian disc test or the diametrical tensile test was considered a reliable and accurate method to determine the strength of brittle and low strength material.

An interesting parameter is the tensile strength, σ_r, which is the maximum nominal tensile stress value of the material. What is defined here is the horizontal tensile stress at the initiation of the large vertical crack at the centre of the disc. For the calculation of the tensile stress in the middle of the disc, Eq. (2) was used, corresponding to a plane stress condition. The usual way in evaluating the tensile strength from diametral compression test is by substituting the maximum load value into Eq. (2).

$$\sigma_r = \frac{P}{S} = \frac{2 \cdot P}{\Pi \cdot D \cdot t} \qquad (2)$$

where P is the maximum applied load, D diameter, t thickness of the disc and σ_r the tensile strength (or mechanical strength).

The essays were realized by means of a device by using "LLOYD EZ50" on cylindrical samples of 6 mm approximately of thickness and 20 mm of diameter. At least six specimens were tested for each test condition. An average of the values was then calculated. The results dispersal is in the order of 15 %.

3. Results and discussion

3.1 Characterization of powder

Table 1 shows the SSA of β-TCP powder, Fap powder and different β-TCP - Fap composites, the average grain size D_{BET} (calculated by equation (1)) and the particle size distribution data (measured by granulometric repartition). The difference between the value deducted by SSA and by granulometric repartition was probably due to the presence of agglomerates in the initial powder. The SSA of different β-TCP - Fap composites increases with the percentages of Fap whereas the average grain size deceases (Table 1). This result shows the effect of Fap additive in tricalcium phosphate matrix. Indeed, the grain of Fap has a dense morphology by report β-TCP, which was responsible of the increasing of different composites SSA.

Compound	SSA (m²/g) ± 1.00	D_{BET} (µm) ± 0.20	D_{50} (µm)[a] ± 0.20	d[b]
Fap	29.00	0.07	6	3.190
β-TCP	0.70	2.80	5	3.070 (β) 2.860 (α)
β-TCP – 13.26 wt% Fap composites	2.90	0.67	-	3.086
β-TCP - 19.90 wt% Fap composites	3.32	0.58	-	3.094
β-TCP - 26.52 wt% Fap composites	3.80	0.51	-	3.101
β-TCP – 33.16 wt% Fap composites	4.37	0.44	-	3.110
β-TCP - 40 wt% Fap composites	5.24	0.36	-	3.118

[a] : Mean diameter, [b] : theoretical density.

Table 1. SSA and average grain size obtained by different analysis of various compounds.

Fig. 1 shows the dilatometric measurements of a different powder used in this study (β-TCP, Fap and different β-TCP-Fap composites). The sintering temperatures began at about

1180°C, 1154°C, 1145°C, 1100°C, 1052°C, 1045°C and 775°C for β-TCP, β-TCP – 13.26 wt% Fap composites, β-TCP - 19.90 wt% Fap composites, , β-TCP – 26.52 wt% Fap composites, β-TCP – 33.16 wt% Fap composites, β-TCP - 40 wt% Fap composites and Fap, respectively. The addition to Fap additive in the matrix of β-TCP decreasing the sintering temperature of the pure tricalcium phosphate (Fig. 1b-1f). Indeed, the sintering temperature of β-TCP decreases when the percentage of Fap increases.

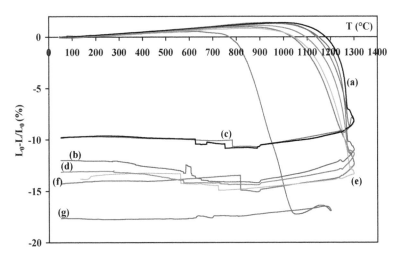

Fig. 1. Linear shrinkage versus temperature of: (a) β-TCP powder, (b) β-TCP - 13.26 wt% Fap composites, (c) β-TCP - 19.9 wt% Fap composites, (d) β-TCP – 26.52 wt% Fap composites, (e) β-TCP - 33.16 wt% Fap composites, (f) β-TCP - 40 wt% Fap composites, (g) Fap powder.

Differential thermal analysis studies detected out a potential phase change during the sintering of a different powder used in this study. Fig. 2a shows DTA curve of β-TCP. The DTA thermogramme shows 2 peaks relatives to the first and second endothermic allotropic transformation of β-TCP (1290°C and 1464°C). The last endothermic peak, at 1278°C, was related to a peritectic between β-TCP and pyrophosphate (β-$Ca_2P_2O_7$). The chemical reaction was illustrated in the following:

$$Ca_3(PO_4)_{2(s)} \quad + \quad Ca_2P_2O_{7(s)} \quad \rightarrow \quad Ca_3(PO_4)_{2(s)} \quad + \quad \text{liquid phase} \tag{3}$$

This was similar to the result previously reported by Destainville and colleagues (Destainville et al., 2003).

Fig. 2b and Fig. 2g show the DTA curves of β-TCP - 13.26 wt% Fap composites and Fap powder, respectively. We notice one endothermic peak around 1180°C. This peak may be due to the formation of a liquid phase, formed from binary eutectic between CaF_2 and Fap (Ben Ayed et al., 2000). We can assume that fluorite (CaF_2) was formed as a second phase during the powder preparation of Fap.

Fig. 2c, 2e and 2f illustrate the DTA curves of β-TCP – 19.9 wt% Fap composites, β-TCP – 33.16 wt% Fap composites and β-TCP - 40 wt% Fap composites, respectively. Fig. 2c and 2f are practically similar with the Fig. 2a. Indeed, DTA thermogramme of composites shows also 2 peaks of tricalcium phosphate allotropic transformations.

The DTA thermogramme of TCP - 26.52 wt% Fap composites shows 3 endothermic peaks at 1268°C, at 1280°C and at 1444°C, which are two peaks related with two allotropic transformations of tricalcium phosphate (Fig. 2d). The temperatures of allotropic transformations have been decreased about 10°C (1280°C in the place 1290°C) and 20°C (1444°C in the place 1464°C) with that of the pure β-TCP. This result has been explained probably by the Fap effect in β-CTCP matrix.

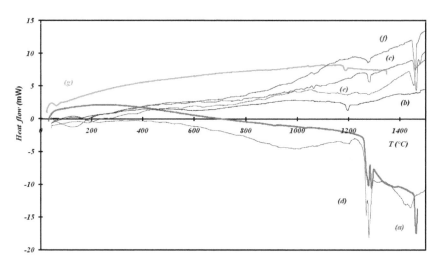

Fig. 2. DTA curves of: (a) β-TCP powder, (b) β-TCP - 13.26 wt% Fap composites, (c) β-TCP - 19.9 wt% Fap composites, (d) β-TCP – 26.52 wt% Fap composites, (e) β-TCP - 33.16 wt% Fap composites, (f) β-TCP - 40 wt% Fap composites, (g) Fap powder.

The DTA thermogrammes of β-TCP - 33.16 wt% Fap composites and β-TCP - 40 wt% Fap composites show only 2 endothermic peaks, which are relative to the allotropic transformations of tricalcium phosphate (Fig. 2e and 2f). In these curves, the endothermic peak relative to a peritectic between β-TCP and β-$Ca_2P_2O_7$ is practically disappeared. This chemical reaction between tricalcium phosphate and pyrophosphate was improved by the addition of 33.16 wt% or more Fap. Also, the temperatures of allotropic transformations decreased when the percentage of Fap increases (Fig. 2c-2f). This result has been explained probably by the Fap effect in β-TCP matrix.

3.2 Sintering and mechanical properties of tricalcium phosphate – fluorapatite composites

This Table 2 summarizes the mechanical strength of different β-TCP sintered at various temperatures (1100°C-1450°C) with different amounts of Fap (13.26 wt%, 19.9 wt%, 26.52 wt, 33.16 wt% and 40 wt%).

Fig. 3 illustrates the evolution of density and rupture strength of the composites sintered for 1h at various temperatures (between 1100°C and 1450°C) at different percentages of Fap (13.26 wt %; 19.9 wt %; 26.52 wt %; 33.16 wt % and 40 wt %). The densification and mechanical resistance varied as a function of Fap in the β-TCP matrix and sintering temperature.

wt% Fap	13.60	19.90	26.52	33.16	40.00
T (°C)			σ_r (MPa)		
1100	0.25	1.68	2.00	2.53	0.23
1150	0.50	2.11	2.20	2.73	0.54
1200	0.90	2.55	3.00	3.01	0.85
1250	1.36	2.86	3.80	6.23	1.33
1300	2.31	3.41	6.00	8.74	1.80
1320	3.80	6.10	8.00	9.80	4.50
1350	6.00	6.45	9.20	13.20	10.10
1400	7.10	7.10	9.60	13.70	11.80
1450	3.60	5.33	8.00	5.70	6.20
Green	0.10	0.20	0.20	0.30	0.20

Table 2. Rupture strength versus temperature of different TCP-Fap composites.

Fig. 3A shows the relative density and mechanical resistance of β-TCP - 13.26 wt% Fap composites. The rupture strength and densification increase with sintering temperature. The optimum densification was obtained at 1350°C (87%). The maximum mechanical resistance reached 7.1MPa at 1400°C (Fig. 3A). This study shows that small additions of a Fap (13.26 wt %) can significantly enhance the sinterability and strength of β-TCP.

The evolution of β-TCP - 19.90 wt% Fap composites of densification and mechanical properties was practically similar to the β-TCP sintered with 13.26 wt% Fap (Fig. 3B). This curve illustrates a maximum densification at about 89% corresponding to the composites sintered at 1350°C. The mechanical resistance was similar to the β-TCP - 13.26 wt% Fap composites in value and sintering temperature (7.1 MPa at 1400°C).

Fig. 3C illustrates the evolution of the composites densification relative to the temperature between 1100°C and 1450°C. The densification was variable as a function of temperature. Between 1100°C and 1200°C, the samples relatives' densities were very small for any samples. An increase of density was shown between 1300°C and 1400°C, where the optimum densities are about 89.1 % at 1350°C. Fig. 3C shows the mechanical properties of the β-TCP - 26.52 wt% Fap composites samples according to the sintering temperature. Between 1100°C and 1250°C, the rupture strength of TCP - 26.52 wt% Fap composites samples was around 2-3 MPa. Above 1250°C, the rupture strength increases and reaches maximum value at 1400°C (9.6 MPa). Above 1400°C, the mechanicals properties of composites decrease during the sintering process. Indeed, the mechanical resistance of TCP - 26.52 wt % Fap composites reaches the 8 MPa at 1450°C. The evolution of mechanical properties of composites was considered a function of a sintering temperature. At 1350°C, the mechanical resistance optimum of β-TCP sintered without Fap additive reached 5.3 MPa (Bouslama et al., 2009), whereas the resistance increases to 9.4 MPa with 26.52 wt% Fap.

Fig. 3D shows the results of densification and mechanical properties of β-TCP - 33.16 wt% Fap composites. The ultimate densification was obtained at 1350°C (93.2%) and the maximum of mechanical resistance was approached at 1400°C (13.7 MPa). The β-TCP - Fap composites ratio was strongly dependent on the percentage Fap addition. The densification and mechanical properties remain low with 13.26 wt% Fap and 19.90 wt% Fap, when β-TCP sintered with 33.16 wt % Fap, the densification and mechanical resistance increase with the sintering temperature and reaches its maximum values.

With 40 wt % Fap, the densification and mechanical resistance decrease slowly with sintering temperature (Fig. 3E). The optimum relative density was obtained at 1350°C (92%) and the maximum mechanical resistance reached at 1400°C (11.8 MPa). As the amount of Fap increased (40 wt %), sinterability and mechanical properties considerably decreased. This result was clarified by the large (increase weight ratio) amounts of Fap used in the prepared composites. In fact the microstructure and thermal properties of Fap weren't stable at high temperature (after 1300°C) (Ben Ayed et al., 2000 and 2001b). Above 1400°C, the densification and the mechanical properties decrease with any β-TCP sintered with different percentages of Fap (Fig. 3).

Table 3 summarizes the optimum values of density and mechanical resistance of β-TCP sintered for 1h with different percentages of Fap. Whatever the content of Fap, the maximum of densification was obtained at 1350°C, whereas the optimum of rupture strength was reached at 1400°C (Table 3). The optimum values were obtained for β-TCP sintered with 33.16 wt% Fap (93.2% and 13.7 MPa).

wt % Fap	13.26	19.90	26.52	33.16	40
Optimum density (%)	87.87	89.00	89.10	93.20	92.62
Optimum strength (MPa)	7.1	7.1	9.6	13.7	11.8

Table 3. Optimum values of density (at 1350°C) and mechanical resistance (at 1400°C) of different β-TCP-Fap composites.

An increase of β-TCP- Fap density was shown between 1300°C and 1400°C, where the optimum densities were obtained at 1350°C with 33.16 wt % Fap. The used temperatures are similar to the densification of tricalcium phosphate when singly used (Ben Ayed et al., 2006a). But, the temperatures are relatively higher in comparison to those used for densification of only Fap (Ben Ayed et al., 2000, 2001b and 2006b). Indeed, Fap presents a good sinterability in the temperature ranging at 900°C-1100°C (Ben Ayed et al., 2000 and 2001b). At 1450°C, the densities decrease at different degrees of Fap additions. These results are similar to the previously reported by Ben Ayed et al. during the study of elaboration and characterization of calcium phosphate biomaterial (Ben Ayed et al., 2007).

This study shows the mechanical properties of the β-TCP-Fap composites samples according to the sintering temperature. At lower temperature (between 1100°C and 1200°C), the rupture strength of β-TCP-Fap composites was around 1 and 3 MPa (Fig. 3). When sintering temperature increase above 1200°C, the rupture strength increases slowly. These results are due to the increase of the densification caused by the growth of the grains size. Beyond 1300°C, the rupture strength increases and reaches maximum value at 1400°C. Between 1350°C and 1400°C, the increase in rupture strength was very clear for all samples with different weight ratio of Fap in the composites specimens. This is attributed to the influence and effect of Fap additive in relative densities and mechanical properties of the sintered composites. In fact, Ben Ayed et al. show that Fap has a good sinterability and mechanical

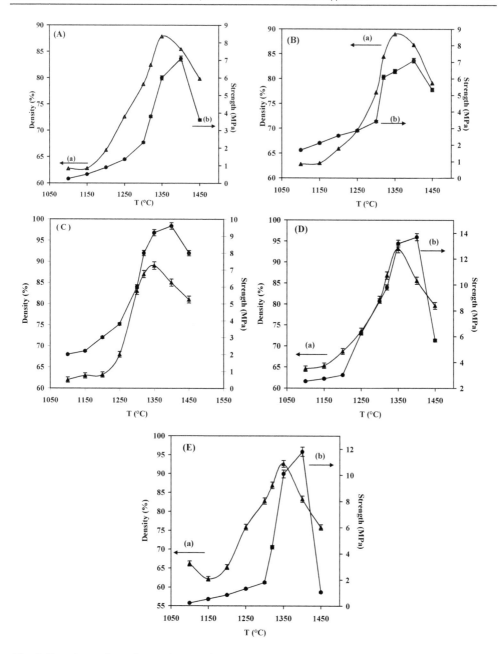

Fig. 3. Density and mechanical strength versus temperature of β-TCP sintered for one hour with different wt % of Fap: (A): β-TCP - 13.26 wt% Fap composites, (B) β-TCP - 19.90 wt% Fap composites, (C) β-TCP – 26.52 wt% Fap composites, (D) β-TCP - 33.16 wt% Fap composites, (E) β-TCP - 40 wt% Fap composites, ((a) : Density, (b): strength).

resistance (Ben Ayed et al., 2006b). It revealed that the mechanical resistance increases with temperature and reaches its maximum value about 14 MPa (Ben Ayed et al., 2006b).

The mechanical properties of the β-TCP and different TCP - Fap composites depend directly on the properties of the departure powder (granulometric, crystallinity of the powder, chemical composition, origin of powder) and on the operative conditions of the sintering process (temperature, heating time, cycle of sintering, atmosphere) (Ben Ayed et al., 2006b). Each of these parameters has a direct effect on the final properties of the composite.

Resorbable beta-tricalcium phosphate bioceramic is known for its excellent biocompatibility. However, it exhibits poor tensile strength. Here, we improved tensile strength of β-TCP bioceramic without altering its biocompatibility by introducing Fap additives, in different quantities (13.26 wt %; 19.9 wt %; 26.52 wt %; 33.16 wt % and 40 wt %). In this study, we showed that the presence of Fap with different amounts in β-TCP matrix improves the mechanical properties of β-TCP-Fap composites. This result was similar to the previous report by Bouslama et al. (Bouslama et al., 2009).

Table 4 displays several examples of dense calcium phosphates composites mechanical properties. The values found by mechanical strength were unequalled with the values in Table 4, because the authors have used different mechanical modes others than the Brazilian test. Indeed, many factors influence the mechanical properties such as: initial materials, the process conditions and annealing treatment. So, it is difficult to compare the results of this study with those found in literature.

Materials	σ_r (MPa)[a]	σ_f (MPa)[b]	σ_c (MPa)[c]	K_{1C} (GPa)[d]	References
Fap	14.00	-	-	4.5	Ben Ayed et al., 2006b
β-CTCP	5.30	-	-	-	Bouslama et al., 2009
β-TCP	-	92.00	-	-	Kalita et al., 2008
β-TCP- Fap composites	9.60	-	-	-	Bouslama et al., 2009
brushite	-	-	32.00	-	Hofmann et al., 2009
Hap	-	-	5.35	3.52	Balcik et al., 2007
TCP-Hap composites (40/60)	-	-	4.89	3.10	Balcik et al., 2007
Hap- poly-L-lactic acid composites	-	-	100.00	6.00	Gay et al., 2009

[a]σ_r: Mechanical strength (Brazilian test), [b]σ_f: Flexural strength, [c]σ_c: Compressive strength, [d]K_{1C}:Young's modulus.

Table 4. Literature examples of dense calcium phosphates composites mechanical properties.

3.3 Characterization of sintered tricalcium phosphate - fluorapatite composites

After sintering, different techniques characterized the samples: X rays diffraction (XRD), ^{31}P MAS-NMR, scanning electronic microscopy (SEM) and infrared spectroscopy (IR).

Composites of β-TCP with Fap were pressureless sintered at temperatures from 1100 to 1450°C. The reactions and transformations of phases were monitored with X-ray diffraction.

The XRD patterns of the β-TCP were sintered at various temperatures (1100°C, 1300°C and 1400°C) with different percentages of Fap (13.26 wt % and 40 wt %) were shown in Fig. 4. These spectra are identical to the initial powder (β-TCP and Fap). We conclude that β-TCP and the Fap were steady during the sintering process. But when left at 1300°C, the thermogrammes indicate the germination of α-TCP phase. This phase is a proof of the fragility of samples because the absolute densities of β-TCP and α-TCP were different. Besides, the incorporation of Fap in tricalcium phosphate does not seem to be in its right decomposition. Indeed, the XRD revealed only phases of departure (β-TCP and Fap) and α-TCP.

The Nuclear magnetic resonance chemical shift spectra of the ^{31}P of the β-TCP as sintered at various temperatures (1100°C; 1300°C and 1400°C) with different percentages of Fap (13.26 wt %; 33.16 wt % and 40 wt %) are shown in Fig. 5. The ^{31}P MAS-NMR solid spectra of composites show the presence of three tetrahedral environments of the phosphorus. Indeed, the ^{31}P MAS - NMR analysis reveals the presence of three tetrahedral P sites for the β-TCP whereas the Fap possesses only one. This was similar to the result previously reported by (Yashima et al., 2003). They show that the phosphorus of tricalcium phosphate is located in three crystallographic sites: $P(1)O_4$, $P(2)O_4$, $P(3)O_4$ (Yashima et al., 2003). There is no record of any phase modification of the β-TCP or Fap, which confirms that the two powders, β-TCP and Fap were thermally steady between 1100°C and 1450°C.

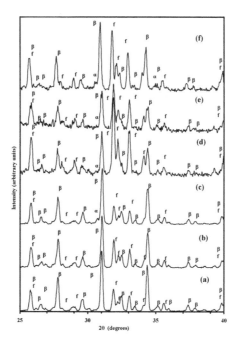

Fig. 4. XRD patterns of β-TCP-Fap composites sintered for one hour at various temperatures with different wt % of Fap: (a) 13.26 wt %, 1100°C; (b) 13.26 wt %, 1300°C; (c) 13.26 wt %, 1400°C; (d) 40 wt %, 1100°C; (e) 40 wt %, 1300°C and (f) 40 wt %, 1400°C (β : β-TCP; f: Fap; α : α-TCP).

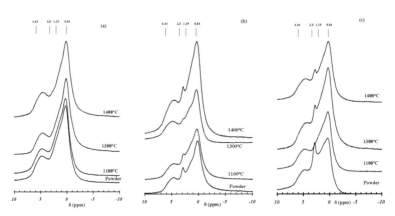

Fig. 5. ^{31}P NMR spectra of β-TCP-Fap composites sintered for one hour at various temperatures with different wt % of Fap : (a) 13.26 wt %; (b) 33.16 wt % and (c) 40 wt % (β : β-TCP; f: Fap).

Fig. 6 and Fig. 7 show the SEM micrographs of β-TCP-Fap composites. This technique helps to investigate the texture and porosity of the biphasic bioceramic. The fracture surfaces clearly show the composite as sintered at various temperatures (1100°C, 1300°C, 1400°C and 1450°C) with different percentages of Fap (13.26 wt %; 19.9 wt %; 33.16 wt % and 40 wt %) revealing the influence of the temperature of microstructural developments during the sintering process (Fig. 6). These effects increased along the sintering temperature and the percentage of Fap.

The results of microstructural investigations of β-TCP-Fap composites sintered at various temperatures (1100°C and 1450°C) with 13.26 wt % Fap are shown in Fig. 6a and b. At 1100°C, the sample presents an important intergranular porosity (Fig. 6a). At higher temperatures (1450°C), the densification was hindered by the formation of large pores (Fig. 6b). During the sintering of Fap and TCP-Fap composites, Ben Ayed et al. also observed the formation of large pores at these temperatures (Ben Ayed et al., 2001b). They are attributed to the hydrolysis of Fap and CaF$_2$, as expressed by the following equation:

$$CaF_{2(s)} \quad + \quad H_2O_{(g)} \quad \rightarrow \quad CaO_{(s)} \quad + \quad 2HF_{(g)} \tag{4}$$

$$Ca_{10}(PO_4)_6F_{2(s)} \quad + \quad xH_2O_{(g)} \quad \rightarrow \quad Ca_{10}(PO_4)_6F_{2-x}(OH)_{x(s)} \quad + \quad xHF_{(g)} \tag{5}$$

The increase of percentages of Fap from 13.26 wt % to 19.9 wt % did not show a significant change of the microstructure. When temperature increases, the microstructure of β-TCP-Fap composites was sintered at 1300°C with different percentages of Fap (19.90 wt %; and 33.16 wt %) reveals moderate grain growth (Fig. 6c and d). Indeed, we notice a partial reduction of the porosity and a presence of some closed pores (Fig. 6c and d).

At 1400°C, the microstructural analysis of β-TCP- 40 wt% Fap composites reveals the creation of a liquid phase (Fig. 6e). The outstanding modification on the microstructure of β-TCP-Fap composites takes place only due to the presence of a liquid phase that can occur by the increase of the sintering temperature or by an increase of the Fap content (Fig. 6e). At this temperature (1400°C), the fragility and the decrease of relative densities and mechanical properties were originated from the micro-crack formed by small expansion of the phase transformation samples due to the low density of α-TCP (2.86 g/cm^3) than that of β-TCP (3.07 g/cm^3) (Fig. 6e). But also from the allotropic transformation from β to α and from α to α'.

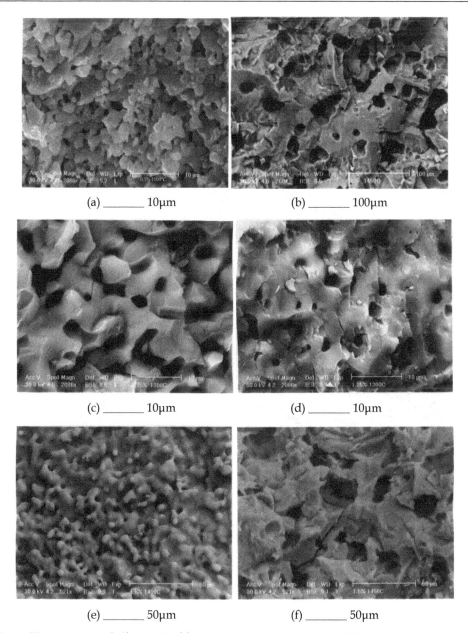

(a) _____ 10µm

(b) _____ 100µm

(c) _____ 10µm

(d) _____ 10µm

(e) _____ 50µm

(f) _____ 50µm

Fig. 6. SEM micrograph (fracture) of β-TCP-Fap composites sintered for one hour at various temperatures with different wt % of Fap: (a) 13.26 wt %, 1100°C; (b) 13.26 wt %, 1450°C; (c) 19.9 wt %, 1300°C; (d) 33.16 wt %, 1300°C; (e) 40 wt %, 1400°C and (f) 40 wt %, 1450°C.

Above 1400°C, the microstructure of β-TCP sintered at 1450°C with different percentages of Fap (13.26 wt % and 40 wt %) was totally modified (Fig. 6b and f). In fact, we observe two types of microstructures; the first was characterized by the creation of large pores engendering the reaction of hydrolysis of Fap and the second was characterized by the exaggerated coarsening of grains (Fig. 6b and f). The microstructural analysis of the composite shows a larger amount of pores, preferentially located on the grain boundaries.

In Fig. 7, the fracture surfaces show clearly the β-TCP–26.52 wt% Fap composite as sintered for 1 h at various temperatures (1300, 1350, 1400 and 1450°C) revealing the influence of the temperature of microstructural developments during the sintering process. These effects increased along the sintering temperature. The results of microstructural investigations of β-TCP–26.52 wt% Fap composites show that the morphology of the samples was completely transformed (Fig. 7a–d).

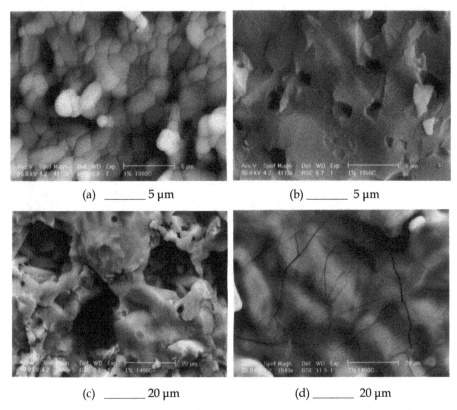

Fig. 7. SEM micrograph of β-TCP - 26.52 wt% Fap composites sintered for 1h at: (a) 1300°C, (b) 1350°C, (c) 1400°C, (d) 1450°C.

At 1300°C, the SEM micrographs of samples show liquid phase relative to the binary peritectic between pyrophosphate with the tricalcium phosphate (Fig. 7a). At 1350°C, one notices a partial reduction of the porosity (Fig. 7b). At higher temperatures (1400–1450°C), the densification was hindered by the formation of both large pores and many cracks (Fig. 7c and d).

Fig. 8A-8B show typical IR spectra of β-TCP sintered for one hour at various temperatures (1100°C, 1200°C, 1300°C and 1400°C) with 13.26 wt% and 40 wt% Fap, respectively. Most bands were characteristic of phosphate group of β-TCP and Fap (at 540-600 cm⁻¹ and 920-1120 cm⁻¹). The peaks at 920 cm⁻¹ and 1120 cm⁻¹ are assigned to the stretching vibration of PO₄³⁻ ions and the peaks at 540 cm⁻¹ and 600 cm⁻¹ are assigned to the deformation vibration of PO₄³⁻ ions. The bands at 3500 cm⁻¹ and at 1640 cm⁻¹ were assigned to the adsorbed water molecule.

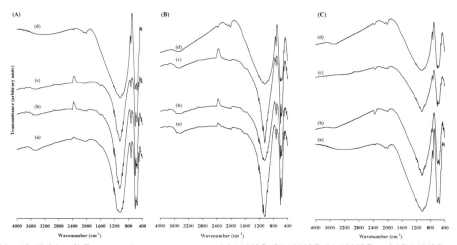

(A) with 13.26 wt% Fap at various temperatures: (a) 1100°C, (b) 1200°C, (c) 1300°C and (d) 1400°C.
(B) with 40 wt% Fap at various temperatures: (a) 1100°C, (b) 1200°C, (c) 1300°C and (d) 1400°C.
(C) at 1400°C with different percentages of Fap: (a) 13.26 wt %; (b) 19.9 wt %; (c) 33.16 wt % and (d) 40 wt%.

Fig. 8. IR of β-TCP sintered for one hour

Figs. 8A(a-b), 8B(a-b) show the IR spectra of the β-TCP-Fap composites sintered at 1100°C and 1200°C, respectively. In the spectra absorptions assigned to OH group, are detected at 3510 cm⁻¹ and 747 cm⁻¹. Those absorptions indicate that the OH group substitutes F in the Fap structure, thus leading to the formation of hydroxyfluorapatite ($Ca_{10}(PO_4)_6F_{2-x}(OH)_x$: FOHap). This compound may be the result of the reaction between Fap and water vapour traces (Ben Ayed et al., 2001b). This reaction was illustrated by Eq. 5. At higher temperature (1300°C-1400°C), the intensity of these bands decreases but does not completely vanish supposing that the amount of fluoride in the samples decreases when the temperature increases (Figs. 8A(c-d), 8B(c-d)). IR spectra of different composites sintered at 1400°C with different percentages of Fap (13.26 wt %; 19.9 wt %; 33.16 wt % and 40 wt %) were practically similar to the Figs. 8A-8B. Indeed, the typical IR spectra show the characteristics of the phosphate group of calcium phosphates (at 500-600 cm⁻¹, 1019 cm⁻¹, 1034 cm⁻¹, 1105 cm⁻¹).

At higher temperatures, the β-TCP-Fap composites densification and mechanical properties were hindered by the exaggerated grain growth, the effect of the allotropic transformation of β-TCP and the formation of intragranular porosity. The SEM micrograph illustrated the change of microstructure at 1450°C. These results are similar to the previously obtained by Ben Ayed et al. during the sintering process of TCP-Fap composites (Ben Ayed et al., 2007).

The preliminary results obtained in this study have shown that the β-TCP-Fap composites have a potential of further development into an alternative system to produce denser β-TCP bodies. Further investigations are still under way to investigate the influence of Fap on the density, microstructure and mechanical properties of β-TCP-Fap biomaterials.

4. Conclusion

Tricalcium phosphate and fluorapatite powder were mixed in order to elaborate biphasic composites. The influence of Fap substitution on the β-TCP matrix was detected in the mechanical properties of the sintered composites. The characterization of samples was investigated by using X-Ray diffraction, differential thermal analysis, scanning electronic microscopy and by an analysis using ^{31}P nuclear magnetic resonance. The sintering of tricalcium phosphate with different percentages of fluorapatite indicates the evolution of the microstructure, densification and mechanical properties. The Brazilian test was used to measure the rupture strength of biphasic composites biomaterials. The mechanical properties increase with the sintered temperature and with fluorapatite additive. The mechanical resistance of β tricalcium phosphate - 33.16 wt % fluorapatite composites reached its maximum value (13.7MPa) at 1400°C, whereas the optimum densification was obtained at 1350°C (93.2%). Above 1400°C, the densification and mechanical properties were hindered by the tricalcium phosphate allotropic transformation and the formation of both intragranular porosity and cracks.

5. Acknowledgment

The author thanks the Professor Bouaziz Jamel and Doctor Bouslama Nadhem for their assistances in this work.

6. References

Aryee, S., Imhoff, A. B., Rose, T., Tischer, T., 2008. *Biomaterials*, 29, 3497-3502.

Agneskirchner, J.D., Freiling, D., Hurschler, C., Lobenhoffer, P., 2006. *Knee Surg. Sports Traumatol Arthrosc*, 14, 291–300.

ASTM C496, *Standard test method for splitting tensile strength of cylindrical concrete specimens Annual Book of ASTM*, Standards, vol. 0.042, ASTM, Philadelphia, 1984, p. 336.

Brouwer, R.W., Bierma-Zeinstra, S.M., van Raaij, T.M., Verhaar, J.A., 2006. J. Bone Joint. Br.,, 454–9.

Bauer, T.W., Muschler, G.F., Bone graft materials. An overview of the basic science, 2000. *Clin. Orthop. Relat. Res.*,10–27.

Balcik, C., Tokdemir, T., Senkoylo, A., Koc, N., Timucin, M., Akin, S., Korkusuz, P., Korkusuz, F., 2007. *Acta Biomaterialia*, 3, 985-996.

Ben Ayed, F., Bouaziz, J., Bouzouita, K., 2000. *J. Eur. Ceram. Soc.* 20 (8), 1069.

Ben Ayed, F., Bouaziz, J., Khattech, I., Bouzouita, K., 2001a. *Ann. Chim. Sci. Mater.* 26 (6), 75.

Ben Ayed, F., Bouaziz, J., Bouzouita, K., 2001b. *J. Alloys Compd.* 322 (1-2), 238.

Ben Ayed, F., Chaari, K., Bouaziz, J., Bouzouita, K., 2006a. *C. R. Physique* 7 (7), 825.

Ben Ayed, F., Bouaziz, J., Bouzouita, K., 2006b. *Ann. Chim. Sci. Mater.* 31 (4), 393.

Ben Ayed, F., Bouaziz, J., 2007. *C. R. Physique* 8 (1), 101-108.

Ben Ayed, F., Bouaziz, J., 2008a. *Ceramics Int.* 34 (8), 1885-1892.

Ben Ayed, F., Bouaziz, J. 2008b. *J. Eur. Ceram. Soc.* 28 (10), 1995-2002.

Brunauer, S., Emmet, P. H., Teller, J., 1938. *Amer. Chem. Soc. J.* 60, 310.

Bouslama, N., Ben Ayed F., Bouaziz, J., 2009. *Ceramics Int.* 35, 1909-1917.

Chaari, K., Ben Ayed F., Bouaziz, J., Bouzouita, K., 2009. *Materials Chemistry and Physics*, 113, 219-226.

Destainville, A., Champion E., Bernache - Assolant, D., 2003. *Mater. Chem. Phys.* 80, 69.

De Long, Jr. W.G., Einhorn, T.A., Koval, K., McKee, M., Smith, W., Sanders, R., 2007. *J. Bone Joint. Am.*, 89, 49–58.

DeSilva, G.L., Fritzler, A., DeSilva, S.P., 2007. *Tech. Hand Up Extrem Surg.*, 11,163–7.

Elliott, J. C., 1994. *Structure and Chemistry of the Apatite and Other Calcium Orthophosphates*, Elsevier Science B.V., Amsterdam.

Fellah, B.H., Gauthier, O., Weiss, P., Chappard, D., Layrolle, P., 2008. *Biomaterials*, 29, 1177–88.

Gaasbeek, R.D., Toonen, H.G., van Heerwaarden, R.J., Buma, P., 2005. *Biomaterials*,26, 6713–9.

Gutierres, M., Dias, A.G., Lopes, M.A., Hussain, N.S., Cabral, A.T., Almeida, L., 2007. *J. Mater. Sci. Mater. Med.*,18, 2377–82.

García-Leiva, M.C., Ocaña, I., Martín-Meizoso, A., Martínez-Esnaola, J.M., 2002. *Engineering Fracture Mechanics* 69, 1007-1013.

Gay, S., Arostegui, S., Lemaitre, J., 2008. *Materials Science and Engineering C*, in press.

Hench, L. L., 1991. *J. Am. Ceram. Soc.* 74 (7), 1487.

Hofmann, M.P., Mohammed, A.R., Perrie, Y., Gbureck, U., Barralet, J.E., 2009. *Acta Biomaterialia*, 5, 43-49.

Hernigou, P., Ma, W., 2001. *Knee*, 8, 103–10.

Hoell, S., Suttmoeller, J., Stoll, V., Fuchs, S., Gosheger, G., 2005. *Arch.Trauma Surg.*, 125, 638–43.

ISRM. Suggested methods for determining tensile strength of rock materials, *Int. J. Rock Mech. Min. Sci. Geomech.* Abstr. 1978. 15, 99.

Jensen, S.S., Broggini, N., Hjorting-Hansen, E., Schenk, R., Buser, D. 2006. *Clin. Oral. Implants Res.*, 17, 237–43.

Kalita, S.J., Fleming, R., Bhatt, H., Schanen, B., Chakrabarti, R., 2008. *Materials Science and Engineering C*, 28, 392 - 398.

Landi, E., Tampieri, A., Celotti, G., Sprio, S., 2000. *J. Eur. Ceram. Soc.* 20, 2377.

Legeros, R. Z., 1993. *Clinical Materials* 14, 65.

Okuda, T., Ioku, K., Yonezawa, I., Minagi, H., Kawachi, G., Gonda, Y., 2007. *Biomaterials*, 28, 2612–21.

Steffen, T., Stoll, T., Arvinte, T., Schenk, R.K., 2001. *Eur. Spine J.*, 10 (Suppl 2),132–40.

Schilling, A.F., Linhart, W., Filke, S., Gebauer, M., Schinke, T., Rueger, J.M., 2004. *Biomaterials*, 25, 3963–72.

Uwe, P., Angela, E., Christian, R., 1993. Mater, *J. Sci.: Mate In Medicine* 4, 292.

Varma, H. K., Sureshbabu, S. 2001. *Materials letters* 49, 83.

Wang, C. X., Zhou, X., Wang, M. 2004. *Materials Characterization* 52, 301.

Yashima, M., Sakai, A. , Kamiyama, T., Hoshikawa, A., 2003. *J. Solid State Chemistry* 175, 272.

10

Fracture Mechanisms of Biodegradable PLA and PLA/PCL Blends

Mitsugu Todo[1] and Tetsuo Takayama[2]
[1]Research Institute for Applied Mechanics, Kyushu University
[2]Graduate School of Science and Engineering, Yamagata University
Japan

1. Introduction

Poly (lactic acid) (PLA), made from natural resources such as starch of plants, is one of typical biodegradable thermoplastic polymers and has extensively been used in medical fields such as orthopedics, neurosurgery and oral surgery as bone fixation devices mainly due to biocompatibility and bioabsorbability (Higashi et al., 1986; Ikada et al., 1996; Middleton & Tipton, 2000; Mohanty, 2000). Its importance has led to many studies on its mechanical properties and fracture behavior which found that the mode I fracture behavior of PLA is relatively brittle in nature (Todo et al., 2002; Park et al., 2004, 2005, 2006). Therefore, blending with a ductile biodegradable and bioabsorbable polymer such as poly (ε-caprolacton) (PCL) has been adopted to improve the fracture energy of brittle PLA (Broz et al., 2003; Dell'Erba et al., 2001; Chen et al., 2003; Todo et al., 2007; Tsuji & Ikada, 1996, 1998; Tsuji & Ishizuka, 2001; Tsuji et al., 2003); however, it was also found that phase separation originated by immiscibility of PLA and PCL tends to degrade the mechanical properties of PLA/PCL blends (Todo et al., 2007). It has recently been found that such phase separation can dramatically be improved by using an isocyanate group, lysine tri-isocyanate (LTI) (Takayama et al., 2006; Takayama & Todo, 2006; Harada et al., 2007, 2008), and the fracture properties of PLA/PCL/LTI are much higher than those of PLA/PCL.

In this chapter, firstly the fracture behavior and micromechanism of pure PLA are summarized (Park et al., 2004, 2005, 2006; Todo et al., 2002). Effects of crystallization behavior and loading-rate on the mode I fracture behavior are discussed. Effect of unidirectional drawing on the fracture energy is also presented as one of the effective ways to improve the brittleness of PLA (Todo, 2007). Secondly, the fracture behavior of PLA/PCL blends is discussed on the basis of the relationship between the microstructure and the fracture property (Todo et al., 2007b). In the third section, improvement of microstructural morphology of PLA/PCL by using LTI is discussed (Takayama et al., 2006; Takayama & Todo, 2006; Todo & Takayama, 2007; Todo et al., 2007a). It has been found that addition of LTI effectively improves the phase morphology of PLA/PCL, resulting in dramatic improvement of fracture energy. Effects of annealing on the mechanical properties of PLA/PCL/LTI blend are discussed in the last section (Takayama et al., 2011). It has been found that a thermal annealing process can effectively improve the mechanical properties of the polymer blend, as a result of strengthened structures due to crystallization of PLA.

2. Fracture behavior of PLA

2.1 Effect of crystallization

The microstructure of crystalline PLA can be changed through annealing process from amorphous to highly crystallized states as shown in Fig.1. 70°C-3h indicates an annealing process that the specimens are kept in an oven at 70°C for three hours.

The annealing process of 70°C-3h results in an amorphous state, on the contrary, the 100°C-3h and 100°C-24h processes create highly crystallized states and longer annealing time tends to increase the size of crystals.

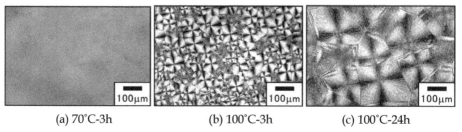

(a) 70°C-3h   (b) 100°C-3h   (c) 100°C-24h

Fig. 1. Polarized micrographs of microstructures of PLA.

It has been found that such microstructure dramatically affects the fracture behavior of PLA. As an example, the critical energy release rate, G_{IC}, measured at a quasi-static rate, 1 mm/min, and an impact rate, 1 m/s, of loading is shown in Fig.2. as a function of crystallinity. At the quasi-static rate, G_{IC} slightly decreases with increase of crystallinity up to 11.6%, and kept constant up to 48.3%. Above 48.3%, G_{IC} rapidly decrease. On the other hand, at the impact rate, G_{IC} tends to increase with increase of crystallinity. As a result, the static G_{IC} is greater than the impact value up to 48.3%, and above 48.3%, on the contrary, the impact value becomes higher than the static value. This result suggests that the fracture mechanism at the static rate is different from that at the impact rate.

Fig. 3 shows polarized microphotographs of arrested cracks in the PLA specimens prepared under different annealing conditions, and tested under static and impact loading rates. For the amorphous specimen with the crystallinity, X_c=2.7%, under the static loading-rate (Fig.3(a)), extensive multiple crazes were generated in the crack-tip region, while only a few crazes were observed under the impact loading-rate (Fig.3(b)). This kind of craze formation in crack-tip region is usually observed in amorphous polymers such as polystyrene in which craze formation is dominant rather than shear plastic deformation (Botosis, 1987). Disappearance of multiple craze formation observed at the impact rate corresponds to the reduction of additional energy dissipation in the crack-tip region compared to the static case where multiple crazes are formed, and therefore results in the decrease of G_{IC} as shown in Fig.2. On the contrary, for the highly crystallized specimen with X_c=55.8% tested at the static rate (Fig.3(c)), a straight single crack without craze formation in the surroundings is observed. This type of crack growth usually corresponds to brittle fracture and lower G_{IC} than the amorphous dominant samples in which crazes are generated in crack-tip region. At the impact rate (Fig.3(d)), the main crack tends to be distorted and branched. These behaviors may be related to the increase of G_{IC} at the impact rate, although the detail of the mechanism is still unclear, and further study will be performed to elucidate such mechanism.

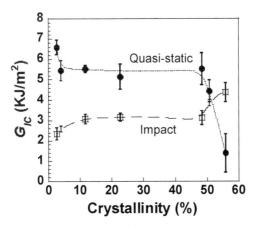

Fig. 2. Dependence of crystallinity on the critical energy release rate under a quasi-static and an impact loading conditions.

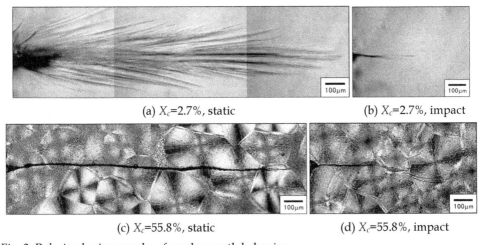

(a) X_c=2.7%, static   (b) X_c=2.7%, impact

(c) X_c=55.8%, static   (d) X_c=55.8%, impact

Fig. 3. Polarized micrographs of crack growth behavior.

Fig.4 shows FE-SEM micrographs of the fracture surfaces of the PLA samples. For the amorphous sample tested at the static rate, the fracture surface exhibits deep concavities and hackles due to multiple craze formation (Fig.4(a)). The fracture surfaces of the crystallized samples (Fig.4(c)) appears to be smoother than the amorphous one, corresponding to the decrease of the toughness values. The impact fracture surface of the amorphous sample (Fig.4(b)) is obviously smoother than the static one, corresponding to the decrease of G_{IC}. It is noted that drawing fibrils are also observed on the impact fracture surface, suggesting that effect of high strain-rate exists. Roughness of the impact fracture surface appears to increase with increase of crystallinity comparing the surfaces shown in Figs.4(b) and (d). For the impact surface of the highly crystallized sample (Fig.4(d)), relatively fine roughness exists suggesting the increase of G_{IC} as crystallinity increases.

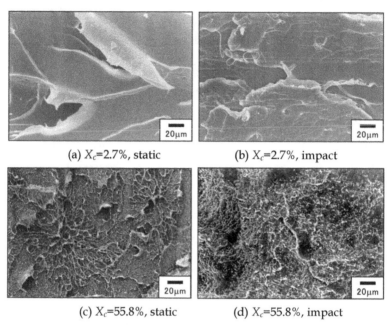

(a) X_c=2.7%, static   (b) X_c=2.7%, impact

(c) X_c=55.8%, static   (d) X_c=55.8%, impact

Fig. 4. FE-SEM micrographs of fracture surfaces.

2.2 Effect of unidirectional drawing process

Drawing process is known to be an effective way to improve the mechanical properties of thermoplastics, and effects of drawing on tensile and fracture properties of thermoplastics such as polypropylene (Mohanraj et al., 2003a, 2003b; Uehara et al., 1996), poly(acrylonitrile) (Sawai et al., 1999; Yamane et al., 1997) and PLA (Todo, 2007) have been studied. PLA is usually draw-processed when it is used for bone fixation devices, and therefore, fundamental effect of drawing on its fracture behavior needs to be characterized. As an example, dependence of draw ratio on the critical J-integral at crack initiation, J_{in}, is shown in Fig.5. In the fracture specimens, the initial notches were introduced in the direction perpendicular or parallel to the drawing direction. Therefore, the two different types of specimens are denoted as 'perpendicular' and 'parallel'. For the parallel, J_{in} decreased with increase of draw ratio, and J_{in} for draw ratio of 2.5 became about one fifth of the original. On the contrary, for the perpendicular, J_{in} increased as draw ratio increased, and J_{in} for draw ratio of 2.5 became five times greater than that of the original. Thus, greater energy is needed for crack propagation in the perpendicular than in the parallel. This is easily understood by considering the effect of drawing on the micromechanism of fracture. In draw-processed polymer, molecules are reoriented in the drawing direction. Therefore, energy dissipation during crack growth by elongation and scission of such oriented molecules is much greater in the perpendicular direction than in the parallel direction where such elongation and scission processes obviously decrease.

FE-SEM micrographs of fracture surfaces are shown in Fig. 6. The perpendicular with draw ratio 2.5 exhibited rougher surface with ductile deformation than the original (without drawing). It is interesting to note that crevices existed on the fracture surfaces that were thought to be cracks transversely propagated between the parallel fibrils reoriented in the drawing direction. It is thus thought that the ductile deformation due to elongation of the oriented molecules and the transverse crack formation are primary mechanisms of toughening in draw-processed PLA. On the other hand, the fracture surface of the parallel were much smoother than that of the original, corresponding to the lower J_{in} value. J_{in} is contributed by energy dissipation through not only creation of fracture surface but also development of process zone. Poralized micrographs of notch-tip regions of the original and the perpendicular are shown in Fig.7. In the original, multiple crazes forming a fan shape were observed. They were initiated from the initial notch-tip and propagated almost perpendicularly to the tensile direction. For the perpendicular with draw ratio 2.5, crazes were much denser and the width of the damage region was much wider than the original. Transverse cracks generated in the drawing direction are observed, and these obviously correspond to the crevices observed on the fracture surfaces as shown in Fig.6(b). Larger damage region consisting of crazes and transverse cracks generated in crack-tip region indicates larger energy dissipation under crack initiation and propagation processes, and therefore, greater J_{in}.

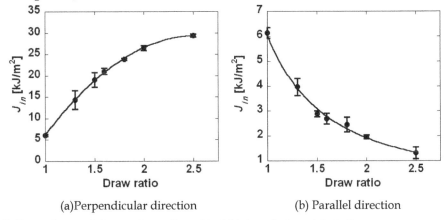

(a)Perpendicular direction   (b) Parallel direction

Fig. 5. Dependence of draw ratio on the critical J-integral at crack initation.

In summary, it was shown that the crystallization behavior greatly affects the fracture behavior of PLA. Microstructure of PLA can easily be changed through annealing process by changing temperature and heating time. The static fracture energy tends to decrease as crystallinity increases, while the impact fracture energy increases.

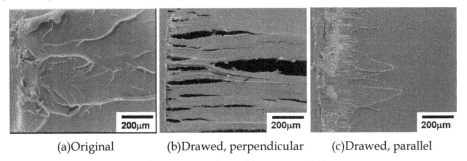

(a)Original   (b)Drawed, perpendicular   (c)Drawed, parallel

Fig. 6. FE-SEM micrographs of fracture surfaces (draw ratio=2.5).

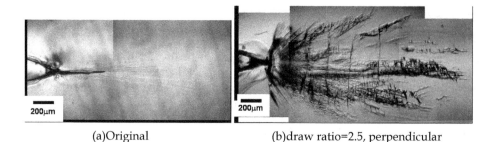

| (a)Original | (b)draw ratio=2.5, perpendicular |

Fig. 7. Poralized micrographs of damage zones.

3. Fracture behavior of PLA/PCL blend

Blending with ductile biodegradable polymers such as PCL (Broz, 2003; Dell'Erba, 2001; Chen, 2003; Todo, 2007; Tsuji, 1996, 1998, 2001, 2003), poly(butylene succinate-co-ε-caprolactone) (PBSC) (Vannaladsaysy, 2010) and poly (butylene succinate-co-L-lactate) (PBSL) (Shibata, 2006, 2007; Vannaladsaysy, 2009; Vilay, 2009) has extensively been investigated in order to improve the fracture energy of PLA. Amoung of them, PCL is known to be bioabsorbable and bioaabsorbable, therefore has been applied in medical fields. In this paragraph, the Mode I fracture behavior of PLA/PCL blend is discussed.

FE-SEM micrographs of the cryo-fracture surfaces of PLA/PCL blends are shown in Fig.8. Phase separations indicated as spherulites of PCL are clearly observed. It is obvious that the size of the PCL spherulites increases with increase of PCL content. It is also seen that voids are created as a result of removal of the dispersed PCL droplets. In general, a blend of immiscible polymers such as PLLA and PCL creates macro-phase separation of the two components due to difference of solubility parameter. This kind of phase separation dramatically affects the physical and mechanical properties of the blend (Dell'Erba et al., 2001; Maglio et al., 1999; Tsuji et al., 2003).

Dependence of PCL content on the critical energy release rate at crack initiation, G_{in}, is shown in Fig.9. G_{in} increases with increase of PCL content up to 5wt%, and G_{in} becomes about 1.5 times greater than that of PLA. G_{in} slightly decreases as PCL content increases above 5wt%; however, G_{in} values of the blends with 10 and 15wt% of PCL are still higher than that of PLA.

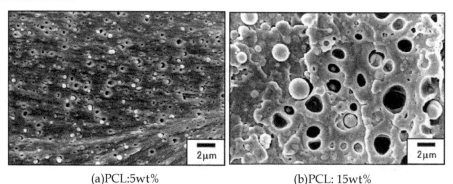

| (a)PCL:5wt% | (b)PCL: 15wt% |

Fig. 8 Morphology of PLA/PCL blends

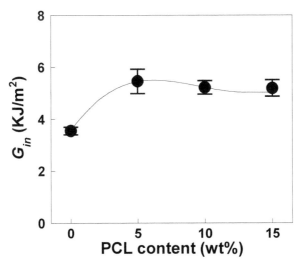

Fig. 9. Dependence of PCL content on the critical energy release rate at crack initiation.

Polarized micrograph of crack-tip region of PLA/PCL (PCL:15wt%) is shown in Fig.10. Craze-like damages similar to neat PLA shown in Fig.7(a) are created in the crack-tip region, and the size of the damage zone is much larger than that of PLA. FE-SEM micrographs of surface on the crack-tip region are also shown in Fig.11. The right-hand figure is a micrograph of the craze-like damages at higher magnification. The micrograph clearly indicates a typical structure of craze, consisting of voids and fibrils. The spherulites of PCL are also seen. The extended fibrils of the matrix PLA were found to be much longer than those of the neat PLA, suggesting that the existence of the dispersed PCL spherulites in PLA tends to enhance ductile deformation of PLA fibrils. It is thus clear that the dispersed PCL droplets play an important role in the formation of the craze-like damages in PLA/PCL blend. FE-SEM micrograph of the fracture surface is shown in Fig.12. Increased ductile deformation of the matrix PLA with appearance of porous structures is clearly observed on the surface. These holes are thought to be created by removal of PCL droplets.

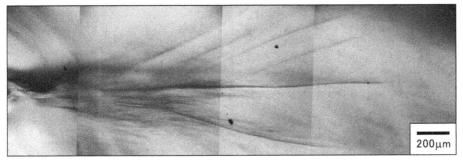

Fig. 10. Polarized micrograph of crack growth behavior in PLA/PCL.

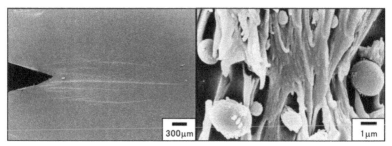

Fig. 11. FE-SEM micrographs of crack-tip region of PLA/PCL.

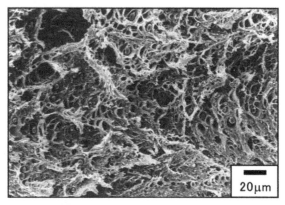

Fig. 12. FE-SEM micrograph of fracture surface of PLA/PCL.

In summary, it was shown that the fracture energy of PLA can be improved by blending with PCL with unchanged biocompatible and bioabsorbable characteristics. This improvement is considered to be achieved by stress relaxation and energy dissipation mechanisms such as extensive multiple craze formation of continuous phase and creation of extended fibril structures of dispersed phase. It is important to note that PLA/PCL exhibited phase separation due to incompatibility of two components, and created voids owing to removal of dispersed PCL phase. Those voids increased with increase of PCL content. PLA/PCL exhibited craze-like deformation of continuous phase similar to neat PLA during mode I fracture process, however, the size of the damage zone was much larger than the PLA, corresponding to the higher G_{in}. PLA/PCL also showed creation of voids by PCL phase separation within the fracture process region, and these voids were likely to be extended at lower stress level, and therefore, decrease G_{in} due to local stress concentration.

4. Toughness improvement of PLA/PCL blend

4.1 Effect of LTI

As shown in the above section, blending with PCL successfully improved the fracture energy of brittle PLA. It was, however, also found that the immiscibility of PLA and PCL causes phase separation, and tends to lower the fracture energy especially when PCL content increases. It has recently been found that the addition of lysine tri-isocyanate (LTI) to PLA/PCL blends effectively improves their immiscibility (Takayama, 2006a, 2006b; Harada, 2007, 2008) and therefore the fracture energy (Takayama, 2006a, 2006b).

FE-SEM micrographs of cryo-fractured surfaces of PLA/PCL and PLA/PCL/LTI are shown in Fig.13. The content of PCL was 15wt% in these materials. Spherical features appeared on the micrograph are thought to be PCL-rich phases. These micrographs clearly showed that the size of the PCL-rich phase dramatically decreases by LTI addition. It is thus presumed that LTI addition effectively improves the miscibility of PLA and PCL. This is thought to be related to the following chemical reaction, that is, the hydroxyl group of PLA and the isocyanate group of LTI creates urethane bond:

$$HO\text{-}R' + R\text{-}N\text{=}C\text{=}O \rightarrow R\text{-}NHCOO\text{-}R'$$

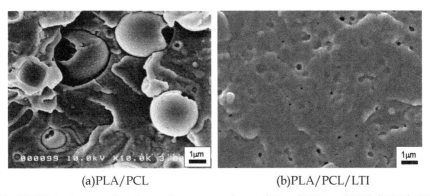

(a)PLA/PCL   (b)PLA/PCL/LTI

Fig. 13. FE-SEM micrographs of cryo-fracture surfaces of PLA/PCL and PLA/PCL/LTI.

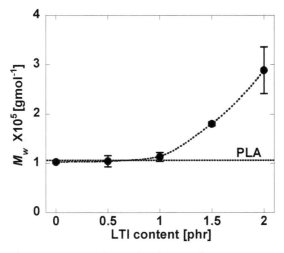

Fig. 14. Dependence of LTI content on the molecular weight.

Dependence of LTI content on the molecular weight, M_w, is shown in Fig.14 · For comparison, M_w of neat PLA is also shown in the figure. M_w values of the blends tend to keep unchanged up to 1phr of LTI content, and then rapidly increase up to 2phr, suggesting that the chemical reaction between the hydroxyl groups of PLA and PCL and the isocyanate groups of LTI was promoted by addition of LTI more than 1 phr. This microstructural change in molecular level due to LTI addition strongly support the macroscopic improvement of the fracture energy.

Dependence of LTI content on the crystallinity of PLA, $x_{c,PLA}$, is shownIn Fig.15. · $x_{c,PLA}$ of PLA/PCL and PLA/PCL/LTI with 0.5phr of LTI are higher than that of PLA. · It is considered that the crystallization of PLA in the blends was progressed actively more than in neat PLA. With increase of LTI content, $x_{c,PLA}$ decreases rapidly at 1 phr and this value is slightly lower than that of neat PLA. These results support that the phase separation between PLA and PCL is improved dramatically at 1 phr. It is presumed that the chemical reation between LTI and PLA/PCL during melt-mixing process results in the improvement of miscibility, and therefore the mobility of PLA and PCL molecules during solidification in cooling process is reduced, resulting in the reduction of crystallization.

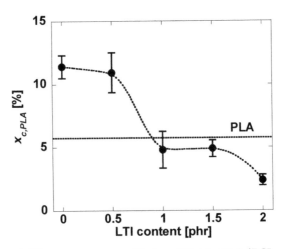

Fig. 15. Dependence of LTI content on crystallinity of PLA in PLA/PCL and PLA/PCL/LTI.

Dependence of LTI content on J_{in} is shown in Fig. 16. It is seen that J_{in} of PLA/PCL is a little larger than that of PLA, indicating the effectiveness of PCL blend on J_{in} is very low. J_{in} of PLA/PCL is effectively improved by LTI addition, and J_{in} increases with increase of LTI content up to 1.5 phr. There is no difference of J_{in} between 2 phr and 1 phr of LTI addition, suggesting that the improvement of J_{in} is saturated with about 1.5 phr of LTI.

Poralized micrographs of crack growth behaviors in PLA/PCL/LTI are shown in Fig.17. In PLA/PCL/LTI with 0.5 phr of LTI, craze-like features are still seen in the crack-tip region as also seen in neat PLA (Fig.7(a)) and PLA/PCL blend (Fig.10). The number of the craze-lines is obviously decreased due to LTI addition. With higher content of LTI, such craze-like feature is no longer generated, and instead, the crack-tip region is plastically deformed, very similar to the crack-tip deformation in ductile plastics and metal. It is known that this kind of plastic deformation dissipates more energy than the craze-like damage, resulting in the greater fracture energy. It is therefore thought that LTI addition to PLA/PCL dramatically changes the crack-tip deformation mechanism; as a result, J_{in} is greatly improved.

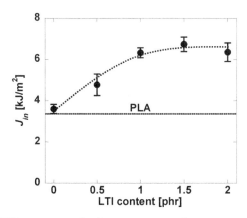

Fig. 16. Dependent of LTI content on the fracture energy, J_{in}.

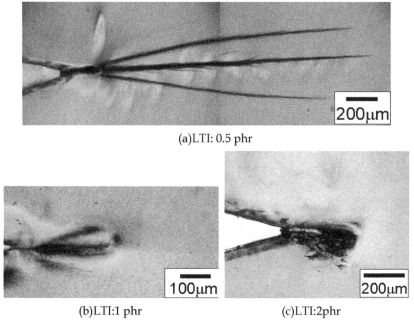

(a)LTI: 0.5 phr

(b)LTI:1 phr   (c)LTI:2phr

Fig. 17. Poralized micrographs of crack growth behaviors in PLA/PCL/LTI.

FE-SEM micrographs of fracture surfaces of PLA, PLA/PCL and PLA/PCL/LTI are shown in Fig. 18. The fracture surface of PLA is very smooth, corresponding to a brittle fracture behavior with low fracture energy. The surface roughness increases with the existence of elongated PCL and cavities by PCL blending. These cavities are thought to be created by debonding of the PCL-rich phases from the surrounding PLA matrix phase and usually cause local stress concentration in the surrounding regions. Thus, this kind of cavitation tends to lower the fracture energy because of the local stress concentration, and compensates to the increase of fracture energy due to the ductile deformation of PCL. This is the reason for the slight improvement of J_{in} in PLA/PCL shown in Fig. 16. It is clearly seen from Fig. 18(c) that cavities do not exist on the fracture surface of PLA/PCL/LTI, indicating that the miscibility of PLA and PCL improves due to LTI addition. In addition, elongated structures are more on PLA/PCL/LTI than PLA/PCL. Thus, extensive ductile deformation associated with disappearance of cavitation is the primary mechanism of the dramatic improvement of J_{in}.

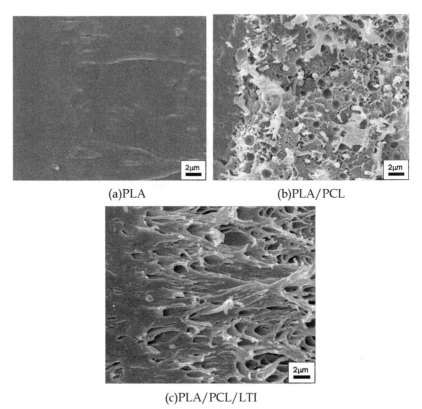

(a)PLA   (b)PLA/PCL

(c)PLA/PCL/LTI

Fig. 18. FE-SEM microgprahs of fracture surfaces of PLA, PLA/PCL and PLA/PCL/LTI.

In summary, the miscibility between PLA and PCL is dramatically improved by introducing LTI as an additive. The increase of molecular weight and the decrease of crystallinity with increase of LTI content clearly indicate that crosslinks are generated by urethane bonds in which the hydroxyl groups at the ends of PLA and PCL molecules react with the isocyanate groups of LTI during molding process. Such microstructural modification results in the dramatic improvement of the macroscopic fracture property, J_{in}.

4.2 Effect of annealing process on PLA/PCL/LTI

As described in the previous section, the immiscibility of PLA/PCL can be improved by adding LTI as a compatibilizer, and as a result, the fracture energy of PLA/PCL is effectively improved. However, blending of ductile PCL with PLA degrades another mechanical properties such as strength and elastic modulus of the base polymer PLA. Recently, Tsuji et al. found that the mechanical properties such as elastic modulus and tensile strength of PLA could be improved by annealing (Tsuji et al., 1995). It is thus expected that annealing process may also affect on the mechanical properties of PLA/PCL/LTI blend.

FE-SEM micrographs of cryo-fracture surfaces of quenched and annealed PLA/PCL and PLA/PCL/LTI are shown in Fig.19. Both the quenched and the annealed PLA/PCL show spherical structures of PCL. It is obviously seen that PCL spherulites in PLA/PCL/LTI are much smaller than those in PLA/PCL. The annealed blends exhibit rougher surfaces with spherical structures than the quenched samples. These spherical structures are thought to be the spherulites of PLA generated by crystallization during annealing process.

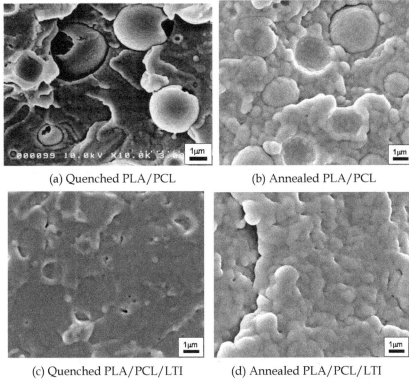

(a) Quenched PLA/PCL   (b) Annealed PLA/PCL

(c) Quenched PLA/PCL/LTI   (d) Annealed PLA/PCL/LTI

Fig. 19. FE-SEM micrographs of microstructures of PLA/PCL and PLA/PCL/LTI.

Effects of annealing on the elastic modulus, E, and the strength, σ_f, under three-point bending condition are shown in Fig.20. Both E and σ_f increase due to annealing. Effects of annealing on the crystallinity, $x_{c,PLA}$, and the molecular weight, M_w, are also shown in Table 1. It is clearly seen that $x_{c,PLA}$ dramatically increases due to annealing. Thus, increasing $x_{c,PLA}$ is likely to strengthen the structure of the polymer, resulting in the increase of E and σ_f. M_w of PLA/PCL increases slightly due to LTI, suggesting that LTI is thought to generate additional polymerization with hydroxyl and carboxyl groups lying in the ends of PLA or PCL molecules. This polymerization results in the improvement of miscibility of PLA and PCL as shown in Fig.19(c). On the other hand, M_w of PLA/PCL decreases slightly due to annealing, indicating progression of thermal degradation in this blend. On the contrary, M_w of PLA/PCL/LTI slightly increases, suggesting that additional polymerization take place during annealing process.

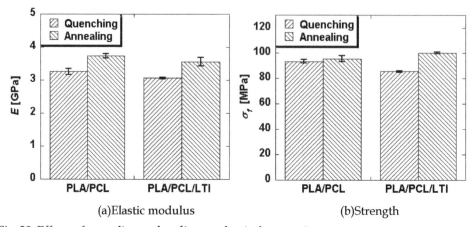

(a)Elastic modulus   (b)Strength

Fig. 20. Effects of annealing on bending mechanical properties.

Blend	$X_{c,PLA}$ (%)	M_w (g/mol)
Quenched PLA/PCL	11.4	1.03×10^5
Annealed PLA/PCL	45.6	9.08×10^4
Quenched PLA/PCL/LTI	4.8	1.13×10^5
Annealed PLA/PCL/LTI	36.6	1.52×10^5

Table 1. Effects of annealing on the crystallinity and molecular weight.

Effects of annealing on the critical J-integral at crack initiation, J_{in}, are shown in Fig.21. It is clearly seen that J_{in} of PLA/PCL/LTI effectively increases due to annealing; on the contrary, PLA/PCL exhibites decrease of J_{in}.

FE-SEM micrographs of the fracture surfaces of the mode I fracture specimens are shown in Fig.22. By comparing Figs.22(a) and (b), it is clearly seen that ductile deformation of spherical PCL phase is suppressed by annealing. Cavities are also observed on the surface of PLA/PCL, as a result of removal of the spherical PCL phases. FE-SEM micrographs at higher magnification show that elongated structures of the spherical PCL phases are observed in the quenched PLA/PCL, while ruptured PLA fibrils and undeformed PCL spherulites are observed in the annealed PLA/PCL. It is also interesting to see in Fig.22(a) that some PCL fibrils are penetrated into the PLA phase and seem to be entangled with PLA fibrils. It is thought that the PLA phase creates a firm structure due to crystallization by annealing and therefore, entangled PCL fibrils with PLA fibrils are firmly trapped in the PLA phase. The PCL spherulites are thus surrounded by such firm crystallized PLA phase with entanglement of PLA and PCL fibrils, resulting in the rupture of the PCL fibrils and the suppression of ductile deformation of the PCL spherulites. This is considered to be the primary reason for the degradation of J_{in} as shown in Fig.21.

It is clearly seen from Figs.22(c) and (d) that in the PLA/PCL/LTI blends, cavity formation is totally suppressed and as a result, ductile deformation is expanded due to the improved miscibility of PLA and PCL by LTI addition. This implies that the miscibility of PLA and PCL is improved by crosslinking of PLA and PCL macromolecules induced by the chemical reaction between the hydroxyl group of PLA and PCL and the isocyanate group of LTI. FE-SEM micrographs at higher magnification show that for both the quenched and annealed PLA/PCL/LTI blends, entangled fibril structures of PLA and PCL are observed. It is thus considered that this kind of structural transformation due to polymerization by LTI blending results in strengthening the structure of the PLA/PCL blends. The microstructure of PLA/PCL/LTI is thought to be further strengthened due to crystallization of PLA by annealing, resulting in the dramatic improvement of the mode I fracture energy J_{in} as shown in Fig.22.

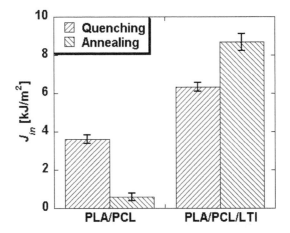

Fig. 21. Effects of annealing on the critical J-integral at crack intiation, J_{in}.

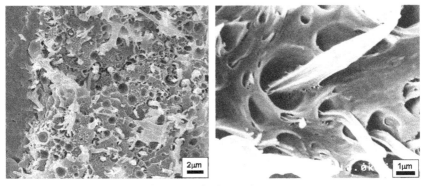

(a) Quenched PLA/PCL

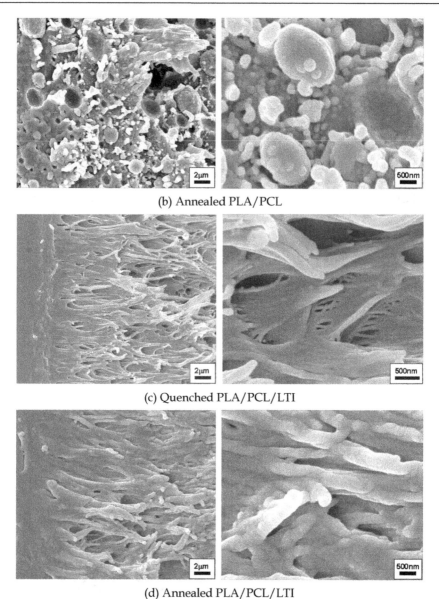

(b) Annealed PLA/PCL

(c) Quenched PLA/PCL/LTI

(d) Annealed PLA/PCL/LTI

Fig. 22. FE-SEM micrographs of fracture surfaces of quenched and annealed samples.

In summary, the bending modulus and strength of both PLA/PCL and PLA/PCL/LTI are effectively improved by annealing. Crystallization of the PLA phase by annealing is thought to strengthen the structure of the PLA/PCL blend, resulting in increase of these properties. The mode I fracture energy of PLA/PCL significantly decreases by annealing mainly owing to embrittlement of the PLA phase. For the case of PLA/PCL/LTI, the structural transformation due to polymerization by LTI addition and crystallization by annealing strengthen the microstructure, resulting in the dramatic improvement of the mode I fracture energy.

5. Conclusion

In this chapter, the fundamental fracture characteristics of bioabsorbable PLA were firstly discussed, and then as examples of toughening, effects of unidirectional drawing and blending with PCL on the fracture behavior were presented. Finally, microstructural modification for PLA/PCL blends using LTI additive was discussed. Thermal processes have great influences on the microstructure and the mechanical properties of PLA mainly due to crystallization behaviour during the heating process. Highly crystallized PLA tends to exhibit very brittle fracture behavior with low fracture energy. Amorphous PLA can generate multiple crazes at crack-tip region to dissipate more energy during fracture process than crystallized materials in which craze formation is suppressed. Drawing process can arrange molecules in one direction so that the fracture resistance in the perpendicular to the drawing direction is greatly improved, while the resistance in the drawing direction tends to degrade. Another effective way to improve the fracture energy is blending with ductile polymer such as PCL. PLA/PCL blends show higher fracture energy with extensive damage formation in crack-tip regions than neat PLA; however, the immiscibility of PLA and PCL results in phase separation morphology in which spherulites of PCL are dispersed in PLA matrix. Such morphological problem can effectively be improved by using LTI as an additive. The phase separation is almost disappeared and the fracture energy is greatly improved. The fracture micromechanism is changed from multiple craze-like damage formation to plastic deformation in crack-tip region. Furthermore, the mechanical properties including elastic modulus, strength and fracture energy of PLA/PCL/LTI blends can effectively be improved by introducing annealing process, although such process tends to degrade the fracture energy of PLA/PCL blends.

6. References

Botsis, J., Chudnovsky, A., Moet, A. (1987). Fatigue crack layer propagation in polystyrene. 1. experimental observations. *International Journal of Fracture*, Vol.33, No.4, pp.263-276.

Broz, M.E., VanderHart, D.L., Washburn, N.R. (2003). Structure and mechanical properties of poly(DmL-lactic acid)/poly(ε-caprolactone) blends. *Biomaterials*, Vol.24, pp.4181-4190.

Chen, C.C., Chueh, J.Y., Tseng, H., Huang, H.M., Lee, S.Y. (2003). Preparation and characterization of PLA plymeric blends. *Biomaterials*, Vol.24, pp.1167-1173.

Dell'Erba, R., Groeninckx, G., Maglio, G., Malinconico, M., Migliozzi, A. (2001). Immiscible polymer blends of semicrystalline biocompatible components: Thermal properties and phase morphology analysis of PLLA/PCL blends. *Polymer*, Vol.42, pp.7831-7840.

Harada, M., Ohya, T., Iida, K., Hayashi, H., Hirano, K., Fukuda, H. (2007). Increased impact strength of biodegradable poly(lactic acid)/poly(butylene succinate) blend composites by using isocyanate as a reactive processing agent, Journal of Applied Polymer Science, Vol.106, No.3, pp.1813-11820.

Harada, M., Iida, K., Okamoto, K., Hayashi, H., Hirano, K. (2008). Reactive compatibilization of biodegradable poly(lactic acid)/poly(epsilon-caprolactone) blends with reactive processing agents, Polymer Engineering and Science, Vol.48, No.7, pp.1359-1368.

Higashi, S., Tamamoto, T., Nakamura, T., Ikada, Y., Hyon, S.H., Jamshidi. K. (1986). Polymer-hydroxyapatite composites for biodegradable bone fillers. *Biomaterials*, Vol.7, pp.183-187.

Ikada, Y., Shikinami, Y., Hara, Y., Tagawa, M., Fukuda, E. (1996). Enhancement of bone formation by drawn poly(L-lactide). *Journal of Biomedical Materials Research*, Vol. 30, pp.553-558.

Maglio, G., Migliozzi, A., Palumbo, R., Immizi, B., Volpe, M.G. (1999). Compatibilized poly(ε-caprolactone)/poly(L-lactide) blends for biomedical uses. *Macromolecules Rapid Communication*, Vol.20, pp.236-238.

Middleton, J.C., Tipton, A.J. (2000). Synthetic biodegradable polymers as orthopedic devices. *Biomaterials*, Vol.21, pp.2335-2346.

Mohanraj, J., Chapleau, N., Ajji, A., Duckett, R.A., Ward, I.M. (2003a). Fracture behavior of die-drawn toughened polypropylenes. Journal of Applied Polymer Science, Vol.88, pp.1336-1345.

Mohanraj, J., Chapleau, N., Ajji, A., Duckett, R.A., Ward, I.M. (2003b). Production, properties and impact toughness of die-drawn toughened polypropylenes. Polymer Engineering Science, Vol.43, No.6, pp.1317-1335.

Mohanty, A.K., Misra, M., Hinrichsen, G. (2000). Biofibres, biodegradable polymers and biocomposites:An overview. *Macromolecular Materrials and Engineering*, Vol.276/277, pp.1-24.

Park, S.D., Todo, M., Arakawa, K. (2004). Effect of annealing on the fracture toughness of poly(lactic acid), *Journal of Materials Science*, Vol.39, pp.1113-1116.

Park, S.D., Todo, M., Arakawa, K. (2005). Effects of isothermal crystallization on fracture toughness and crack growth behavior of poly(lactic acid) , *Journal of Materials Science*, Vol.40, pp.1055-1058.

Park, S.D., Todo, M., Arakawa, K., Koganemaru, M. (2006). Effect of crystallinity and loading-rate on mode I fracture behavior of poly(lactic acid). *Polymer*, Vol.47, pp.1357-1363.

Sawai, D., Yamane, A., Kameda, T., Kanamoto, T., Ito, M., Yamazaki, M., Hisatani, K. (1999). Uniaxial drawing of isotactic poly(acrylonitrile): development of oriented structure and tensile proeprties. Macromolecules, Vol.32, No.17, pp.5622-5630.

Shibata, M., Teramoto, N., Inoue Y. (2006). Mechanical properties, morphologies, and crystallization behavior of blends of poly(L-lactide) with poly(butylene succinate-co-L-lactate) aaand poly(butylene succinate). *Polymer*, Vol.47, pp.3557-3564.

Shibata, M., Teramoto, N., Inoue Y. (2007). Mechanical properties, morphologies, and crystallization behavior of plasticized poly(L-lactide)/poly(butylene succinate-co-L-lactate) blends. *Polymer*, Vol.48, pp.2768-2777.

Takayama, T., Todo, M. (2006). Improvement of impact fracture properties of PLA/PCL polymer blend due to LTI addition. *Journal of Materials Science*, Vol.41, No.15, pp.4989-4992.

Takayama, T., Todo, M., Tsuji, H., Arakawa, K. (2006). Effect of LTI content on impact fracture property of PLA/PCL/LTI polymer blends. *Journal of Materials Science*, Vol.41, No.19, pp.6501-6504.

Takayama, T., Todo, M., Tsuji, H. (2011). Effect of annealing on the mechanical properties of PLA/PCL and PLA/PCL/LTI polymer blends. *Journal of the Mechanical Behavior of Biomedical Materials*, Vol.4, pp.255-260.

Todo , M., Shinohara , N., Arakawa, K. (2002). Effects of crystallization and loading-rate on the mode I fracture toughness of biodegradable poly(lactic acid). *Journal of Materials Science Letters*, Vol.21, pp.1203-1206.

Todo, M., Kagawa, T., Takenoshita, Y., Myoui, A. (2008). Effect of press processing on fracture behavior of HA/PLLA biocomposite material. *Journal of Solid Mechanics and Materials Engineering*, Vol.2, No.1, pp.1-7.

Todo, M., Takayama, T. (2007). Improvement of mechanical properties of poly(l-lactic acid) by blending of lysine triisocyanate. *Journal of Materials Science*, Vol.42, pp.4721-4724.

Todo, M., Takayama, T., Tsuji, H., Arakawa, K. (2007a). Effect of LTI blending on fracture proeprties of PLA/PCL polymer blend. *Journal of Solid Mechanics and Materials Engineering*, Vol.1, No.9, pp.1157-1164.

Todo, M., Park, S.D., Takayama, T., Arakawa, K. (2007b). Fracture micromechanisms of bioabsorbable PLLA/PCL polymer blends. *Engineering Fracture Mechanics*, Vol.74, pp. 1872-1883.

Todo, M. (2007). Effect of unidirectional drawing process on fracture behavior of poly(L-lactide). *Journal of Materials Science*, Vol.42, No.4, pp.1393-1396.

Tsuji, H., Ikada, Y. (1995). Properties and morphologies of poly (L-lactide): 1. Annealing condition effects on properties and morphologies of poly (L-lactide). *Polymer*, Vol.36, pp.2709-2716.

Tsuji, H., Ikada, Y. (1996). Blends of aliphatic polyesters. I. Physical proeprties and morphologies of solution-cast blends from poly(DL-lactide) and poly(ε-caprolactone). *Journal of Apllied Polymer Science*, Vol.60, pp.2367-2375.

Tsuji, H., Ikada, Y. (1997). Blends of aliphatic polyesters. II. Hydrolysis of solution-cast blends from poly(L-lactide) and poly(ε-caprolactone) in phosphate-buffered solution. *Journal of Apllied Polymer Science*, Vol.67, pp.405-415.

Tsuji, H., Ishizuka, T. (2001). Porous biodegradable polyesters, 3 Preparation of porous poly(ε-caprolactone) films from blends by selective enzymatic removal of poly(L-lactide). *Macromolecular Bioscience*, Vol.1, No.2, pp.59-65.

Tsuji, H., Yamada, T., Suzuki, M., Itsuno, S. (2003). Blends of aliphatic polyesters. part 7. effects of poly(L-lactide-co-ε-caprolactone) on morphology, structure, crystallization, and physical properties of blends of poly(L-lactide) and poly(ε-caprolactone). *Polymer Journal*, Vol.52, pp.269-275.

Uehara, H., Yamazaki, Y., Kanamoto, T. (1996). Tensile properties of highly syndiotactic polypropylene. *Polymer*, Vol.37, No.1, pp.57-64.

Vannaladsaysy, V., Todo, M., Takayama, T., Jaafar, M., Zulkifli, A., Pasomsouk, K. (2009). Effect of lysine triisocyanate on the mode I fracture behavior of polymer blend of poly(L-lactic acid) and poly(butylene succinate-co-L-lactide). *Journal of Materials Science*, Vol.44, No.11, pp.3006-3009.

Vannaladsaysy,V., Todo, M., Jaafar, M., Ahmad, Z. (2010). Characterization of microstructure and mechanical properties of biodegradable polymer blends of poly(L-lactic acid) and poly(butylene succinate-co-ε-caprolactone) with lysine triisocyanate. *Polymer Engineering and Science,* Vol.50, No.7, pp.1485-1491.

Vilay, V., Mariatti, M., Ahmad, Z., Pasomsouk, K., Todo, M. (2010). Improvement of microstructures and properties of biodegradable PLLA and PCL blends compatibilized with a triblock copolymer. *Materials Science and Engineering A,* Vol.527, No.26, pp.6930 – 6937.

Vilay, V., Mariatti, M., Ahmad, Z., Pasomsouk, K., Todo, M. (2009). Characterization of the mechanical and thermal properties and morphological behavior of biodegradable poly(L-lactide)/poly(ε-caprolactone) and poly(L-lactide)/poly(butylenes succinate-co-L-lactide) polymer blends. *Journal of Applied Polymer Science,* Vol.114, pp.1784-1792.

Vilay, V., Mariatti, M., Ahmad, Z., Pasomsouk, K., Todo, M.(2010). Improvement of microstructure and fractured property of poly(L-lactic acid) and poly(butylene succinate-co-ε-caprolactone) blend compatibilized with lysine triisocyanate. *Engineering Letters,* Vol.18, No.3, pp.303-307.

Yamane, A., Sawai, D., Kameda, T., Kanamoto, T., Ito, M., Porter, R.S. (1997). Development of high ductility and tensile properties upon two-stage draw of ultrahigh molecular weight poly(acrylonitrile). *Macromolecules,* Vol.30, No.14, pp.4170-4178.

Part 2

Evaluation of the Interaction and Compatibility of Biomaterials with Biological Media

11

A Preliminary In Vivo Study on the Histocompatibility of Silk Fibroin

Lu Yan[1], Zhao Xia[1], Shao Zhengzhong[2], Cao Zhengbing[2] and Cai Lihui[1]
[1]Department of Otorhinolaryngology, Huashan Hospital, Fudan University, Shanghai
[2]Department of Macromolecule Science, Fudan University, Shanghai, China

1. Introduction

Biomaterials used for tissue engineering should have the property of good histocompatibility, superb plasticity and desired degradability, so that it can be extensively applied for defect tissue repairing with excellent clinical outcome. In the past decade, silk fibroin has become one of the most favored biomaterials for its wide availability, superb performance and readiness to be shaped for different purposes in tissue engineering[1-14]. Porous scaffolds made by silk fibroin can be made into different pore size and porosity to serve for different needs of tissue repairing. The porous structure may contribute to the mass exchange in the scaffolds. However, the implanted protein scaffolds will degrade and can hardly be separated from host tissues. Therefore, little has been reported on histocompatibility experiment in vivo for silk fibroin[15-22].

In this chapter, progress in study of silk fibroin scaffold in tissue engineering application and biocompatibility research will be introduced, then our histocompatibility experiment of porous scaffolds will be reported. In our experiment, porous scaffolds were made platy and buried in subcutaneous part of the back of SD rat. Tissue reaction was observed, and the value of silk fibroin as tissue engineering scaffold material was discussed.

2. The application of silk fibroin in tissue engineering

It is known to all that the application of silk production as un-absorb suture has many years of history. Along with the progress of tissue engineering techniques, scientists can extract natural polymeric materials- fibroin from silk and make it into different forms to fit various needs of tissue engineering[1]. Vitro studies show silk fibroin is biodegradable. The speed and degree of degradation can be adjusted through changing physicochemical property[2-4]. The product of degradation is mainly free amino acid and has no toxic side-effect on tissue.

Silk fibroin can be easily made into different forms to serve for various needs such as membrane, gelatum, knitting scaffold, porous scaffold and electrospinning scaffold. Therefore, it has been gradually utilized in various medical field such as drug delivery[5,6], nerve regeneration[7-9], dermis healing[10], artificial ligament repair[11,12], bone or cartilage healing[13,14], vascular tissue engineering[15,16], otology application[17] and so on.

3. Biocompatibilty research of silk fibroin

Silk fibroin has been utilized as natural biomaterial in tissue engineering applications. Many researches have been made in vitro to reveal its attractive properties such as slow-controllable degradation, mechanical robustness, and inherent biocompatibility[18-21].

For the silk fibroin scaffolds are made up of proteins, those materials which have been implanted will degrade and could hardly be separated from host tissues, that made the study of histocompatibility difficult. Therefore, few report can be found about research on biocompatibility and degradation of porous fibroin scaffolds in vivo [4,22].

4. Our in vivo research on the histocompatibility of silk fibroin

4.1 Materials and methods

Laboratory animals: 8 healty female Sprague-Dawley rat, body weight 220g~250g, get from department of laboratory animal medicine, Fudan University

Preparation method of porous silk fibroin scaffold: After degummed, the raw silk was dissolved into aqueous solution of LiBr (lithium bromide). The silk fibroin solution was adjusted to the concentration of 20% after filtration and concentration. Then n-butyl alcohol was added into the solution as volume ratio of 2: 1. Mixed solution was added into self-made mould after low-speed agitation, and white porous scaffold was obtained after freeze drying. The average pore size of the scaffold was between 10 and 20um.

Material prepration: We cut the porous silk fibroin scaffolds to 1cm×1cm squares (about 1 mm thick), infused the material into 75% ethanol solution for 0.5 hour, then immersed the materials into 0.9% NaCl water for 24 hours.

Technique: After peritoneal injection anesthesia, we longitudinal incised the skin of back open, blunt dissected surrounding tissues, and buried the porous silk fibroin scaffolds subcutaeous. Then we sew up the incision and raised the animals in different captivities.

Histological examination The rats were executed 2,4,6,8 weeks after operation. General observation and hematoxylin and eosin stain histological examination was performed.

4.2 Result and discussion

General observation Animals recovered well after operation. Throughout the period of implantation, all scaffolds were well tolerated by the host animals and immune responses to the implants were mild. No obvious systemic reaction, abnormal eating and sleeping behaviors were found. The implant sample in animals' back hunched mildly. The wounds and surrouding skin healed well when drawing materials at different times (figure 1). No obvious hyperemia, seepage or purulent exudates signs were found in surrounding skin and muscular tissue. The surface of the scaffolds was wrapped up by the thin, semitransparent fibrous membranes. There was no obvious change in shape and appearance after implanting(figure 2). The inflammatory reaction surrounding the tissue was slight. And no obvious granulation was found.(table 1)

The second week: Fibroin scaffold is eosinophilic staining. The surface is not regular while there are big or small pores inside. Hematoxylin-eosin staining microscopic examination(table 2). Fibroblast grows adherently on the surface of the material. The surrounding tissue hyperplasticly changes, inflammatory cell infiltrate is visible, most of which are lymphocytes and macrophages. No signs of tissue necrosis are found.(figure 3,4)

Fig. 1. The wound healed up well while taking samples

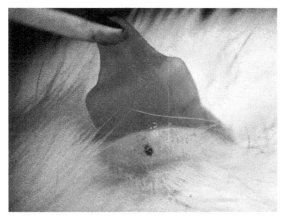

Fig. 2. Thin connective tissue wrap could be found on the scaffold, and the inflammatory reaction surrounding the tissue was slight.

Time(week)	Section number	inflammatory reaction	fibrous encapsulation	granulation formation
2	2	slight	yes	no
4	2	slight	yes	no
6	2	slight	yes	no
8	2	slight	yes	no

Table 1. general observation

The fourth week: There is few parts with eosinophilic staining inside the material disappears 4 weeks after operation, which shows crumbling phenomenon, while it's general structure remains(figure 5). Fibroblasts grow along the surface of the scaffold or into the pores. There are lymphocytes and macrophages infiltrate. Vessel hyperplasia is visible, while new capillary vessel inside the material is obvious. No signs of tissue necrosis are (figure 6).

The sixth week: There are a few parts with eosinophilic staining inside the material disappears 6 weeks after operation, which shows crumbling phenomenon, while it's general structure remains.Inflammatory cell infiltrate is visible 6 weeks after operation. Histocytes gradually grow deep into the scaffold. No signs of tissue necrosis are.(figure 7,8)

The eighth week: Histocytes grow more deeply into the scaffold. Inflammatory cell infiltrate is visible, most of which are macrophages. Vessel hyperplasia reduces inside the tissue, while hyperblastosis still obvious. No signs of tissue necrosis are.(figure 9,10)

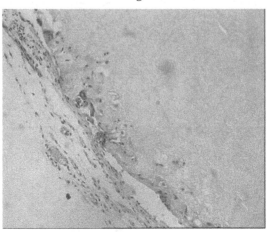

Fig. 3. Fibroblast grows adherently on the surface of the material two weeks after operation. The surrounding tissue hyperplasticly changes, inflammatory cell infiltrate is visible, most of which are lymphocytes and macrophages. No signs of tissue necrosis are found. （×200）

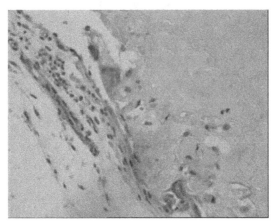

Fig. 4. Fibroin scaffold is eosinophilic staining. The surface is not regular while there are big or small pores inside. Fibroblasts grow adherently on the surface of the material. Inflammatory cell infiltrate is visible, most of which are lymphocytes and macrophages. （×400）

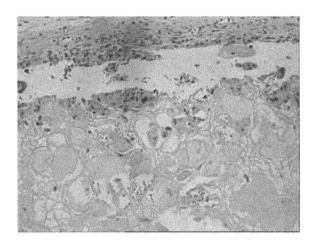

Fig. 5. There is few parts with eosinophilic staining inside the material disappears 4 weeks after operation, which shows disaggregation phenomenon, while it's general structure remains. (×200)

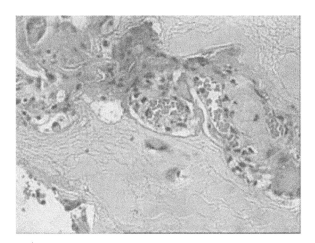

Fig. 6. Fibroblasts grow along the surface of the scaffold or into the pores. There are lymphocytes and macrophages infiltrate. Vessel hyperplasia is visible, while new capillary vessel inside the material is obvious. No signs of tissue necrosis are found. (×400)

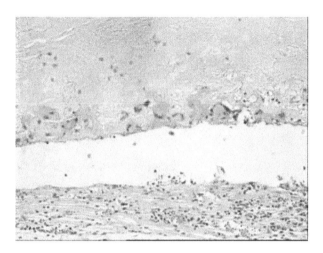

Fig. 7. Inflammatory cell infiltrate is visible 6 weeks after operation. Histocytes gradually grow deep into the scaffold. No signs of tissue necrosis are found. (×200)

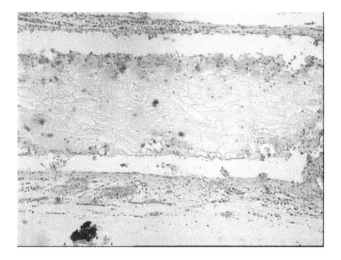

Fig. 8. There are a few parts with eosinophilic staining inside the material disappears 6 weeks postoperatively, which shows disaggregation phenomenon, while it's general structure remains. (×200)

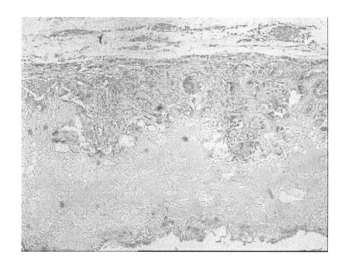

Fig. 9. Histocytes grow more deeply into the scaffold. Inflammatory cell infiltrate is visible, most of which are macrophages. Vessel hyperplasia reduces inside the tissue, while hyperblastosis still obvious. No sign of tissue necrosis is found. (×200)

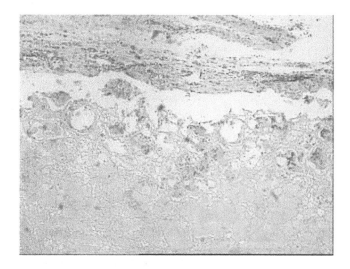

Fig. 10. The border of the scaffold break down at some part 8 weeks postoperatively, while the inside part remains as it was. (×200)

Time (week)	section number	fibration	inflammatory cell type	Scaffold disaggregation	tissue necrosis	tissue ingrowth
2	2	yes	lymphocyte and macrophage	no	no	yes
4	2	yes	lymphocyte and macrophage	yes	no	yes
6	2	yes	lymphocyte and macrophage	yes	no	yes
8	2	yes	mosly macrophage	yes	no	yes

Table 2. histopathology observation

Porous scaffols made by silk fibroin can be formed into different pore size and porosity, which could meet the needs of different histocytes during repairing process. Porous structure is also helpful to the exchange of nutrient substance among different parts of scaffolds. So we choose the porous silk fibroin scaffold as the research object.

Biological material will be treated as foreign matter when implanted in the body, even if its biocompatibility is perfectly good. Therefore, foreign body reaction will be inevitable surrounding the material. The strength and extent of foreign body reaction are closely related to its biocompatibility and biodegradablity. So, we evaluate the tissue reaction and material change after implantation by fibrous encapsulation, granulation formation, tissue fibrosis, type of inflammantary cells, tissue necrosis and tissue ingrowth, morphological changes and disaggregation of the scaffolds.

We found that wet porous fibroin scaffold was soft in texture, and could be easily cut into different forms to fit the need of implantation while the operation was performed. Tissue reaction around the materials was slight when we removed the scaffold, and the surface of the material was wrapped up by semitransparent fibrous membranes. The porous fibroin scaffolds were eosinophilic stained by HE. Its surface was irregular. And there are lacunes of different size inside the material. Chronic inflammation reaction occurred, mainly lymphocytes and macrophages could be found around the material. And fibrous capsule could be observed. After two weeks of implantation, blood capillary could be found proliferate into the lacuna of the material, and the fibroblasts also attached to the irregular surface of the material. These phenomena became more and more obvious as time went by, which showed that the surface texture of porous fibroin was favourable for interstitial cell to grow on. Moreover, the porous structure of the scaffold made the material permeable, which could probably cause the exchange of the nutrition and metabolite through the scaffold that lead to active tissue ingrowth. This may also be a plus factor for tissue regeneration.

Because the fibroin scaffolds were formed by protein and could hardly be separated from surrounding tissues, histopathology observation was performed after experiment instead of quantitative determination. During eight-weeks' observation, the scaffold disintegration started from the fourth week. But there was still no obvious change inside the scaffold structure until the eighth week, which indicated to us that we should set the observation period much longer to get more information about the degradation of fibroin porous scaffold in further research. We didn't make effective statistical analysis about the quantity and type of inflammatory cells surrounding the scaffold for the number of samples was too small. Moreover, according to the requirements of different tissue repair, the pore size and porosity of porous scaffolds should be different. And materials with different structure may cause different result in tissue reaction, tissue regeneration and material degradation. These are also challenging questions that need to be considered in further experiment.

5. Conclusion

To sum up, porous silk fibroin scaffold shows good histocyte attachment and has good histocompatibility. The Porous silk fibroin scaffold can degradate in vivo, but more study should be made on the mechanism and degradation products.

6. Acknowledgment

The chapter publication was funded by Shanghai Science and Technology Committee, 1052nm03801.

7. References

Murphy AR, Kaplan DL. Biomedical applications of chemically-modified silk fibroin. J Mater Chem. 2009 Jun 23;19(36):6443-6450.

Rebecca L. Horan, Kathryn Antle, Adam L. Collette, etc.In vitro degradation of silk fibroin. Biomaterials,Volume 26, Issue 17, June 2005, Pages 3385-3393

Cao Y, Wang B. Biodegradation of silk biomaterials. Int J Mol Sci. 2009 Mar 31;10(4):1514-24.

Wang Y, Rudym DD, Walsh A,etc. In vivo degradation of three-dimensional silk fibroin scaffolds. Biomaterials. 2008 Aug-Sep;29(24-25):3415-28.

Lammel AS, Hu X, Park SH, etc. Controlling silk fibroin particle features for drug delivery. Biomaterials. 2010 Jun;31(16):4583-91.

Wang X, Yucel T, Lu Q, etc. Silk nanospheres and microspheres from silk/pva blend films for drug delivery.Biomaterials. 2010 Feb;31(6):1025-35.

Lu yan, Chi fang-lu, Zhao Xia, etc. Experimental study on facial nerve regeneration by porous silk fibroin conduit. Zhonghua Er Bi Yan Hou Tou Jing Wai Ke Za Zhi. 2006, 41(8),603-606.

Yang Y, Ding F, Wu J, Development and evaluation of silk fibroin-based nerve grafts used for peripheral nerve regeneration. Biomaterials. 2007 Dec;28(36):5526-35.

Tang X, Ding F, Yang Y,etc. Evaluation on in vitro biocompatibility of silk fibroin-based biomaterials with primarily cultured hippocampal neurons. J Biomed Mater Res A. 2009 Oct;91(1):166-74.

Guan G, Bai L, Zuo B, etc. Promoted dermis healing from full-thickness skin defect by porous silk fibroin scaffolds (PSFSs).Biomed Mater Eng. 2010;20(5):295-308.

Bosetti M, Boccafoschi F, Calarco A,etc. Behaviour of human mesenchymal stem cells on a polyelectrolyte-modified HEMA hydrogel for silk-based ligament tissue engineering. J Biomater Sci Polym Ed. 2008;19(9):1111-23.

Altman GH, Horan RL, Lu HH, et al . Silk matrix for tissue engineered anterior cruciate ligaments. Biomaterials, 2002, 23(20): 4131-4141.

Meinel L, Fajardo R, Hofmanna S, et al . Silk implants for the healing of critical size bone defects. Bone, 2005, 37(5): 688-698.

Hofmann S,Hagenmuller H,Koch AM, et al .Control of in vitro tissue-engineered bone-like structures using human mesenchymal stemcells and porous silk scaffolds. Biomaterials, 2007, 28(6): 1152-1162.

Zhou J, Cao C, Ma X,etc. Electrospinning of silk fibroin and collagen for vascular tissue engineering.Int J Biol Macromol. 2010 Nov 1;47(4):514-9.

Tamada Y. Sulfation of silk fibroin by chlorosulfonic acid and the anticoagulant activity. Biomaterials, 2004, 25(3): 377-383.

Levin B, Redmond SL, Rajkhowa R, etc. Preliminary results of the application of a silk fibroin scaffold to otology. Otolaryngol Head Neck Surg. 2010 Mar;142(3 Suppl 1):S33-5.

Ghanaati S, Orth C, Unger RE,etc. Fine-tuning scaffolds for tissue regeneration: effects of formic acid processing on tissue reaction to silk fibroin. J Tissue Eng Regen Med. 2010 Aug;4(6):464-72.

Mandal BB, Das T, Kundu SC. Non-bioengineered silk gland fibroin micromolded matrices to study cell-surface interactions. Biomed Microdevices. 2009 Apr;11(2):467-76.

Acharya C, Ghosh SK, Kundu SC. Silk fibroin protein from mulberry and non-mulberry silkworms: cytotoxicity, biocompatibility and kinetics of L929 murine fibroblast adhesion. J Mater Sci Mater Med. 2008 Aug;19(8):2827-36.

Yang Y, Chen X, Ding F, etc. Biocompatibility evaluation of silk fibroin with peripheral nerve tissues and cells in vitro. Biomaterials. 2007 Mar;28(9):1643-52.

Etienne O, Schneider A, Kluge JA, etc. Soft tissue augmentation using silk gels: an in vitro and in vivo study. J Periodontol. 2009 Nov;80(11):1852-8.

12

Histopatological Effect Characteristics of Various Biomaterials and Monomers Used in Polymeric Biomaterial Production

Serpil Ünver Saraydin[1] and Dursun Saraydin[2]
[1]Cumhuriyet University, Faculty of Medicine, Department of
Histology and Embryology Sivas,
[2]Cumhuriyet University, Faculty of Science, Department of Chemistry Sivas,
Turkey

1. Introduction

When a synthetic material is placed within the human body, tissue reacts towards the implant in a variety of ways depending on the material type. The mechanism of tissue interaction (if any) depends on the tissue response to the implant surface. In general, there are three terms in which a biomaterial may be described in or classified into representing the tissues responses. These are bioinert, bioresorbable, and bioactive.

Biomaterials are often used and/or adapted for a medical application, thus comprises whole or part of living structures or biomedical devices which performs, augments, or replaces biological functions. Biomaterials are used in dental and surgical applications, in controlled drug delivery applications. A biomaterial may be an autograft, an allograft or a xenograft used as a transplant material.

Biomaterials are mostly polymers produced by monomers, and are used in artificial organ production in contemporary medicine. They are prepared by the polymerization reaction of certain monomers.

In several previous studies, we investigated whether acrylamide, methacrylamide, N-isopropylacrylamide, acrylic acid, 2-hydroxyethyl methacrylate, 1-vinyl-2-pyrrolidone and ethylene glycol had cytotoxic effects and induced apoptosis or not in spinal cord. Immunolocalization of glial fibrillary acidic protein (GFAP) was also determined, and it was evaluated by using semi-quantitative morphometrical techniques. The cytotoxicity of monomers on cultured fibroblastic cell lines was also examined *in vitro*.

Acrylic acid had the most cytotoxic effect when compared to the methacrylamide and the ethylene glycol groups. GFAP immunoreactivity was found to be rather stronger in the methacrylamide than the other monomers application groups. The methacrylamide, acrylic acid, N-vynil pyrrolidine, acrylamide, N-isopropylacrylamide and 2-hydroxyethyl methacrylate application groups had TUNEL positive cells when compared to the other groups. While some monomers used in biomaterial production seemed not to affect the cell viability and GFAP immunoreactivity, some other monomers had adverse effects on those features. This in turn may contribute to the pathological changes associated to the monomer type.

In our previous other works, *in vitro* swelling and *in vivo* biocompatibility of radiation crosslinked acrylamide and its co-polymers such as acrylamide (AAm) and acrylamide/crotonic acid (AAm/CA), acrylamide/itaconic acid (AAm/IA), and acrylamide/maleic acid (AAm/MA) hydrogels were investigated.

The radiation crosslinked AAm, AAm/CA, AAm/IA and (AAm/MA) co-polymers were found to be well tolerated, non-toxic and highly biocompatible.

On the other hand, calcium phosphate ceramics and xenografts have been used in different fields of medicine and dentistry. We demonstrated the effects of calcium phosphate ceramics (Ceraform) and xenograft (Unilab Surgibone) in the field of experimentally created critical size parietal bone defects in rats. Although Ceraform was less resorptive and not osteoconductive properties, it could be considered as a biocompatible bone defect filling material having a limited application alternative in dentistry and medicine. However, xenograft seems biocompatible, osteoconductive, and could be used in a limited manner as a filling material in osseous defects in clinical practice.

2.Toxicological effect of the water-soluble monomers

2.1 Monomers

Monomer is a molecule of any of a class of compounds, mostly organic, that can react with other molecules of the same or other compound to form very large molecules, or polymers. The essential feature of a monomer is polyfunctionality, the capacity to form chemical bonds to at least two other monomer molecules. Bifunctional monomers can form only linear, chainlike polymers, but monomers of higher functionality yield cross-linked, network polymeric products. Toxicological effects of the monomers are changing from very low (zero) to very high.

Some polymeric biomaterials such as hydrogels are produced by the effect of initiator such as chemical initiator, heat, light or high energy radiation from the water soluble-monomers.

2.2 Cytotoxic effects

Biomaterial suitable for a biomedical application must be biocompatible at least on its surface. In several previous studies, we investigated whether acrylamide, methacrylamide, N-isopropylacrylamide, acrylic acid, 2-hydroxyethyl methacrylate, 1-vinyl-2-pyrrolidone and ethylene glycol used in polimeric biomaterial production had cytotoxic effects (Unver Saraydin et al., 2011). The cytotoxicity of xenograft (one of the alternative graft materials) was also examined in vitro (Unver Saraydin et al., 2011).

The viability of cultured fibroblastic cell lines following all monomer applications except for the ethylene glycol group were found to be decreased in all time intervals (Figure 1, 2), and differences were statistically significant ($p<0.05$). In addition, the cell viability was significantly ($p<0.05$) lower in the acrylamid application group when compared to the control group. Acrylic acid demonstrated the maximum cytotoxic effect when compared to the methacrylamide and ethylene glycol groups. On the other hand, the ethylene glycol group showed no cytotoxicity for cells (Graphic 1).

In our study of the xenograft cytotoxic activities, the xenograft showed no cytotoxicity for the cells (Figure 3). There was no decolorization zone around the samples. Although the cells were directly in contact with the xenograft in the culture media, they did not show any signs of injury and preserved their morphological characteristics and wholeness like those seen in the controls.

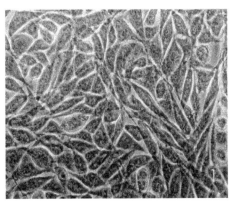

Fig. 1. Fibroblast viability %100 after 12 h incubation period with ethylene glycol

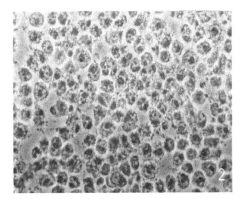

Fig. 2. Fibroblast viability % 0 after 12h incubation period with N-isopropyl acrylamide

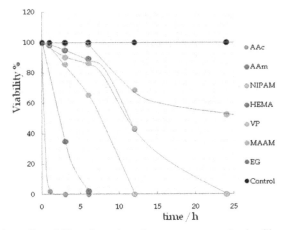

Graphic 1. Shows the cell viability alterations between groups in the fibroblastic cell lines by the time.

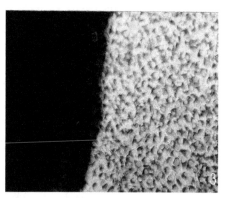

Fig. 3. There is no cytotoxicity for the cells.

2.3 Neurotoxic effects

Several studies revealed neurotoxic effects as well as ataxi and muscle weakness caused by biomaterials on humans and on laboratory animals. It has been suggested that they cause axonal degeneration in central and peripheric nervous system (Barber et al., 2001).

Astrocytes are the stellate glial cells in the central nervous system, which play a major role in supporting neurons, scar formation and development and maintenance of the blood-brain barrier. The physiological and metabolic properties of astrocytes indicate that those cells are involved in the regulation of water, ions, neurotransmitters, and pH of the neuronal milieu (Montgomery 1994). They are also implicated in protection against toxic insults such as excitotoxicity and oxidative stress (Lamigeon et al., 2001). Glial fibrillary acidic protein (GFAP) is an intermediate filament protein found predominantly in astrocytes (McLendon 1994). Therefore it is important to determine the glial fibrillary acidic protein (GFAP) immunoreactivity in astrocytes for the evaluation of biomateials.

In our study, immunolocalization of glial fibrillary acidic protein (GFAP) was determined, and it was evaluated by using semi-quantitative morphometrical techniques (Unver Saraydin et al., 2011). GFAP immunoreactivity was found to be very strong in the methacrylamide, N-isopropylacrilamid, ethylene glycol and N-vinyl pyrrolidine application groups whereas it was weak in acrylic acid, acrylamide and 2-hydroxyethyl metacrylad applied groups (Table 1, Figure 4-10). Changes in GFAP immunoreactivity could be due to following conditions; astrocyte dysfunction, astrocyte loss accompanied by astroglial cell proliferation, de-differentiation, and changes in functional state of neuronal cell types, thus altering the neuron-glial homeostasis. The over-expression of GFAP could probably indicate the protective strategy of these tissues.

Although the neurotoxicity of acrylamide and many monomers has been known since 1950s, its' mechanisms have remained obscure (Lee et al., 2005, Gold and Schaumburg, 2000). Acrylamide increases p53 protein (Okuno et al., 2006), recent studies indicate that it plays a role in apoptotic cell death in neurons (Morrison et al., 2003). Acrylamide can activate caspase- 3 and cause apoptosis in neuronal cells (Sumizawa and Igisu, 2007). The cellular process of apoptosis is an important component of tissue and organ development as well as the natural response to disease and injury (David et al., 2003). DNA fragmentation in neurons was characterized by double staining with terminal deoxynucleotidyl transferase-mediated deoxyuridine triphosphate-biotin nick end labeling (TUNEL) (Bao and Liu, 2004). To our knowledge, however, it has not been determined whether acrylamide and other

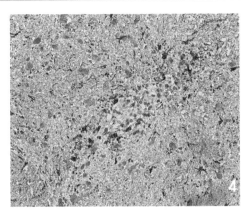

Fig. 4. Control group GFAP immunoreactivity. GFAP40X

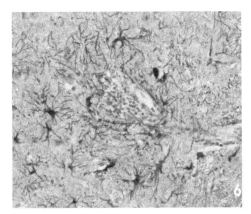

Fig. 5. GFAP immunoreactivity 6 week after Acrylic acid exposure. GFAP 40X

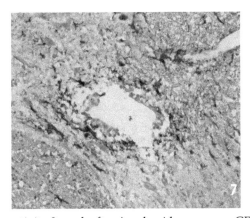

Fig. 6. GFAP immunoreactivity 2 week after Acrylamide exposure. GFAP 40X

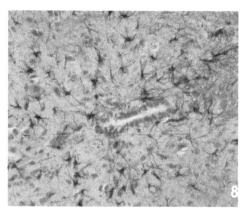

Fig. 7. GFAP immunoreactivity 6 week after 2-hydroxyethyl methacrylate exposure GFAP 40X

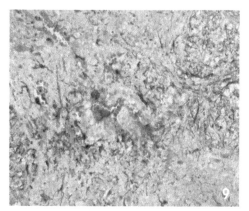

Fig. 8. GFAP immunoreactivity 4 week after methacrylamide exposure GFAP 40X

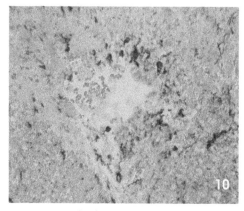

Fig. 9. GFAP immunoreactivity 6 week after N-isopropylacrylamide exposure GFAP 40X

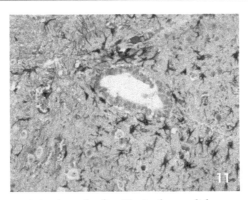

Fig. 10. GFAP immunoreactivity 4 week after N-vinyl pyrrolidone exposure. GFAP 40X

Monomer	1st week	2nd week	4th week	6 th week	12 th week
Ethylene glycol	++	++	+++	++	++
N-vinyl pyrrolidone	+++	+++	+++	+++	++
2-hydroxyethyl methacrylate	+	++	+++	++	++
Acrylamide	++	++			
Methacrylamide	+++	+++	++	++	+
N-isopropylacrylamide	+++	+++	+++	+++	
Acrylic acid	+	++		++	++
Control	+++	+++	+++	+++	+++

Table 1. Demonstrates the semi-quantitative scoring findings of GFAP immunolocalization in rat medullaspinalis following 1, 2, 4, 6 and 12 weeks of particular monomer applications

monomers cause apoptosis in neuronal cells. We therefore examined apoptosis by using terminal deoxynucleotydil transferase dUTP nick and labelling (TUNEL) method in spinal cord (Unver Saraydin et al., 2011).

While TUNEL positive cells has been detected rarely in the control and in the ethylen glycol application groups, numerous TUNEL positive cells were intensively observed in the spinal cord of the methacrylamide, acrylic acid, N-vinyl pyrrolidine, acrylamide, N-isopropylacrylamide and 2-hydroxyethyl metacrylate application groups (Figure 11-14).

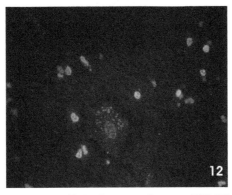

Fig. 11. TUNEL-positive apoptotic cells in the control group. TUNEL 100X

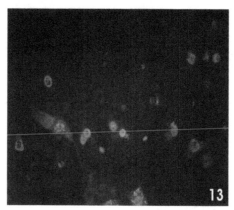

Fig. 12. TUNEL positive cells 6 week after 2-hydroxyethyl methacrylate. TUNEL 100X

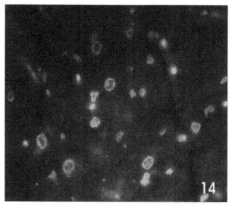

Fig. 13. TUNEL positive cells 6 week after N-isopropylacrylamide. TUNEL 100X

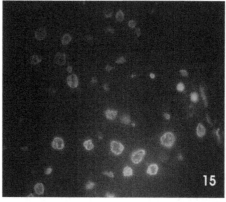

Fig. 14. TUNEL positive cells 6 week after N-vinyl pyrrolidine. TUNEL 100X

3. Polymeric biomaterials

Some polymeric biomaterials such as hydrogels are made of water-soluble molecules, connected usually by covalent bonds, forming a three-dimensional insoluble network. The space between chains is accessible for diffusion of solutes and this space is controllable by the level of cross-linked (connected) molecules. They usually show good biocompatibility in contact with blood, body fluids, and tissues. Therefore, they are very often used as biomaterials for medical purposes, for instance contact lenses, coating of catheters, etc.

Biomaterials are defined as materials that can be interfaced with biological systems in order to evaluate, treat, augment, or replace any tissue, organ, or function of the body.

The clinical application of a biomaterial should not cause any adverse reaction in the organism and should not endanger the life of the patient; any material to be used as part of a biomaterial device has to be biocompatible. The definition of biocompatibility includes that the material has to be nontoxic, non-allergenic, noncarcinogenic, and non-mutagenic, and that it does not influence the fertility of a given patient. Preliminary use of in vitro methods is encouraged as screening tests prior to animal testing. In order to reduce the number of animals used, these standards use a step-wise approach with review and analysis of test results at each stage. Appropriate in vitro investigations can be used for screening prospective biomaterials for estimations of toxic effect. Cytotoxicity in vitro assay is the first test to evaluate the biocompatibility of any material for use in biomedical devices (Rogero et.al. 2003).

Hydrogels can be synthesized by accomplishing crosslinking via γ-irradiation (Guven, O; et.al. 1999, Saraydın et.al. 1995, 2002, Karadağ et. al. 2004). However, little work is done on the biomedical applications of the hydrogels prepared by crosslinking of a homo- or copolymer in solution with γ-irradiation. It is well known that the presence of an initiator and a crosslinking agent affects the macromolecular structure and phase behavior of hydrophilic polymers in solution and contributes to inhomogeneity of the network structure. It is argued that more homogeneous network structures can be synthesized, if crosslinking is accomplished with γ-irradiation in the absence of an initiator and a crosslinking agent. The structural homogeneity of the network affects the swelling behavior and mechanical properties that improved the biological response of materials and subsequently the performance of many medical devices (Benson 2002). Thus, looking to the significant consequences of biocompatibility of biomaterials, we, in the present study, are reporting the results on the biocompatibility with the copolymeric hydrogels prepared with acrylamide (AAm) and crotonic acid (CA) or itaconic acid (IA) or maleic acid (MA) via radiation technique. The selection of AAm as a hydrophilic monomer for synthesizing hydrogel rests upon the fact that it has low cost, water soluble, neutral and biocompatible, and has been extensively employed in biotechnical and biomedical fields. On the other hand, CA monomer consists of single carboxyl group, while IA and MA monomers are consisting of double carboxyl groups. These carboxylic acids could provide the different functional characteristics to acrylamide-based hydrogels. So, these monomers were selected for the preparation of the hydrogels and their biocompatibility studies.

In our previous other works, *in vitro* swelling and biocompatibility of blood *in vivo* biocompatibility of radiation crosslinked acrylamide co-polymers such as acrylamide (AAm), acrylamide/crotonic acid (AAm/CA), acrylamide/itaconic acid (AAm/IA) and acrylamide/maleic acid (AAm/MA) hydrogels were investigated (Saraydin et al., 1995, Karadağ et. al. 1996, Saraydin et al., 2001, 2004).

3.1 In vitro swelling of the hydrogels in the simulated body fluids
In this stage of the study, the swelling of the hydrogels in the simulated physiological body fluids was investigated (Saraydin et al., 1995, Karadağ et. al. 1996).
The phosphate buffer at pH 7.4 (pH of cell fluid, plasma, edema fluid, synovial fluid, cerebrospinal fluid, aqueous humour, tears, gastric mucus, and jejunal fluid), glycine-HCl buffer at pH 1.1 (pH of gastric juice), human sera, physiological saline and distilled water intake of initially dry hydrogels were followed for a long time until equilibrium (Saraydin et al., 2001, 2002).
The fluid absorbed by the gel network is quantitatively represented by the EFC (equilibrium body fluids content), where: EFC% = [mass of fluid in the gel/mass of hydrogel] x 100. EFCs of the hydrogels for all physiologically fluids were calculated. The values of EFC% of the hydrogels are tabulated in Table 2.

Simulated body fluid	AAm	AAm /CA	AAm /MA	AAm /IA
Distilled Water	86.3	93.9	94,7	92.0
Isoosmotic phosphate buffer	87.5	93.8	89,7	92.2
Gastric fluid	87.7	93.6	92,4	88.7
physiological saline	87.8	92.9	89,7	88.7
Human Sera	88.6	92.5	89.8	86.4
In rat	89.0	93.1	91.9	91.7

Table 2. EFC values of the hydrogels

All EFC values of the hydrogels were greater than the percent water content values of the body about 60%. Thus, the AAm and AAm/CA, AAm/MA and AAm/IA hydrogels were exhibit similarity of the fluid contents with those of living tissues.

3.2 In vitro blood biocompatibility
In the second stage of this study, the biocompatibility of the hydrogels was investigated against some biochemical parameters of human sera at 25 °C.
The mean and standard deviation values of control and test groups for biochemical parameters of human sera are listed in Table 3.
Table 3 shows that the values of means of control and test groups are in the range of normal values and there is no significant difference in values before and after contacting these sera with the hydrogels. On the other hand, Student's t-test is applied to control and test groups. No significant difference in values of biochemical parameters was found.

3.3 In vivo tissue biocompatibility
In this part, hydrogels based on copolymer of AAm, AAm/MA, AAm/CA and AAm/IA with capacity of absorbing a high water content in biocompatibility with subcutaneous tissues of rats were examined. After one week implantation, no pathology such as necrosis, tumorigenesis or infection were observed in the excised tissue surrounding the hydrogels and in skin, superficial fascia and muscle tissues in distant sites. After 2–4 weeks, thin fibrous capsules were thickened. A few macrophage and lymphocyte were observed in

Biochemical parameters of human serum / Unit	Normal values	Control	AAm	AAm/CA	AAm/MA	AAm/IA
Glucose/mg dl⁻¹	70-110	87.0±8.2	91.0± 6.1	88.1±3.99	88.8±6.0	88.4±5.30
Triglyceride/mg dl⁻¹	40-160	127.3±24.6	127.0±25.8	130.6±19.9	127.2±25.1	125.6±20.7
Cholesterol/mg dl⁻¹	125-350	158.6±10.9	160.6±14.3	159.8±11.3	157.8±10.8	160.6±14.3
BUN/mg dl⁻¹	8-25	14.8±1.27	15.2±4.56	14.6±3.73	15.2±4.10	15.6±3.84
Creatinin/mg dl⁻¹	0.8-1.6	0.98±0.14	1.06±0.17	1.02±0.14	0.98±0.18	1.00±0.18
Total protein/g dl⁻¹	6.0-8.4	6.52±0.15	6.72±0.15	6.70±0.13	6.60±0.22	6.48±0.30
Albumin/mg dl⁻¹	3.5-5.6	4.02±0.15	3.88±0.15	3.98±0.18	3.96±0.20	3.94±0.10
Alkaline phosphatase/U	35-125	53.6± 13.1	54.5±12.3	54.0±14.9	52.6±12.6	52.6± 10.3
Alanine transaminase/U	7-56	14.6±2.12	16.0±2.63	15.7±3.23	15.9±2.47	16.0±2.63
Aspartate transaminase/U	5-40	16.2±5.33	15.2±3.19	16.5±3.03	17.2±5.16	15.2±3.19
Direct bilirubin/mg dl⁻¹	0.0-0.3	0.12±0.04	0.12±0.04	0.11±0.03	0.11±0.03	0.12±0.04
Indirect bilirubin/mg dl⁻¹	0.1-1.1	0.45±0.05	0.35±0.09	0.35±0.09	0.45±0.05	0.40±0.07
Chlorine/meq dl⁻¹	95-107	98.5±2.17	98.8±2.3	98.6±2.12	97.8±1.75	98.2±2.10
Sodium/meq dl⁻¹	137-146	142.7±1.4	142.8± 0.9	143.0±1.6	142.7±1.4	142.0±1.6
Potassium/meq dl⁻¹	3.5-5.5	4.80±0.28	4.68±0.36	4.94±0.39	4.87±0.35	4.70±0.35
Calcium/mg dl⁻¹	8.5-10.8	9.40±0.39	9.47±0.28	9.42±0.28	9.63±0.42	9.47±0.28
Phosphorus/mg dl⁻¹	2.5-4.5	3.60±0.41	3.60±0.32	3.68±0.42	3.60±0.36	3.56±0.38

Table 3. Means and standard deviations of biochemical parameters of human sera

these fibrous capsules consisting of fibroblasts, and a grouped mast cells and lymphocyte were observed between tissues and capsule in the some samples (Figure 15, 16).

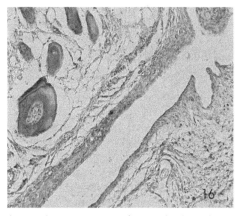

Fig. 15. After one week, the implan-tation site of AAm hydrogel, H-E, 20X

After 6–10 weeks, the adverse tissue reaction, giant cells and necrosis of cells, inflammatory reaction such as deposition of foamed macrophage were not observed in the implant site, however, it is observed to increase in the collagen fibrils due to proliferation and activation of fibroblasts (Fig. 17). No chronic and acute inflammation, adverse tissue reaction were observed in the all test groups. It is no determination related to the loss of activation and liveliness of cells in the capsule cells and in distant sites. No pathology were observed in the skin and the tissues of straight muscle in the close to implant sites.

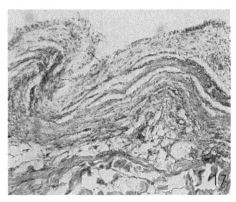

Fig. 16. After 4 week, the implantation site of AAm hydrogel, H-E, 20X

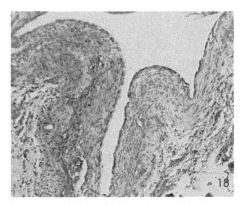

Fig. 17. 10 week postimplantation of AAm/CA hydrogel. H-E, 20X

The thickness of the fibrous capsules were measured in the optical microscope using a micrometer scale. The means of five measurements for each the sample and each time point were calculated. The thickness of fibrous capsules are gradually increased to 6 weeks, and then these values are becomed a constant value. The thickness of fibrous capsule occurred due to AAm/CA, AAm/MA and AAm/IA hydrogels implant are high from the values of AAm and hydrogels. The carboxyl groups on the chemical structure and ionogenic character of AAm/CA, AAm/MA and AAm/IA hydrogels can be caused to the high thickness of the fibrous capsule (Smetana et al., 1990). The thickness of the fibrous capsules were measured in the optical microscope using a micrometer scale. The means of five measurements for each the sample and each time point were calculated and shown in Graphic 2. The thickness of fibrous capsules are gradually increased to 6 weeks, and then these values are becomed a constant value. The thickness of fibrous capsule occurred due to AAm/CA, AAm/MA and AAm/IA hydrogels implant are high from the values of AAm and hydrogels. The carboxyl groups on the chemical structure and ionogenic character of AAm/CA, AAm/MA and AAm/IA hydrogels can be caused to the high thickness of the fibrous capsule (Smetana et al., 1990). On the other hand, Student's t test was applied to the all constant values of thickness of fibrous capsules of the hydrogels, and no significant differences ($p > 0.05$) was found. These thickness of fibrous capsule indicated well within the critical tissue tolerance range. It was given by the some reporters that the threshold capsule thickness should not exceed 200–250 μm for an implanted biomaterial (Jeyanthi and Rao, 1990). Our results clearly indicated that the capsule thickness of the excised tissue were well within these stipulated threshold limits. On the basis of the findings we can conclude that the biological response against the tested hydrogels was very similar to the biocompatibility of very low swollen of poly(2-hydroxyethyl methacrylate) hydrogel, which considered as a biologically inert polymer (Smetana et al., 1990). However, it is important that the swelling of acrylamide based hydrogels are very high than the swelling of poly(2-hydroxyethyl methacrylate) hydrogels for the biomedical uses.

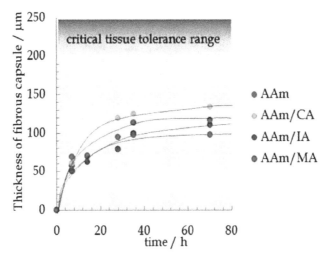

Graphic 2. The curves of thickness of fibrous capsule — implantation time.

4. Bioactive ceramic biomaterials

Bioactive refers to a material, which upon being placed within the human body interacts with the surrounding bone and in some cases, even soft tissue. This occurs through a time – dependent kinetic modification of the surface, triggered by their implantation within the living bone. An ion – exchange reaction between the bioactive implant and surrounding body fluids – results in the formation of a biologically active carbonate apatite (CHAp) layer on the implant that is chemically and crystallographically equivalent to the mineral phase in bone. Prime examples of these materials are synthetic hydroxyapatite, glass ceramic and bioglass.

Calcium phosphate ceramics and xenografts have been used in different fields of medicine and dentistry. We demonstrated the effects of calcium phosphate ceramics (Ceraform) and xenograft (Unilab Surgibone) in the field of experimentally created critical size parietal and mandibular bone defects in rats (Develioglu et al., 2006, 2007, 2009, 2010).

Many researches are currently conducted to find out the ideal material to support bone repair or regeneration. The limitations of autogenous grafts and allogeneic bankbone have led to a search for synthetic alloplast alternatives. Calcium phosphate ceramics have been widely used because the mineral composition of these implants materials does fully biocompatible (Rey C. 1990, LeGeros. 2002). The porous structure of the ceramics is claimed to enhance bone deposition and implant stabilization in the recipient bone. The optimal pore size is still debated to be ranging from 50 and 565 µm (Gauthier et al., 1998, Chang et al., 2000). However, porosity of the material is inversely proportional to the mechanical stability of these calcium phosphate based ceramics (Le Huec et al., 1995). This loss of stability is often cited as a limitation in the use of calcium phosphate-based ceramics in clinical practice. A convenient compromise to overcome this problem is to use a biphasic ceramic, which maintains its mechanical resistance until the resorption is achieved (Gauthier et al., 1998).

Various types of xenografts are used in medicine, dentistry, and also in periodontology. One of the xenografts is Unilab Surgibone, which is currently being used succesfully in medicine and implantology. Moreover, osteoconductive properties are also known (Zhao et al., 1999). Unilab Surgibone is obtained from freshly sacrificed calves which is partially deproteinized and processed by the manufacturers. It is available in varius shapes like tapered pins, blocks, cubes, granules, circular discs and pegs (Balakrishnan et al., 2000). Xenograft materials, bovine bones have been the most preferred ones, basically because they are easily obtainable and there are no great ethical considerations. Additionally they have the great advantage of practically unlimited availability of source/raw material. Partially deproteinized and defatted preparations (e.g.Unilab Surgibone) was indicated reduce antigenity and mild immune response (William et al., 2008).

Generally, xenografts are one of the alternative graft materials used in different fields for filling osseous defects Slotte and Lundgren, 1999, Salama 1983). Nonetheless, an interesting alternative to xenografts is Biocoral® (natural coral), which has been shown to exhibit osteoconductive and biocompatible properties whereby gradual replacement with newly formed bone occurred after its resorption (Guillemin et al.,1989, Doherty et al., 1994, Yılmaz and Kuru, 1996, Yukna Ra and Yukna CN, 1998).

Another xenogeneic, bone-derived implant material is Bio-Oss, which is similar to the xenograft investigated in our studies (Develioglu et al.2009, 2010). Bio-Oss has been proposed as a biocompatible graft material for bony defects for it has shown osteoconductive properties — that is, it was replaced with newly formed bone after grafting (Yıldırım et al, 2001, Sculean et al., 2002, Carmagnola et al., 2002). However, regarding the resorption of Bio-Oss, contradicting reports have emerged. On one hand, a previous study revealed that the bovine bone mineral underwent resorption (Pinholt et al., 1991). On the other hand, numerous researchers claimed that the resorption process of Bio- Oss® was very slow (Skoglund et al., 1997, Jensen et al 1996, Klinge et al., 1992).

In our previous studies with Ceraform (calcium phosphate ceramics) and xenograft (Unilab Surgibone), multinuclear giant cells (MNGC) were observed in the implantation region on 1st, 3rd, 6th ve 18th months.

The observed MNGCs are featured morphologic characteristics of foreign body giant cell (FBGC). These cells are osteoclast-like cells. Both cell types develop from a common precursor (Anderson, 2000) Since foreign body giant cell (FBGC) are the fusion products of monocytic precursors, which are also the precursors of macrophages, (Brodbeck at al., 2002, Matheson et al., 2004) the presence of such leukocytes in the wound healing compartment may be of central importance in driving the tissue reaction to the material. No necrosis, tumorigenesis, or infection was observed at the implant site up to 18 months (Figure 18-20).

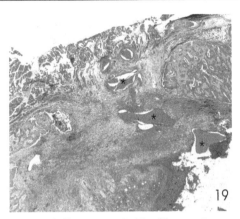

Fig. 18. Remnants of the Xenograft (*) surrounded by fibrous tissue at 30 days. A-T 4X.

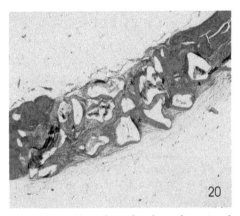

Fig. 19. A dense, fibrovascular tissue (*) in the side of ceraform implantation at 12th month M-T, 4X.

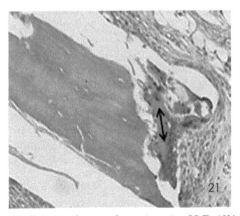

Fig. 20. Multinuclear giant cell (↔)in the implantation site. H-E, 40X

A long-term study would be useful to evaluate the biological degradation behavior of the material utilized in this study. BCP ceramics are well known to be biodegradable due both to body fluid dissolution and bio resorption cellular activity Nery et al., 1990, Piatelli et al., 1996). It might indicate that the implants utilized in our studies are progressively resorbed, but the size of the particle might be big (Handschel et al., 2002) The studies reveal that Ceraform and xenograft are biocompatible. However, the materials did not promote bone formation.

5. Conclusion

In conclusion, while some vinyl monomers had cytotoxic effects on tissues, their polymers, Ceraform and Unilab Surgibone were found to be biocompatible in soft and hard tissues and they seem only to be beneficial bone filler materials in treatment of the bone defects. Unilab Surgibone and xenograft could be used as bone filler materials in the treatment of traumatic and post-traumatic skeletal complications (e.g delayed unions, non-unions), defects due to bone removal (e.g. bone tumors, congenital diseases) or low bone quality (e.g. osteoporosis, osteopenia) and in other medical fields.

6. References

Anderson JM. (2000). Multinucleated giant cells. *Current Opinion in Hematology*, Vol. 7,pp. 40–47, ISSN 1065-6251.

Balakrishnan, M,; Agarwal DC, Kumar SA (2000). Study of efficacy of heterogeneous bone graft (Surgibone) in orthopaedic surgery. *Medical Journal Armed Forces India*, Vol. 56, pp. 21-23, ISSN 0377-1237.

Bao F., Liu D. (2004). Hydroxyl radicals generated in the rat spinal cord at the level produced by impact injury induce cell death by necrosis and apoptosis: Protection by a metalloporphyrin. *Neuroscience,* Vol. 126, pp. 285-295, ISSN 0306-4522.

Barber DS, Hunt JR, Ehrich MF, Lehning EJ, Lo-Pachin RM (2001). Metabolism, Toxicokinetics and Hemoglobin Adduct Formation in Rats Folowing Subacute and Subchronic Acrylamide Dosing. *Neuro Toxicology*, Vol. 22, pp. 341-353, ISSN 0161-813.

Benson RS. (2002) . Use of radiation in biomaterials science. *Nuclear Instruments & Methods In Physics Research Section B–Beam Interactions With Materials And Atoms;* Vol. 191, pp. 752-757, ISSN 0168-583.

Brodbeck WG, Nakayama Y, Matsuda T, Colton E, Ziats NP, Anderson JM. (2002). Biomaterial surface chemistry dictates adherent monocyte/macrophage cytokine expression *in vitro. Cytokine*, Vol. 18, pp. 311–319, ISSN 1043-4666.

Carmagnola D, Berglundh T, Lindhe J. (2002). The effect of a fibrin glue on the integration of Bio-Oss with bone tissue. An experimental study in Labrador dogs. *Journal of Clinical Periodontology*, Vol. 29, pp. 377-383, ISSN 1600-051.

Chang BS, Lee CK, Hong KS, Youn HJ, Ryu HS, Chung SS, Park KW. (2000). Osteoconduction at porous hydroxyapatite with various pore configurations. *Biomaterials*, Vol. 21, pp. 1291–1298, ISSN 0142-9612.

David H. Kim, Alexander R. Vaccaro, Fraser C. Henderson, Edward C. Benzel. (2003). Molecular biology of cervical myelopathy and spinal cord injury: role of oligodendrocyte apoptosis. *The Spine Journal*. Vol. 3, pp. 510-519, ISSN 1529-9430.

Develioğlu H, Ünver Saraydın S, G. Bolayır, L. Dupoirieux. (2006). Assessment of the effect of a biphasic ceramic on bone response in a rat calvarial defect model. *Journal of Biomedical Materials Research,* Vol. 77, pp. 627-631, ISSN 1549-3296.

Develioğlu H, Ünver Saraydın S, Laurent Duopoirieux, Zeynep Deniz Şahin. (2007). Histological findings of long-term healing of the experimental defects by application of a synthetic biphasic ceramic in rats. *Journal of Biomedical Materials Research,* Vol. 80, pp. 505-508, ISSN 1549-3296.

Develioğlu H, Ünver Saraydın S, Ünal Kartal. (2009). The bone healing effect of a xenograft in arat calvarial defect model. *Dental Materials Journal,* Vol. 28, pp. 396-400, ISSN 0287-4547.

Develioğlu H, Ünver Saraydın S, Ünal Kartal, Levent Taner. (2010). Evaluation of the long term results of rat cranial bone repair using a particular xenograft. *Journal of Oral Implantology.* Vol. 36, pp. 167-173, ISSN 0160-6972.

Doherty MJ, Schlag G, Schwarz N, Mollan RA, Nolan PC, Wilson DJ. (1994). Biocompatibility of xenogeneic bone, commercially available coral, a bioceramic and tissue sealant for human osteoblasts. *Biomaterials* Vol. 15, pp. 601-608, ISSN 0142-9612.

Gauthier O, Bouler JM, Aguado E, Pilet P, Daculsi G. (1998). Macroporous biphasic calcium phosphate ceramics: Influence of macropore diameter and macroporosity percentage on bone ingrowth. *Biomaterials;* Vol. 19, pp. 133–139, ISSN 0142-9612.

Gold BG, Schaumburg HH. (2000). Acrylamide. In: Spencer PS, Schaumburg HH (eds) *Experimental and clinical neurotoxicology,* 2nd edn. ISBN 0195084772, Oxford University Pres, New York pp 124-132.

Guillemin G, Meunier A, Dallant P, Christel P, Pouliquet JC, Sedel L. (1989). Comparison of corals resorption and bone apposition with two natural corals of different porosities. *Journal of Biomedical Materials Research,* Vol. 23, pp. 767-779, ISSN 1549-3296.

Guven O, Sen M, Karadag E, Saraydın D. (1999). A review on the radiation synthesis of copolymeric hydrogels for adsorption and separation purposes. *Radiation Physics And Chemistry,* Vol. 56, No. 4, pp. 381–386, ISSN 0969-806.

Handschel J, Wiesmann HP, Sratmann U, Kleinheinz J, Meyer U, Joos U. (2002). TCP is hardly resorbed and not osteoconductive in a nonloading calvarial model. *Biomaterials;* Vol. 23, pp. 1689 –1695, ISSN 0142-9612.

Jensen SS, Aaboe M, Pinholt EM, Hjorting-Hansen E, Melsen F, Ruyter IE. (1996). Tissue reaction and material characteristics of four bone substitutes. *The International Journal of Oral & Maxillofacial Implants,* Vol. 11, pp. 55-56, ISSN 0882-2786.

Jeyanthi R., Rao KP. (1990). *In vivo* biocompatibility of collagen poly(hydroxyethyl methacrylate) hydrogels. *Biomaterials,* Vol. 11, pp. 238-243, ISSN 0142-9612.

Karadag, E; Saraydın, D; Cetinkaya, S; Guven, O. (1996). In vitro swelling studies and preliminary biocompatibility evaluation of acrylamide–based hydrogels *Biomaterials,* Vol. 17, pp. 67–70, ISSN 0142-9612.

Karadag, E; Saraydın, D; Guven, O. (2004). Water absorbency studies of gamma–radiation crosslinked poly (acrylamide-co-2,3–dihydroxybutanedioic acid) hydrogels. *Nuclear Instruments &Methods In Physics Research Section B–Beam Interactions With Materials And Atoms.* Vol. 225, No. 4, pp. 489–496, ISSN 0168-583.

Klinge B, Alberius P, Isaksson S, Jönsson J. (1992). Osseous response to implanted natural bone mineral and synthetic hydroxyapatite ceramic in the repair of experimental skull bone defects. *Journal Oral Maxillofacial Surgery* Vol. 50, pp. 241-249, ISSN 02782391.

Lamigeon C, Bellier JP, Sacchettoni S, Rujano M, Jacquemont B. (2001). Enhanced neuronal protection from oxidative stress by coculture with glutamic acid decarboxylase-expressing astrocytes. *Journal of Neurochemistry*, Vol. 77, pp. 598-606, ISSN 1471-4159.

LeGeros RZ. (2002). Properties of osteoconductive biomaterials: Calcium phosphates. *Clinical Orthopaedics*, Vol. 395, pp. 81-98, ISSN 0009-921.

Le Huec JC, Schaeverbeke T, Clement D, Faber J, Le Rebeller A. (1995). Influence of porosity on the mechanical resistance of hydroxyapatite ceramics under compressive stress. *Biomaterials*, Vol. 16, pp. 113-118, ISSN 0142-9612.

Lee K-Y, Shibutani M, Kuroiwa K, Takagi H, Inoue K, Nishikawa H, Miki T, Hirose M. (2005). Chemoprevention of acrylamide toxicity by antioxidative agents in rats-effective suppression of testicular toxicity by phenylethyl isothiocyanate. *Arch Toxicol*, Vol. 79, pp. 531-541, ISSN 0340-5761.

Matheson MA, Santerre JP, Labow RS. (2004). Changes in macrophage function and morphology due to biomedical polyurethane surface undergoing biodegradation. *Journal of Cell Physiology*, Vol. 199, pp. 8 -19, ISSN 0021-9541.

McLendon, R.E, Bigner, D.D. (1994). Immunohistochemistry of the glial fibrillary acidic protein: basic and applied considerations. *Brain Pathology*, Vol. 4, pp. 221-228, ISSN 1015-6305.

Montgomery D. L. (1994). Astrocytes: form, functions, and roles in disease. *Veterinary Pathology*, Vol. 31, pp. 145-167, ISSN 0300-9858.

Morrison RS, Kinoshita Y, Johnson MD, Guo W, Garden GA. (2003). P53-Dependent cell death signaling in neurons. *Neurochemistry Research*, Vol. 28, pp. 15-27, ISSN 0364-3190.

Nery EB, Eslami A, Van SR. (1990). Biphasic calcium phosphate ceramic combined with fibrillar collagen with and without citric acid conditioning in the treatment of periodontal osseous defects. *Journal of Periodontology*, Vol. 61, pp. 166 -172, ISSN 0022-3492.

Okuno T, Matsuoka M, Sumizawa T, Igisu H. (2006). Involvement of the extracellular signal-regulated protein kinase pathway in phosphorylation of p53 protein and exerting cytotoxicity in human neuroblastoma cells (SH-SY5Y) exposed to acrylamide. *Archives of Toxicology*, Vol. 80, pp. 146-153, ISSN 0340-5761.

Piatelli A, Scarano A, Mangano C. (1996). Clinical and histologic aspects of biphasic calcium phosphate ceramic (BCP) used in connection with implant placement. *Biomaterials*, Vol. 17, pp. 1767-1770, ISSN 0142-9612.

Pinholt EM, Bang G, Haanes HR. (1991). Alveolar ridge augmentation in rats by Bio-Oss. *Scandinavian Journal of Dental Research*, Vol. 99, pp. 154-161, ISSN 0029-845.

Rey C. (1990). Calcium phosphate biomaterials and bone mineral. Differences in composition, structures and properties. *Biomaterials*, Vol. 11, pp. 13-15, ISSN 0142-9612.

Rogero S. O., Malmonge S. M., Lugão A. B., Ikeda T. I., Miyamaru L., and Cruz Á. S.(2003). Biocompatibility Study of Polymeric Biomaterials. *Artificial Organs*, Vol. 27, No. 5, pp. 424–427, ISSN 1525-1594.

Salama R. (1983). Xenogenic bone grafting in humans. *Clinical Orthopaedics and Related Research*, Vol. 174, pp. 113-121, ISSN 0009-921.

Saraydın, D; Karadag, E; Guven, O. (1995). Acrylamide/maleic acid hydrogels. *Polymers For Advanced Technologies*, Vol. 6, No. 12, pp. 719–726, ISSN 1042-7147.

Saraydın, D; Karadag, E; Çetinkaya, S; Guven, O. (1995). Preparation of acrylamide maleic-acid hydrogels and their biocompatibility with some biochemical parameters of human serum. *Radiation Physics And Chemistry*, Vol. 46, No. 4-6, pp. 1049–1052, ISSN 969-806.

Saraydın, D; Koptagel, E; Unver–Saraydin, S; Karadag, E; Guven, O. (2001). In vivo biocompatibility of radiation induced acrylamide and acrylamide/maleic acid hydrogels. *Journal of Materials Science* Vol. 36, No. 10, pp. 2473–2481, ISSN 0022-2461.

Saraydın, D; Caldiran, Y. (2001). In vitro dynamic swelling behaviors of polyhydroxamic acid hydrogels in the simulated physiological body fluids. *Polymer Bulletin* Vol. 46, No. 1, pp. 91–98, ISSN 0170-0839.

Saraydın, D; Isikver, Y; Karadag, E; Sahiner, N; Guven, O. (2002). In vitro dynamic swelling behaviors of radiation synthesized polyacrylamide with crosslinkers in the simulated physiological body fluids. *Nuclear Instruments & Methods In Physics Research Section B–Beam Interactions With Materials And Atoms*, Vol. 187, No. 3, pp. 340–344, ISSN 0168-583.

Saraydın, D; Karadag, E; Sahiner, N; Guven, O. (2002). Incorporation of malonic acid into acrylamide hydrogel by radiation technique and its effect on swelling behaviour. *Journal Of Materials Science* Vol. 37, No. 15, pp. 3217–3223, ISSN 0022-2461.

Saraydın, D; Unver–Saraydın, S; Karadag, E; Koptagel, E; Guven, O. (2004). In vivo biocompatibility of radiation crosslinked acrylamide copolymers. *Nuclear Instruments & Methods In Physics Research Section B–Beam Interactions With Materials And Atoms* Vol. 217, No. 2, pp. 281–292, ISSN 0168-583.

Sculean A, Chiantella GC, Windisch P, Gera I, Reich E. (2002). Clinical evaluation of an enamel matrix protein derivate (Emdogain®) combined with a bovine-derived xenograft (Bio-Oss®) for the treatment of intrabony periodontal defects in humans. *International Journal of Periodontics Restorative Dentistry*, Vol. 22, pp. 259-267, ISSN 0198-7569.

Skoglund A, Hising P, Young C. (1997). A clinical and histologic examination in humans of the osseous response to implanted natural bone mineral. *International Journal of Oral Maxillofacial Implants* Vol. 12, pp. 194-199, ISSN; 0882-2786.

Slotte C, Lundgren D. (1999). Augmentation of calvarial tissue using non-permeable silicone domes and bovine bone mineral. An experimental study in the rat. *Clinical Oral Implants Research*, Vol. 10, pp. 468-476, ISSN 0905-7161.

Smetana Jr. K, Vacík J., Součková D., Krčová Z., Šulc J. (1990). The influence of hydrogel functional groups on cell behavior *Journal of Biomedical Materials Research*, Vol. 24, pp. 463-470, ISSN 1549-3296.

Sumizawa T, Igisu H. (2007). Apoptosis induced by acrylamide in SH-SY5Y cells. *Archives of Toxicology*, Vol. 81, pp. 279-282, ISSN 0340-5761.

Ünver Saraydın S, Develioğlu H. (2011). Evaluation of the bone repair capacity and the cytotoxic properties of a particular xenograft: An Experimental study in rats Turkiye Klinikleri Journal of Medical Sciences, Vol.31, No.3, pp.541-547, ISSN 1300-0292.

Ünver Saraydın S, H. Eray Bulut, Ünal Özüm, Z. Deniz Şahin İnan, Zübeyde Akın Polat, Yücel Yalman, Dursun Saraydın. (2011). Evaluation of the Cytotoxic effects of various monomers, In vitro also their effects on Apoptosis and GFAP immunolocalization in rat spinal cord In vivo. *HealthMED*. Vol. 5, No. 1, pp. 17-28, ISSN 1840-2291.

William S. Pietrzak, Charles A. Vacanti, (2008). *Musculoskeletal Tissue Regeneration: Biological Materials and Methods*, Publisher: Totowa, NJ : Humana Press, p:121, ISSN 0895-8696.

Yılmaz S, Kuru B. (1996). A regenerative approach to the treatment of severe osseous defects: report of an early onset periodontitis case. *Periodontal Clinical Investigation* Vol. 18, pp. 13-16, ISSN 1065-2418.

Yukna RA, Yukna CN. (1998). A 5-year follow-up of 16 patients treated with coralline calcium carbonate (BIOCORAL®) bone replacement grafts in infrabony defects. *Journal of Clinical Periodontology*, Vol. 25, pp. 1036-1040, ISSN 0303-6979.

Yıldırım M, Spiekermann H, Handt S, Edelhoff D. (2001). Maxillary sinus augmentation with the xenograft Bio-Oss® and autogenous intraoral bone for qualitative improvement of the implant site: a histologic and histomorphometric clinical study in humans. *International Journal of Oral Maxillofacial Implants*, Vol. 16, pp. 23-33, ISSN 0882-2786.

Zhao YF, Mendes M, Symington JM. Listrom RD, Pritzker KP. (1999) Experimental study of bone growth around a dental implant after Surgibone grafting. *International Journal of Oral Maxillofacial Implants*, Vol. 14, pp. 889-897, ISSN 0882-2786.

13

Cell Adhesion and Spreading on an Intrinsically Anti-Adhesive PEG Biomaterial

Marga C. Lensen[1,2], Vera A. Schulte[1] and Mar Diez[1]
[1]*DWI e.V. and Institute of Technical and Macromolecular Chemistry, RWTH Aachen*
[2]*Technische Universität Berlin, Institut für Chemie, Nanostrukturierte Biomaterialien*
Germany

1. Introduction

This Chapter deals with bulk hydrogels consisting of a widely used biomaterial: poly(ethylene) glycol (PEG). PEG is renown for its bio-inertness; it is very effective in suppressing non-specific protein adsorption (NSPA) and thereby preventing cell adhesion. However, we have observed unexpected adhesion of fibroblast cells to the surface of bulk PEG hydrogels when the surface was decorated with micrometer-sized, topographic patterns. This Chapter describes the aim of our investigations to unravel the biophysical, biochemical and biomechanical reasons why these cells do adhere to the intrinsically anti-adhesive PEG material when it is topographically patterned.

1.1 Application of hydrogels in biomaterial science

Amongst the different classes of materials which find use in the field of medicine and biology, hydrophilic polymers have demonstrated great potential. Networks formed from hydrophilic polymer often exhibit a high affinity for water, yet they do not dissolve due to their chemically or physically crosslinked network. Water can penetrate in between the chains of the polymer network, leading to swelling and the formation of a hydrogel (Langer & Peppas, 2003; Peppas et al., 2000; Wichterle & Lim, 1960). Generally such polymer networks can be formed via chemical bonds, ionic interactions, hydrogen bonds, hydrophobic interactions, or physical bonds (Hoffman, 2002; Peppas, 1986). Hydrogels have found numerous applications in drug delivery as well as in tissue engineering where they are used as scaffolds for the cultivation of cells to enable the formation of new tissues (Jen et al. 1996; Krsko & Libera, 2005; Langer & Tirrell, 2004; Peppas et al., 2006). Hydrogels are especially attractive for this purpose as they meet numerous characteristics of the architecture and mechanics of most soft tissues and many are considered biocompatible (Jhon & Andrade, 1973; Saha et al., 2007). Furthermore, concerning the intended purpose of cell encapsulation and delivery, hydrogels support sufficient transport of oxygen, nutrients and wastes (Fedorovich et al., 2007; Lee & Mooney, 2001; Nguyen & West, 2002).

In general, hydrogel matrices can be prepared from a variety of naturally derived proteins and polysaccharides, as well as from synthetic polymers (Peppas et al., 2006). Depending on their origin and composition, natural polymers have specific utilities and properties. Hydrogels from natural sources have for example been fabricated from collagen, hyaluronic acid (HA), fibrin, alginate and agarose (Hoffman, 2002). Collagen, HA and fibrin are components which are in vivo present in the extracellular matrix (ECM) of mammalian cells. Since they are derived from natural sources, hydrogels formed from these polymers are inherently cytocompatible and bioactive. They can promote many cellular functions due to a diversity of endogenous factors present. However, scaffolds fabricated from natural sources are rather complex and often ill-defined, making it difficult to determine exactly which signals are promoting the cellular outcome (Cushing & Anseth, 2007). Furthermore they can possess an inherent batch-to-batch variability which can affect sensitive cells in their viability, proliferation, and development (Cushing & Anseth, 2007). Due to these limitations of gels formed from natural polymers, a wide range of synthetic polymers has been found suitable regarding their chemical and physical properties (Hoffman, 2002). The advantages of synthetic gels include their consistent composition and predictable manipulation of properties.

A few examples of synthetic hydrogel building blocks are given in **Figure 1**, including neutral (upper row) and ionic (bottom row) monomers (Peppas et al., 2006).

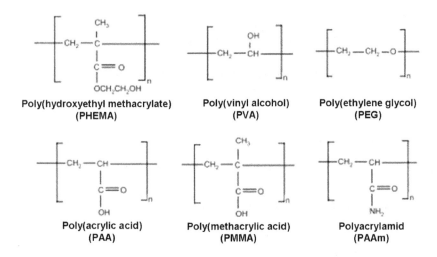

Fig. 1. Some examples of synthetic hydrogels that are used in biomedical applications. Reproduced with permission from Peppas et al., *Adv. Mater.*, *18*, 1345-60. Copyright 2006 John Wiley and Sons.

Proteins are molecules, which often adsorb unspecifically from solution at biomaterial interfaces, a phenomenon that has been documented in a wealth of publications, e.g. references: (Andrade & Hlady, 1986; Andrade et al., 1992; Wahlgren & Arnebrant, 1991). Almost any material, when exposed to a physiological, protein-containing solution, becomes coated with proteins within seconds. As widely recognized, this adsorption of proteins to synthetic material surfaces is of great importance in the field of biomaterials as it plays a determining role for the subsequent cellular responses. Failure of most implant materials stems from an inability to predict and control the process of protein adsorption and cell interaction, resulting in an inappropriate host response to the material (Castner & Ratner, 2002; Hlady & Buijs, 1996; Tsai et al., 2002). Biomaterial surface-induced thrombosis, for example, one of the major problems in clinical applications of materials in contact with circulating blood, begins with the unspecific adsorption of plasma proteins (Andrade & Hlady, 1986; Harris, 1992; Horbett, 1993).

Not only with regards to tissue engineering and implant design unspecific protein adsorption is a highly critical process, also different devices in diagnostics (e.g. protein arrays) and biosensors are based on specific receptor-ligand binding, demanding a non-interacting background. Therefore, much effort has been focused on the development of inert, protein resistant materials and coatings (Chapman et al., 2000; Elbert & Hubbell, 1996). Many synthetic hydrophilic polymers, including PAA, PHEMA, PVA, PEG and poly(ethylene oxide) (PEO) have been applied for this purpose (see Figure 1) (Castillo et al., 1985).

1.2 Biomedical applications of PEG- or PEO-based hydrogels

Some of the earliest work on the use of PEG and PEO as hydrophilic biomaterials showed that PEO adsorption onto glass surfaces prevented protein adsorption (Merrill et al., 1982). Several subsequent studies confirmed that PEO, or its low molecular weight (Mw<10 kDa) equivalent, PEG, were showing the most effective protein-repellent properties (Harris, 1992). PEG-modified surfaces are non-permissive to protein adsorption, bacterial adhesion and eukaryotic cell adhesion (Zhang et al., 1998; Desai et al., 1992; Drumheller et al., 1995; Krsko et al., 2009.

Based on these properties, PEG hydrogels are one of the most widely studied and used materials for a variety of biomedical applications such as tissue engineering, coating of implants, biosensors, and drug delivery systems (Langer & Peppas, 2003; Langer & Tirell, 2004; Krsko & Libera, 2005; Tessmar & Gopferich, 2007; Veronese & Mero, 2008; Harris & Zalipsky, 1997). PEG substrates have also been used to generate patterns of proteins or cells using for example the technique of microcontact printing (Whitesides et al., 2001; Mrksich & Whitesides, 1996; Mrksich et al., 1997). PEG hydrogels are approved by the US Food and Drug Administration (FDA) for oral and topical application; they are little immunogenic and non-toxic at molecular weights above 400 Da. Since PEG itself is not degradable by simple hydrolysis and undergoes only limited metabolism in the body, the whole polymer chains are eliminated through the kidneys or eventually through the liver (Mw < 30 kDa) (Harris, 1992; Knauf et al., 1988).

Many groups have investigated surface coverings of PEG or PEO in order to try to elucidate why PEG has such remarkably effective properties and different theories have been proposed (Jeon et al., 1991; Prime & Whitesides, 1993). First, there are generally only weak attractive interactions between the PEG-coatings and a wide range of proteins, as protein adsorption is generally known to be more pronounced on hydrophobic surfaces in comparison to hydrophilic ones (Morra, 2000). Furthermore, as the interaction between water and PEG via hydrogen bonds is more favorable and surpasses possible attractive interactions of proteins with the surfaces, a repulsion force is created. Therefore the hydration of the layer, i.e. the binding of interfacial water is of high relevance for the exclusion of other molecules coming near the polymer surface (Harris, 1992; Harder et al., 1998). Additionally, molecules approaching the rather flexible, loosely crosslinked PEG hydrogel from the surrounding medium initiate the compression of the extended PEG molecules inducing a steric repulsion effect (Jeon et al., 1991; Morra, 2000). More specifically, a loosely crosslinked gel has relatively long segments between the crosslinks, which can take a relatively large number of conformations. The number of segment conformations would be substantially restricted by the binding of a protein molecule to the gel surface. This would lead to a relatively large unfavorable entropic change, making the process of protein adsorption very unfavorable for thermodynamic reasons. Additionally, the high mobility of PEG chains allows little time for proteins to form durable attachments.

Many techniques have been developed to create PEG or PEO-bearing surfaces, e.g. exploiting physical adsorption, chemical coupling, and graft polymerization (Harris, 1992; Harris & Zalipsky, 1997; Prime & Whitesides, 1993; Fujimoto et al., 1993; Prime & Whitesides, 1991). Whitesides and co-workers have studied covalent coatings of oligo(ethylene glycol)s, so-called self-assembled monolayers (SAMs) and found that the resistance to protein adsorption increased with the chain length of the oligomers (Prime & Whitesides, 1991 and 1993). Furthermore, it has been demonstrated that the adhesion resistance of PEG increases with chain packing density (Sofia et al., 1998; Malmsten et al., 1998).

In recent years the versatility of star-shaped PEG molecules has been recognized, as they present a high number of end-groups per molecule allowing interconnectivity and functionalization (Groll et al., 2005a & 2005b; Lutolf et al., 2003). Some star polymers have been shown to achieve a high surface coverage and localization of the end-groups near the top of the star polymer (Irvine et al., 1996). Therefore, star-shaped PEG molecules are an interesting and promising alternative to linear PEG.

1.3 PEG-based hydrogels formed by UV-curing: patternable biomaterials

We have been using PEG hydrogels that are prepared by UV-based radical crosslinking of six-armed star-shaped macromonomers via acrylate (Acr) end-groups. The polymer backbone consists of a statistical copolymer of 85 % ethylene oxide and 15 % propylene oxide (P(EO-stat-PO)) and each star molecule bears 6 reactive Acr end-groups. The formal notation of the precursor polymer would thus be Acr sP(EO stat PO). Nevertheless, in the following the resulting, crosslinked network will be denoted PEG-based (hydro)gel, even though the arms of the precursors do not consist of pure PEG, but contain a fraction (15%) of propylene glycol units in the copolymer. These PO-units give the prepolymer its unique and very useful property of being a liquid at room temperature, before crosslinking. The crosslinking reaction was initiated by a UV-based radical reaction with benzoin methyl ether as photoinitiator (PI) and an additional crosslinker (CL) (pentaerythritol triacrylate). Further experimental details concerning the synthesis and the curing conditions can be found in our recent publications (Lensen et al., 2007; Diez et al., 2009).

The hydrogel substrates were applied as free-standing bulk gels for 2D cell culture studies. Due to the fact that the prepolymer Acr-sP(EO-stat-PO) is liquid before crosslinking, the precursor mixture can be molded in any shape, which has enabled us to imprint desired micro- and nanometer topographic patterns into the hydrogel surface (Lensen et al., 2007; Diez et al., 2009). In the following, the properties of this hydrogel system in view of its use in biomedical applications will be evaluated, e.g. the cytotoxicity and cytocompatibility will be assessed, and the cell behavior on the surface of the hydrogels will be demonstrated. Finally, the remarkable effect of surface topography and substrate elasticity on protein adsorption, cell adhesion and cell spreading will be discussed.

2. Fabrication and properties of PEG-based substrates

2.1 Synthesis of PEG-based hydrogels from Acr-sP(EO-stat-PO) macromonomers

Hydrogels fabricated for the application in cell culture studies were crosslinked from Acr-sP(EO-stat-PO) prepolymers. UV-irradiation was used to initiate radical polymerization of the macromonomer mixture with added photoinitiator (PI) and crosslinking agent (CL) **(Figure 2)**.

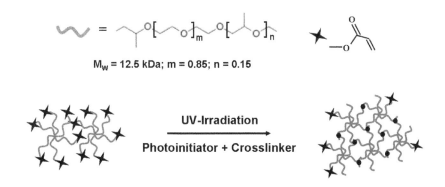

M_W = 12.5 kDa; m = 0.85; n = 0.15

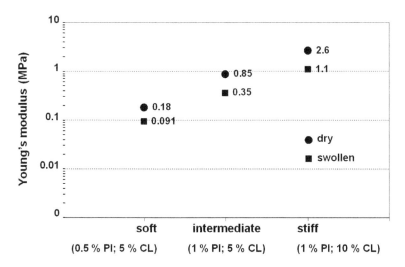

Fig. 2. Fabrication of bulk PEG-based hydrogels by means of UV-curing.

To ensure a complete reaction of the acrylate end-groups which could otherwise undergo undesired reactions with the biological system, the curing kinetics of the system were monitored. It was confirmed that after 10 min more than 90 % of the C-C double bonds of the acrylate end-groups had been consumed. After 60 min only 2.3 % of unreacted end-groups were left. Based on these observations it was decided to apply 60 min of UV-irradiation to the samples in order to achieve virtually complete crosslinking. Bulk PEG-based substrates were fabricated by casting the prepolymer mixture against a smooth silicon surface.

Fig. 3. Young's modulus (MPa) of bulk PEG-based hydrogel samples in dry and swollen state; gels were fabricated from three precursor mixtures with different percentages (w/v) of photoinitiator (PI) and crosslinking agent (CL). Reprinted with permission from: Schulte et al. *Biomacromolecules, 11,* 3375-83. Copyright 2010 American Chemical Society.

The UV-photopolymerization via the acrylate end-groups on the sP(EO-stat-PO) arms does not only allow topographic patterning of the hydrogel's surface, but also enables tuning of the crosslinking density, hence stiffness. Thus, varying the amount of added photoinitiator (PI) and crosslinker (CL) represents a practical approach to controlling the mechanical properties of PEG-based hydrogels (**Figure 3**). This is of high relevance for biomedical applications, as it is well known that cells feel and respond to the stiffness of the underlying substrate (Discher et al., 2005; Engler et al., 2006; Yeung et al., 2005). PEG-based hydrogels with distinctly different mechanical properties were fabricated; the resulting hydrogels from 3 different formulations are denoted as soft, intermediate and stiff (Schulte et al., 2010; Diez et al., 2011).

The stiffness in the swollen state, which is obviously the most relevant for cell culture, was shown to be approximately half of that measured in the dry state, ranging from ~100 kPa for the softest to 1 MPa for the stiffest, thus covering one order of magnitude in elastic modulus (Figure 3).

2.2 Cytotoxicity assessment of PEG-based substrates

As the material has not been used in this exact composition before, cytotoxicity tests were conducted to prove that possible traces of unreacted acrylate end-groups, photoinitiator or crosslinker did not interfere with cell viability. The impact of the material on the viability of L929 fibroblasts was tested with an indirect cell test. Cell membrane integrity as an important indicator for cell viability was tested with a commercially available enzymatic assay. Values shown in **Figure 4** were derived by comparison with a control sample of mortalized cells (incubation with DMSO), which was set to 100 % cytotoxicity. It should be noted that no direct test was possible as PEG is known to be anti-adhesive and the majority of the seeded cells would not stay attached to the PEG-substrate and could therefore not be used for quantitative studies.

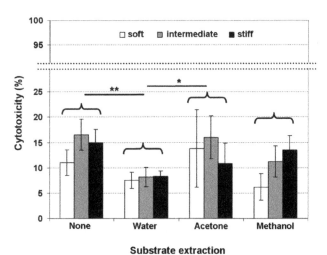

Fig. 4. Cytotoxicity of bulk PEG samples; shown are the results of an indirect cytotoxicity test with L929 cultured for 24 h in PEG conditioned (72 h) medium. The PEG samples had before been extracted in water, acetone or methanol for 24 h. The test was performed in triplicate with substrates of three different crosslinking densities. Statistical significance indicated by **: $p < 0.01$; *: $p < 0.05$.

The percentage of dead cells after 24 h in the indirect cell test with medium that had been conditioned with PEG samples gave values between 7 % and 16 % depending on the sample composition (variations in the amount of PI and CL) and the solvent used for previous extraction of the samples. No significant difference in cytotoxicity could be observed comparing soft, intermediate and stiff samples. Looking at the impact of sample extraction medium, substrates which had been incubated in water for 24 h prior to the test showed the lowest cytotoxic effect. Compared with results gained by extraction with organic solvents (acetone) or without any treatment, the cytotoxicity level detected for the water washed samples was significantly lower ($p<0.05$ and $p< 0.01$, respectively). As a consequence of these test results PEG samples were stored in sterile water for at least 24 h after the crosslinking process prior to in vitro application.

2.3 Fibroblast culture on smooth PEG hydrogel substrates

In order to assess the cytocompatibility of the PEG-based hydrogels in direct contact with cells, L929 cells were cultured directly on the surface of bulk free-standing substrates. Samples of the three introduced elasticities were applied and cell morphology was documented by live imaging after 24 h and 48 h in culture (**Figure 5**). Subsequently, fixed and dried cells were further evaluated by means of electron microscopy (FESEM).

As can be seen in Figure 5, the cell morphology on the three different PEG substrates did not vary significantly, all cells displayed a round shape and only little or no stable cell adhesion was evident. This clearly confirms the cell-repulsive properties of the PEG-based material. Electron microscopy further showed that the rinsing and fixation of the samples led to the removal of a large number of cells. This observation underlines that only a negligible amount of cells was able to establish a contact to the surface. The majority of the cells tended to form aggregates after longer cultivation times. The number of cells visible on the images of the different samples cannot be compared directly as a slight shaking of the medium led to an immediate re-distribution of the cells above the surface.

Based on the observation that the cells were not able to build stable contacts to the surface the consequences for intracellular processes were studied as well. Many anchorage-dependent cell types stop to proliferate or can undergo a programmed cell death (apoptosis) without the presence of integrin-mediated cell-surface contacts (Frisch & Francis, 1994; Gilmore, 2005). The amount of the apoptotic markers caspase-3 and caspase-7 after 48 h of cultivation time on the three different PEG substrates and polystyrene (PS) was assessed with a commercially available assay and compared to a culture where apoptosis was specifically induced by staurosporin addition (values set to 100 %). The results are depicted in **Figure 6**.

As seen in Figure 6, cells cultured on the three different sP(EO-stat-PO) hydrogels did not show an enhanced level of apoptotic activity compared to those seeded on the control substrate PS. This observation is in accordance with results from other groups showing that fibroblasts are not very sensitive to lack of adhesions to a solid substrate if serum is present in the medium (Ishizaki et al., 1995; McGill et al., 1997). There was also no significant difference between the samples with the three different crosslinking degrees. Cell adhesion to PS after 48 h was confirmed by light microscopy.

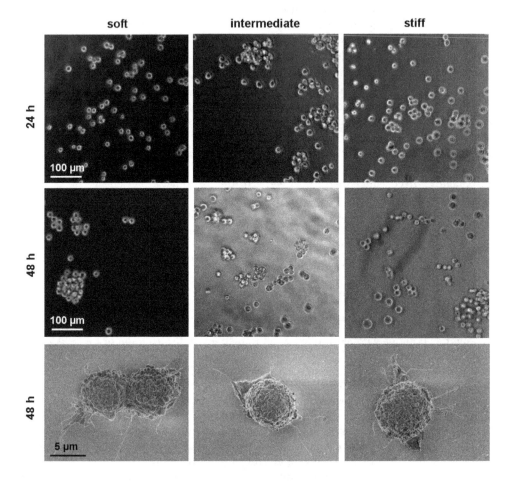

Fig. 5. Microscopic investigation of cytocompatibility of PEG-based hydrogels. L929 fibroblasts were cultured on smooth, bulk PEG samples that were fabricated with different percentages (w/v) of photoinitiator (PI) and crosslinking agent (CL); resulting in different mechanical properties. Cell morphology was monitored by light microscopy at different time points (24 h and 48 h) after initial cell seeding. Electron microscopy images (FESEM) of cells which were fixed and dried after 48 h are shown in the bottom row.

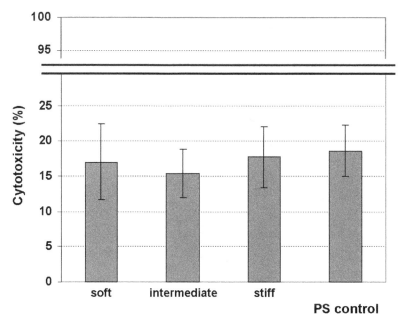

Fig. 6. Apoptotic level of fibroblasts cultured on PEG-based samples. The induction of apoptosis in L929 cells after 48 h cultivation time on smooth PEG samples was quantified with a caspase-3/-7 assay. The samples had before been extracted in water for 24 h. The test was performed in triplicate with substrates of three different crosslinking densities.

2.4 Fibroblast culture on micropatterned PEG hydrogel substrates

While no cell adhesion was observed on the smooth surface of the PEG-based hydrogels, we have discovered that cells do adhere to bulk PEG hydrogels when they are topographically patterned (Lensen et al., 2008; Schulte et al., 2009). **Figure 7** depicts some representative images of fibroblasts adhering to the micropatterned surface of the PEG hydrogels.

In order to explain this observation we have considered that proteins may adsorb to the physically patterned gels and/or that the cells themselves 'feel' the physical pattern and respond on the level of cytoskeletal adaptations. In two follow-up studies we investigated those two (biochemical and biophysical) arguments in more detail. Thus, we examined the possibility of protein adsorption to the gels and we explored the effect of systematic variations of geometry and stiffness of the hydrogels on the enabled cell adhesion.

The first investigation concerned the systematic variation of the deformability of the topographic structures that the cells are assumed to perceive. Although it is unlikely that the cells are able to really deform the micrometer-sized bars, we did envisage that the deformability of the surface structures to depend on the groove width, on the aspect ratio of the bars, and on the inherent mechanical properties of the gels. Thus we prepared line patterns with different groove width (5, 10, 25 and 50 μm) and depth (5, 10 and 15 μm). Also we prepared hydrogel formulations resulting in gels with three different stiffnesses, denoted soft (~90 kPa), intermediate (~350 kPa) and stiff (~1 MPa).

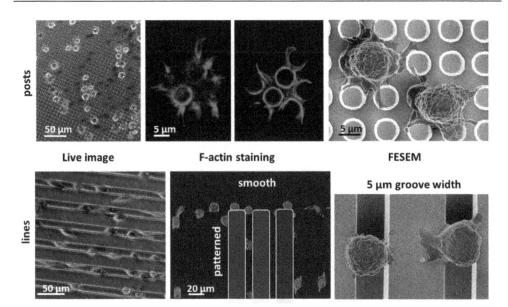

Fig. 7. Reprinted with permission from: Schulte et al., *Biomacromolecules*, 10, 2795–2801. Copyright 2009 American Chemical Society.

We have examined the effect of geometric parameters and of mechanical properties of the hydrogels and found that cells prefer to bind inside of grooves that are of comparable size as their cell body, i.e. 10 μm wide and notably when those 10 μm wide grooves were shallow (5 μm deep). The effect of stiffness variations was only evident in combination with topography, and increased cell adhesion and spreading was observed on the softer gels (**Figure 8**; Schulte et al., 2010).

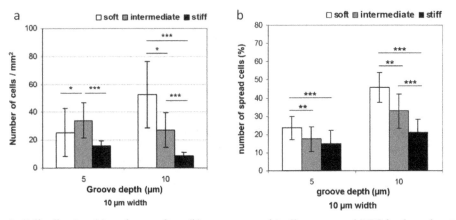

Fig. 8. Cell adhesion (a) and spreading (b) on topographically patterned PEG hydrogels with varied pattern geometries and stiffness. Reprinted with permission from: Schulte et al. *Biomacromolecules*, 11, 3375-83. Copyright 2010 American Chemical Society.

The cells were found to adhere inside the grooves and form adhesion contacts with the side walls as well as to the bottom of the shallower grooves. They were able to adhere to the narrower (5 μm wide) grooves as well, but had to undergo tremendous shape adaptations, including deformation of the cell nucleus, as observed from fluorescence microscopy using selective staining agents for the actin cytoskeleton and for the nucleus (**Figure 9**, right image). The cells apparently like to snug into well-fitting grooves in a more compliant hydrogel material. No visual deformation of the hydrogels was found; the cells rather adapted their shape to fit into too narrow grooves (i.e. 5 μm wide; Schulte et al., 2010).

Second, we investigated whether proteins could adsorb non-specifically to the gels and if so, if they would adsorb differently on the topographic structures, e.g. preferentially on the walls of the grooves, or on the convex (outer) or concave (inner) corners, since the eventually adherent cells were found to be located inside of the grooves and aligned along the ridges. We incubated the patterned hydrogel samples in protein solutions of three selected extracellular matrix (ECM) proteins, i.e. Fibronectin (FN), Vitronectin (VN) and bovine serum albumin (BSA). The former two are cell adhesion mediating proteins, while BSA does not facilitate cell adhesion. BSA is the most abundant protein in serum, and besides its abundance it is also a small protein, which diffuses fast and reaches the surface faster than the larger proteins FN and VN and finally the cells.

With help of immunological staining using fluorescently labeled antibodies we could demonstrate that from pure protein solutions all three proteins were able to adsorb to the PEG surfaces to a certain extent (Schulte et al., 2011). The fluorescence was homogeneously distributed over the surface; there was no detectable difference between for example the vertical walls or the horizontal planes between the grooves. Finally, it seemed that BSA was able to diffuse into the PEG hydrogels, since the fluorescence was not restricted to the surface (Schulte et al., 2011).

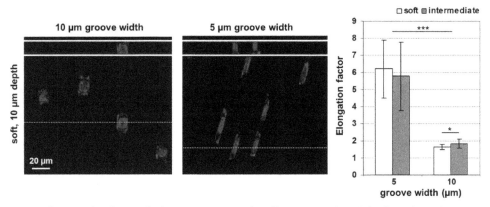

Fig. 9. Elongated cell morphology on topographically patterned PEG hydrogels with 10 or 5 μm wide grooves. Reprinted with permission from: Schulte et al. *Biomacromolecules, 11*, 3375-83. Copyright 2010 American Chemical Society.

Notwithstanding the ability of all three proteins to adsorb to the PEG hydrogel, only VN was observed to adsorb in a detectable amount under competitive conditions, i.e. from a mixture of VN and FN and when serum, a complex mixture of proteins, was supplemented (**Table 1**; Schulte et al., 2011). This result was rather unexpected, since BSA is usually observed to adsorb to virtually any surface, and because FN is generally considered to be the most important cell adhesion mediating protein in serum and consequently has been much more studied than VN.

Solution for Incubation		Protein detected by antibody staining	
		VN	FN
FBS	10 % in PBS	+ + +	
FBS	100 %	+ + +	
VN FN	5 µg/ml in PBS 50 µg/ml in PBS	+	+
VN FN	5 µg/ml in PBS 20 µg/ml in PBS	+ +	+
VN FN	10 µg/ml in PBS 20 µg/ml in PBS	+ +	
VN FN	10 µg/ml in PBS 50 µg/ml in PBS	+	

Table 1. Protein adsorption to the surface of the (topographically patterned) PEG-based hydrogel; qualitative results are given for samples that were incubated with serum (100% or 10% in buffer solution) or with buffer solutions containing a mixture of the pure proteins Vitronectin (VN) and Fibronectin (FN) in various ratios. Reproduced from: Schulte et al., (2011) *Macromol. Biosci. (in press)*. Copyright 2011 John Wiley and Sons.

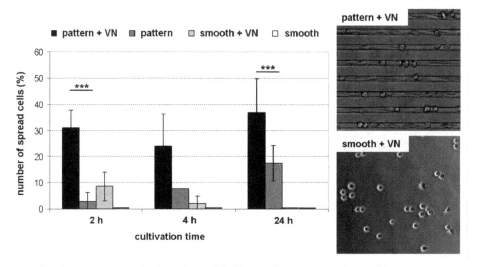

Fig. 10. Cell adhesion on PEG hydrogels enabled by surface topography and/or pre-incubation with the cell adhesion-mediating protein VN. Reproduced from: Schulte et al., *Macromol. Biosci. (in press)*. Copyright 2011 John Wiley and Sons.

In order to verify whether this small but significant amount of adsorption of VN to the PEG surface is responsible for the observed cell adhesion on topographically patterned hydrogels we investigated the pre-incubated hydrogels (both smooth and patterned) in cell culture.

We analyzed the number of adherent cells and found an increased number of cells on the VN-incubated hydrogels. This effect was also found for the unpatterned, smooth hydrogels; after VN-incubation a very small but significant number of cells were able to adhere to the PEG surface. In **Figure 10** these results are depicted; comparing smooth and patterned hydrogels with and without VN-incubation.

It can be seen that compared to the effect of topography alone, the VN-incubation alone was less effective in enabling cell adhesion. Remarkably, the effect of VN on cell adhesion was only evident at early time points; after 24 hours the enabling effect was completely lost. Finally, a striking synergistic effect was observed from the combination of VN-incubation and topography; the number of adherent and spread cells was larger than the sum of the individual contributions (Schulte et al., 2011). Taking into account the apparent difference in the effect of topography and VN with time, we tentatively conclude that the cell adhesion protein VN facilitates the initial cell adhesion, while the adhesion-enabling effect of surface topography becomes dominant at longer times and is necessary for the development of durable and stable adhesion complexes.

3. Conclusion

Hydrogels are of high relevance for several biomedical applications. We have described the fabrication of a hydrogel system based on poly(ethylene glycol) and evaluated the potential of this PEG-based gel as a patternable biomaterial. PEG-based polymers are of great importance as biomaterials for applications in cell and tissue engineering, as coating of implants or biosensors, and as drug delivery systems. In particular, PEG coatings have been used to minimize surface biofouling by plasma proteins to create surfaces that are "invisible" to cells. Cell biological studies with murine fibroblasts (NIH L929) confirmed the expected non-adhesive nature of the smooth hydrogel surfaces and furthermore ruled out any toxic effect of the material. Alterations of the mechanical properties could easily be achieved by varying the crosslinking density.

The most striking result from our studies is that the very popular and versatile PEG biomaterial is not cell-repellent per se. Only when the surface of the bulk PEG hydrogels is smooth it is anti-adhesive to cells, and this applies to all hydrogels we have investigated with a stiffness ranging from 0.1 to 1 MPa. However, we have discovered that on the same PEG hydrogels when decorated with micropatterns of topography, cells are able to adhere and spread. We have explored several underlying biochemical, biophysical and biomechanical factors that could attribute to this phenomenon and found that these factors do have an effect indeed, and notably the combination of these parameters, e.g. protein adsorption, surface topography and substrate compliance, work together to enable cell adhesion to the intrinsically anti-adhesive PEG biomaterial.

More specifically, three investigated PEG-based hydrogels with different stiffness were all cell anti-adhesive when smooth. However, in combination with topography, the softer gels were clearly more attractive for the cells; on softer gels with the same pattern geometry, significantly more cells adhered and spread than on the intermediate or stiffer gels. It seems that the compliance of the softer gels enables the cells to 'squeeze' into the grooves, although the cells apparently deform their own cytoskeleton rather than the topographic features.

We also discovered that a slight but significant amount of the ECM-protein Vitronectin is able to adsorb to the PEG surface and that this leads to an increase in initial cell adhesion during the first 4 hours of cell culture. However, this effect rapidly falls off. The effect of topography is a more durable effect that dominates at longer time scales, suggesting its role in the enabling of stable adhesion complexes, which is a process that occurs during more than several hours. We consider that the topographic features may provide shelter to the cells and prolong their residence time inside of the grooves. Added to this, the 'pulling' of the cells on the weakly bound cell adhesion proteins may be less effective when they are confined between the vertical walls than when the surface is smooth; as a consequence focal adhesion contacts can develop into stable focal adhesion complexes.

Thus, the combination of different (bio)chemical, physical and mechanical properties of the PEG hydrogels results in the observed cell adhesion on this intrinsically anti-adhesive biomaterial. The effects are difficult to disentangle, and marked synergistic effects were observed for example when using topographically patterned hydrogels that were incubated with Vitronectin prior to cell culture.

As PEG is generally well known for its anti-adhesive properties and is widely applied in biomedical applications, it is important to take into consideration what our study has shown: physical and mechanical surface properties can impede the anti-adhesive characteristics of PEG. On the other hand it also opens new opportunities for biomimetic material design which does not rely on complicated and expensive biochemical surface functionalization for manipulating cellular responses.

4. Acknowledgment

The author gratefully acknowledge funding from the Alexander von Humboldt Foundation and the Federal Ministry for Education and Research (BMBF) in the form of a Sofja Kovalevskaja Award (M. C. L.) and from the DFG Graduate School "Biointerface" (GRK 1035).

5. References

Andrade, J. D.& Hlady, V. (1986) Protein adsorption and materials biocompatibility - a tutorial review and suggested hypotheses. *Adv. Polym. Sci., 79*, 1-63.

Andrade, J. D.; Hlady, V.; Wei, A. P.; Ho, C. H.; Lea, A. S.; Jeon, S. I.; Lin, Y. S.& Stroup, E. (1992) Proteins at interfaces principles multivariate aspects protein resistant surfaces and direct imaging and manipulation of adsorbed proteins. *Clin. Mater., 11*, 67-84.

Baroli, B. (2006) Photopolymerization of biomaterials: Issues and potentialities in drug delivery, tissue engineering, and cell encapsulation applications. *J. Chem. Technol. Biot., 81*, 491-9.

Castner, D. G.& Ratner, B. D. (2002) Biomedical surface science: Foundations to frontiers. *Surf. Sci., 500*, 28-60.

Castillo, E. J.; Koenig, J. L.; Anderson, J. M.& Lo, J. (1985) Protein adsorption on hydrogels. 2. Reversible and irreversible interactions between lysozyme and soft contact-lens surfaces. *Biomaterials, 6*, 338-45.

Chapman, R. G.; Ostuni, E.; Takayama, S.; Holmlin, R. E.; Yan, L.& Whitesides, G. M. (2000) Surveying for surfaces that resist the adsorption of proteins. *J. Am. Chem. Soc., 122*, 8303-4.

Cushing, M. C.& Anseth, K. S. (2007) Hydrogel cell cultures. *Science, 316*, 1133-4.

Desai, N. P.; Hossainy, S. F. A. & Hubbell, J. A. (1992) Surface-immobilized polyethylene oxide for bacterial repellence. *Biomaterials, 13,* 417-20.

Diez, M.; Mela, P.; Seshan, V.; Möller, M.; Lensen, M. C. Nanomolding of PEG-based hydrogels with sub-10-nm resolution. *Small* 2009, 5, 2756-60.

Diez, M.; Schulte, V. A.; Stefanoni, F.; Natale, C. F.; Mollica, F.; Cesa, C. M.; Chen, Y.; Möller, M.; Netti, P. A.; Ventre, M. & Lensen, M. C. (2011) Molding Micropatterns of Elasticity on PEG-based Hydrogels to Control Cell Adhesion and Migration *Adv. Eng. Mater. (Adv. Biomater.), 13* (4): n/a. doi: 10.1002/adem.201080122, in press.

Discher, D. E.; Janmey, P.& Wang, Y. L. (2005) Tissue cells feel and respond to the stiffness of their substrate. *Science, 310,* 1139-43.

Drumheller, P. D.& Hubbell, J. A. (1995) Densely cross-linked polymer networks of poly(ethylene glycol) in trimethylolpropane triacrylate for cell-adhesion-resistant surfaces. *J. Biomed. Mater. Res., 29,* 207-15.

Elbert, D. L.& Hubbell, J. A. (1996) Surface treatments of polymers for biocompatibility. *Annu. Rev. Mater. Sci., 26,* 365-94.

Engler, A. J.; Sen, S.; Sweeney, H. L.& Discher, D. E. (2006) Matrix elasticity directs stem cell lineage specification. *Cell, 126,* 677-89.

Fedorovich, N. E.; Alblas, J.; de Wijn, J. R.; Hennink, W. E.; Verbout, A. J.& Dhert, W. J. A. (2007) Hydrogels as extracellular matrices for skeletal tissue engineering: State-of-the-art and novel application in organ printing. *Tissue Eng., 13,* 1905-25.

Fisher, J. P.; Dean, D.; Engel, P. S.& Mikos, A. G. (2001) Photoinitiated polymerization of biomaterials. *Ann. Rev. Mater. Res., 31,* 171-81.

Frisch, S. M.& Francis, H. (1994) Disruption of epithelial cell-matrix interactions induces apoptosis. *J. Cell Biol. 124,* 619-26.

Fujimoto, K.; Inoue, H.& Ikada, Y. (1993) Protein adsorption and platelet-adhesion onto polyurethane grafted with methoxy-poly(ethylene glycol) methacrylate by plasma technique. *J. Biomed. Mater. Res., 27,* 1559-67.

Gilmore, A. P. (2005) Anoikis. *Cell Death Differ. 12,* 1473-7.

Groll, J.; Ameringer, T.; Spatz, J. P.& Möller, M. (2005a) Ultrathin coatings from isocyanate-terminated star PEG prepolymers: Layer formation and characterization. *Langmuir, 21,* 1991-9.

Groll, J.; Fiedler, J.; Engelhard, E.; Ameringer, T.; Tugulu, S.; Klok, H. A.; Brenner, R. E.& Möller, M. (2005b) A novel star PEG-derived surface coating for specific cell adhesion. *J. Biomed. Mater. Res. Part A, 74A,* 607-17.

Harder, P.; Grunze, M.; Dahint, R.; Whitesides, G. M.& Laibinis, P. E. (1998) Molecular conformation in oligo(ethylene glycol)-terminated self-assembled monolayers on gold and silver surfaces determines their ability to resist protein adsorption. *J. Phys. Chem. B, 102,* 426-36.

Harris, J. H. (1992) *Poly(ethylene glycol) chemistry: Biotechnical and biomedical applications.* Plenum Press: New York.

Harris, J. H.& Zalipsky, S. (1997) *Poly(ethylene glycol) chemistry: Chemsitry and biomedical applications.* Am. Chem. Soc.: Washington D.C.

Hlady, V.& Buijs, J. (1996) Protein adsorption on solid surfaces. *Curr. Opin. Biotech., 7,* 72-7.

Hoffman, A. S. (2002) Hydrogels for biomedical applications. *Adv. Drug Deliv. Rev., 54,* 3-12.

Horbett, T. A. (1993) Principles underlying the role of adsorbed plasma-proteins in blood interactions with foreign materials. *Cardiovasc. Pathol.*, *2*, S137-S48.

Irvine, D. J.; Mayes, A. M.& GriffithCima, L. (1996) Self-consistent field analysis of grafted star polymers. *Macromolecules*, *29*, 6037-43.

Ishizaki, Y.; Cheng, L.; Mudge, A. W.& Raff, M. C. (1995) Programmed cell-death by default in embryonic-cells, fibroblasts, and cancer-cells. *Mol. Biol. Cell*, *6*, 1443-58.

Jen, A. C.; Wake, M. C.& Mikos, A. G. (1996) Review: Hydrogels for cell immobilization. *Biotechnol. Bioeng.*, *50*, 357-64.

Jeon, S. I.; Lee, J. H.; Andrade, J. D.& Degennes, P. G. (1991)Protein surface interactions in the presence of polyethylene oxide .1. Simplified theory. *J. Colloid Interface Sci.*, *142*, 149-58.

Jhon, M. S.& Andrade, J. D. (1973) Water and hydrogels. *J. Biomed. Mater. Res.*, *7*, 509-22.

Knauf, M. J.; Bell, D. P.; Hirtzer, P.; Luo, Z. P.; Young, J. D.& Katre, N. V. (1988) Relationship of effective molecular-size to systemic clearance in rats of recombinant interleukin-2 chemically modified with water-soluble polymers. *J. Biol. Chem.*, *263*, 15064-70.

Krsko, P.& Libera, M. (2005) Biointeractive hydrogels. *Materials today*, *8*, 36-44.

Krsko, P.; Kaplan, J. B.& Libera, M. (2009) Spatially controlled bacterial adhesion using surface-patterned poly(ethylene glycol) hydrogels. *Acta Biomater.*, *5*, 589-96.

Langer, R. & Peppas, N. A. (2003) Advances in biomaterials, drug delivery, and bionanotechnology. *AIChE J.*, *49*, 2990-3006.

Langer, R. & Tirrell, D. A. (2004) Designing materials for biology and medicine. *Nature*, *428*, 487-92.

Lee, K. Y.& Mooney, D. J. (2001) Hydrogels for tissue engineering. *Chem. Rev.*, *101*, 1869-79.

Lensen, M. C.; Mela, P.; Mourran, A.; Groll, J.; Heuts, J.; Rong, H. T.& Möller, M. (2007) Micro- and nanopatterned star poly(ethylene glycol) (PEG) materials prepared by UV-based imprint lithography. *Langmuir*, *23*, 7841-6.

Lensen, M. C.; Schulte, V. A.; Salber, J. ; Díez, M.; Menges, F. R. & Möller, M. (2008) Cellular responses to novel, micro-patterned biomaterials *Pure Appl.Chem. 11*, 2479–2487.

Lutolf, M. P.; Raeber, G. P.; Zisch, A. H.; Tirelli, N.& Hubbell, J. A. (2003) Cell-responsive synthetic hydrogels. *Adv. Mater.*, *15*, 888-92.

Malmsten, M.; Emoto, K.& Van Alstine, J. M. (1998) Effect of chain density on inhibition of protein adsorption by poly(ethylene glycol) based coatings. *J. Colloid Interface Sci.*, *202*, 507-17.

McGill, G.; Shimamura, A.; Bates, R. C.; Savage, R. E.& Fisher, D. E. (1997) Loss of matrix adhesion triggers rapid transformation-selective apoptosis in fibroblasts. *J. Cell Biol.*, *138*, 901-11.

Merrill, E. W.; Salzman, E. W.; Wan, S.; Mahmud, N.; Kushner, L.; Lindon, J. N.& Curme, J. (1982) Platelet-compatible hydrophilic segmented polyurethanes from polyethylene glycols and cyclohexane diisocyanate. *Trans. Am. Soc. Artif. Intern. Organs.*, *28*, 482-7.

Morra, M. (2000) On the molecular basis of fouling resistance. *J. Biomater. Sci.-Polym. E.*, *11*, 547-69.

Mrksich, M.& Whitesides, G. M. (1996)Using self-assembled monolayers to understand the interactions of man-made surfaces with proteins and cells. *Annu. Rev. Biophys. Biomolec. Struct.*, *25*, 55-78.

Mrksich, M.; Dike, L. E.; Tien, J.; Ingber, D. E.& Whitesides, G. M. (1997) Using microcontact printing to pattern the attachment of mammalian cells to self-assembled monolayers of alkanethiolates on transparent films of gold and silver. *Exp. Cell Res.*, *235*, 305-13.

Nguyen, K. T.& West, J. L. (2002) Photopolymerizable hydrogels for tissue engineering applications. *Biomaterials*, *23*, 4307-14.

Peppas, N. A. (1986) *Hydrogels in medicine and pharmacy: Fundamentals*. CRC Press: Boca Raton.

Peppas, N. A.; Bures, P.; Leobandung, W.& Ichikawa, H. (2000) Hydrogels in pharmaceutical formulations. *Euro. J. Pharm. Biopharm.*, *50*, 27-46.

Peppas, N. A.; Hilt, J. Z.; Khademhosseini, A.& Langer, R. (2006) Hydrogels in biology and medicine: From molecular principles to bionanotechnology. *Adv. Mater.*, *18*, 1345-60.

Prime, K. L.& Whitesides, G. M. (1991) Self-assembled organic monolayers - model systems for studying adsorption of proteins at surfaces. *Science*, *252*, 1164-7.

Prime, K. L.& Whitesides, G. M. (1993) Adsorption of proteins onto surfaces containing end-attached oligo(ethylene oxide) - a model system using self-assembled monolayers. *J. Am. Chem. Soc.*, *115*, 10714-21.

Saha, K.; Pollock, J. F.; Schaffer, D. V.& Healy, K. E. (2007) Designing synthetic materials to control stem cell phenotype. *Curr. Opin. Chem. Biol.*, *11*, 381-7.

Schulte, V. A.; Diez, M.; Möller, M. & Lensen, M. C. (2009) Surface Topography induces Fibroblast Adhesion on intrinsically non-adhesive Poly(ethylene Glycol) Substrates *Biomacromolecules*, *10*, 2795–2801.

Schulte, V. A.; Diez, M.; Hu, Y.; Möller, M. & Lensen, M. C. (2010) The combined influence of substrate elasticity and surface topography on the anti-adhesive properties of poly(ethylene glycol). *Biomacromolecules*, *11*, 3375-83.

Schulte, V. A.; Diez, M.; Hu, Y.; Möller, M. & Lensen, M. C. (2011) Topography induced cell adhesion to Acr-sP(EO-stat-PO) hydrogels: the role of protein adsorption *Macromolecular Bioscience* in press

Sofia, S. J.; Premnath, V.& Merrill, E. W. (1998) Poly(ethylene oxide) grafted to silicon surfaces: Grafting density and protein adsorption. *Macromolecules*, *31*, 5059-70.

Tessmar, J. K.& Gopferich, A. M. (2007) Customized PEG-derived copolymers for tissue-engineering applications. *Macromol. Biosci.*, *7*, 23-39.

Tsai, W.-B.; Grunkemeier, J. M.; McFarland, C. D.& Horbett, T. A. (2002) Platelet adhesion to polystyrene-based surfaces pre-adsorbed with plasmas selectively depleted in fibrinogen, fibronectin, vitronectin, or von willebrand's factor. *J. Biomed. Mater. Res.*, *60*, 348-59.

Wahlgren, M.& Arnebrant, T. (1991) Protein adsorption to solid-surfaces. *Trends Biotechnol.*, *9*, 201-8.

Veronese, F. M.& Mero, A. (2008) The impact of pegylation on biological therapies. *Biodrugs*, *22*, 315-29.

Whitesides, G. M.; Ostuni, E.; Takayama, S.; Jiang, X. Y.& Ingber, D. E. (2001) Soft lithography in biology and biochemistry. *Annu. Rev. Biomed. Eng.*, *3*, 335-73.

Wichterle, O.& Lim, D. (1960) Hydrophilic gels for biological use. *Nature*, *185*, 117-8.

Yeung, T.; Georges, P. C.; Flanagan, L. A.; Marg, B.; Ortiz, M.; Funaki, M.; Zahir, N.; Ming, W. Y.; Weaver, V.& Janmey, P. A. (2005) Effects of substrate stiffness on cell morphology, cytoskeletal structure, and adhesion. *Cell Motil. Cytoskeleton*, *60*, 24-34.

Zhang, M. Q.; Desai, T.& Ferrari, M. (1998) Proteins and cells on PEG immobilized silicon surfaces. *Biomaterials*, *19*, 953-60.

14

Facial Remodelling and Biomaterial

G. Fini, L.M. Moricca, A. Leonardi, S. Buonaccorsi and V. Pellacchia
La Sapienza/ Roma
Italy

1. Introduction

Facial remodelling comprises all the surgical techniques able to reconstruct the correct proportion between soft and hard tissues of the face. In order to obtain facial harmony maxillo-facial surgeons have at disposal many surgical techniques such as reconstructive, orthognatic, aesthetic surgery and camouflage. In the study of a patient who presents facial asymmetry two base evaluations are necessary: aesthetic analysis and cephalometric analysis. The first is an evaluation of the skeleton in association to the evaluation of the soft tissues according to the harmonic proportions of the face, while the second consists in the evaluation of the skeletal relationships with respect to the basicranium through the identification of specific craniometrical points. It is possible to establish a specific therapeutic plan for the specific facial
asymmetry by means of the combination of these analyses and the addition of the radiographies study. There are various pathologies that need facial remodelling: acquired syndromes, congenital syndrome such as the Hemifacial Microsomiae, autoimmune disorder or atrophic disease such as Perry-Romberg Syndrome, traumas, demolitive surgery and infectious pathologies. In these clinical conditions the choice of a therapeutic treatment of camouflage instead of corollary surgery and conventional aesthetics techniques is made when there is a specific request from patients. This decision is also made when there is an increased operative risk and the deficits to fill are not massive. The described surgical treatments present the advantage of being not invasive, easy to position, not much traumatic and with immediate results. The complications can be the following: shifting of the biomaterial , chronic inflammation, quick reabsorption of the used materials, infections and reject. The camoufflage is growing as a surgical technique for the continuos scientific studies, on the new bio-materials. The studied filling bio material are the porous polyethylene and the bio-bone for the hard tissues, the Polyalkylimide and Polyacrylamide for the soft tissues. Some representative clinical cases are presented.
During our experience, patients have been treated with the following bio materials: porous polyethylene, bio-bone, polialkylimide, polyacrylamide, and with a combined treatment with polyacrylamide and porous polyethylene. The patients treated with porous polyethylene presented pathologies deriving 50% from traumas, 40% from malformation and a 10% from congenital asymmetries. The patients treated with bio-bone (7% of the total patients) of the total were presented in all cases the bony atrophy of the jaw.
Patients suffering from infectious pathologies (HIV) were included among the patients treated with polialkylimide, others with autoimmune pathologies (PRS) and with malformative syndromes were included too. The treatment with polyacrylamide was

carried out in patients with autoimmune syndrome (Scl and PRS), LPS results people, HIV affected, and patients with congenital malformation (HMF). We have a diagnostic and therapeutic procedure uniform for all the patients; the first clinical evaluation concerns radiographic and laboratory examination, such as head and neck Dimensional Computed Tomography, Magnetic Resonance, Ultrasonography, Orthopanoramic x-rays, searches for ANA-ENA-Anti Cardiolipina anticorpal; specialized infective and immunologic consulence relating to the single patient has been committed. The 7% of the patients has been treated with a replenishment composed of the combination with porous polyethylene and polyacrylamide because of the wide loss of both skeleton and soft tissues. Some other representative clinical cases are presented:

2. Biomaterials

Polyacrylamide - Clinical Case 1: CC, a female of 17 years old, came in our center for hemifacial hasimmetry. This anomaly wascharacterized by a progressive atrophy of the left hemiface. The patient did not present diplopy. The objective examination we observed the presence of slight atrophy of the left middle and inferior third of the face, including nasolabial region and the omolateral upper lip. A nasal pyramid deviation also appeared and confirmed by the anterior rhynoscopy, which also showed a left-convex deviation of the nasal sept. A Dimensional Computed Tomography scan was performed. This exam confirmed a maxillo- mandibular fault. Then an orthotic evaluation was performed to value muscular structures. We adopted a surgical approach which primarily provided an improvement of the respiratory activity. Then, we planned a filling of the atrophic soft tissues through infiltrations of Polyacrylamide. The infiltrations have been performed after 2 months from the septorhynoplastic surgery. Biomaterial (5ml) in the left nasolabial fold was filled. After 1 month from the beginning of the treatment, we noticed an evident reduction of the pre-existing deficit of the soft tissues. The clinical and Ultrasonography checks after 2,6, 12 and 24 months confirmed a correct integration of the used biomaterial.

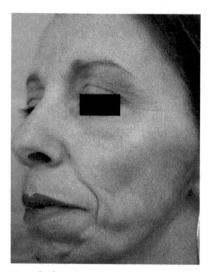

Fig. 1. Lateral view of the patient before treatment

Clinical case 2: E.M, female of 45 years referred about the appearance of a progressive facial asymmetry alterations interesting the lower third of the face in particular in the last 3 years. The tissue deficit was becoming clearer without any symptoms or alterations in the facial motility. (FIGURE 1) Therefore a Computed Tomography Scan of the face was performed, in order to evaluate the entity of such deficit. It shows marked deficit of the soft tissue, which was extending partially to the skeletal structures". During the objective exam conduced even with a photographic study, the loss of the symmetry was appreciating, in particular to the third lower of the hemiface. Problems related to the function of the facial nerve, were not noticed, and the patient did not refer facial hypoesthesia.

Therefore the patient underwent to fill of the facial soft tissues with biomaterial. The biomaterial was implanted in the left middle and inferior third of the hemiface; moreover it has been noticed a partial resorption of the biomaterial at the end of treatment. A total of 3 infiltration has been performed for a tot of 8ml of infiltrated biomaterial. (FIGURE 2)

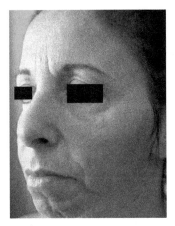

Fig. 2. Lateral view of the patient after treatmet

Clinical Case 3: E.D.E, male of 55 years old, immuno-compromised patient, affected by HIV from 20 year. He had underwent to the Highly Active Anti-Retroviral Therapy treatment (HAART) from about ten years. The patient referred the appearence of an atrophy in the middle third of face, he had developed a lypodistrophy lesion of soft tissues. (FIGURE 3)

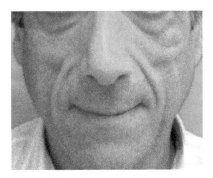

Fig. 3. Particular of the lypodistrophy area

The patient was stable from clinical and infective point of view, confirmed by hematologic exams; so that we decided to underwent the patient to biomaterial infiltration with Polyacrylamide. The sites of treatment were the areas where the atrophy and the lypodistrophy are happened. Clinical-aesthetic and infective results in six years of follow-up were good. (FIGURE 4)

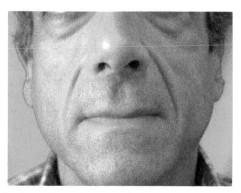

Fig. 4. Particular of the the lypodistrophy area after treatment

Polyalkylimide - Clinical case 4: D.L.B, a male of 30 years old, showed an atrophy of the middle and inferior third of the face. The patient eight years before, referred a not treated facial trauma. The objective exam showed the presence of a moderate atrophy of the rigth hemiface, associated to aesthetic and functional alterations. In particular, a combination of the right orbital-malar asymmetry complex, and a rigth orbital oenophtalmo was noticed as well. 7. The patient referred about the appearance of dyplopia. It has not been well clarifyied. The patient underwent before further clinical and radiological checks exams to study the soft tissues, and to evaluate the ocular motility, so as a Perry Romberg Syndrome was suggested for diagnosis. A surgical treatment was planned to correct the oenophtalmo and to restore the ocular motility correcting the dyplopia disfunction. On the other hand, this surgery has been performed with the goal of resolving the atrophy and the face's deformity through the use of porous polyetilene. Afterwards, a treatment of polyackylimide infiltrations has been planned. The patient, after 1 week from the treatment with biomaterial, presented a good tolerance and a total restoration of the facial eurytmia. An ultrasonographic evaluation was performed to value the integration of the biomaterial after 6,12 and 24 months; this exam showed a compartment of the biomaterial implanted, associated to fibrotyc branches compatible with the basal pathology.

Porous polyethylene - Clinical Case 5: S.B, a female of 40 years old, referred facial asymmetry. The objective exam showed a skletal deformity of mandibular and maxillary component, associated to mycrogenia.(FIGURE 5)

The patient presented a congenital nose deformity too. The surgical treatment was planned with mandibular promotion by means graft of porous polietylene. Then in the same surgical step a graft "on lay" of septal cartilage was positionated on the nasal dorsum. Ultrasonographic checks after 1 month, 23 months and 6 months were optimal; radiologic exams were performed to check the planted biomaterial. These investigations verified a good tolerance and a well fixed of the biomaterial. (FIGURE 6)

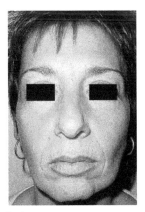

Fig. 5. Frontal view of the patient before treatment

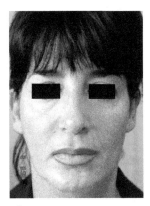

Fig. 6. Frontal view of the patient after treatment

Bio-bone - Clinical case 6: ML, 19 years old female, came up to our attention for a mandibular lesion. It was an occasionally finding of a Magnetic Risonance carried out previously after a lipothymia event. This mandibular 8 lesion was an osteolytic one, positioned in the left ramous under 3.6 to 3.8 roots. Patient had not simpthoms, and the clinical exam didn't show any visible or touchable lesion in the fornix gum, or at the corresponding teeth. Patient only referred a mononucleosis infection six months before. Computed Tomography of the Head and Neck and Orthopanoramic x-rays were carried out (FIGURE 7) .

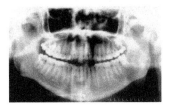

Fig. 7. Pre operative Orthopanoramic x-rays

These images were likely suggestive for adamantinoma. That hypothesis led to a particular operative intervention with the aim of a definitive diagnosis and treatment. In fact the histologic exam would have led, or not, to a mandibular resection. So left mucous fornix section was performed in order to uncover the mandibular bone from 3.6 to 3.8 dental elements and to dissect the bone through osteotribe, as long as the lesion was found. Strangely enough the surgical finding was a rarefaction of the bone, no capsular structure or any other elements that could help with the diagnosis were observed as well. Consequently a conservative surgical technique was carried on, such as cutting out the bone box with 3.7 and 3.8 because of their roots inclusion in the osteolitic lesion as well. The missing bony part was filled with a demineralized bone matrix, in order to prevent iatrogenic fracture. So before performing a mandibular ramous resection, we have been waiting for the definitive histologic diagnosis. Unfortunately, against every expectation, it resulted as a follicular cyst within Candida A. yeasts. It was performed a batteriologic exam that resulted positive for C. Albicans too and for Hafnia alveii. Antibiotic and antimicotic therapies were carried out for a long period. After three months patient underwent to a Orthopanoramic x-rays, that revealed the biomaterial integration but a surgical interventation of removing bio-bone it was necessary in order to assure the complete eradication of the Candida infection. (FIGURE 8)

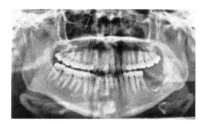

Fig. 8. Post operative Orthopanoramic x-rays

Porouse polyethylene and Polyacrylamide - Clinical case 7: R.A, a male of 40 years old, affected by the Goldenhar syndrome, he underwent to differentreconstructive surgical treatments, to restore the normal symmetry of the face soft tissues. The patient showed a facial asymmetry characterized by an atrophy of the right hemifacial soft tissues, associated to auricular agenesys, and a behind-positioning of the left auricle.(FIGURE 9)

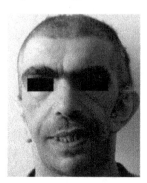

Fig. 9. Frontal view of the patient before treatment

Radiological and clinical exams with Computer Tomography Dental Scan and Telecranium x ray in two projections with cefalometric study were performed to evaluet bony and soft tissues. After 1 month surgery was performed: two fixures with abutment have been positioned in the right mastoid bone, Then the left auricular was positionated to restablish the normal structures of the face. In the same surgical time, two porous polyetylene prosthesis were implanted in the malar region, to restore the sagittal diameter of the middle third of the face; other two porous polyetylene prosthesis were implanted on the mandibular angle and one more prosthesis was implated on the sinphisis, to restore the transversal and sagittal diameter of the thrid inferior of face. After three mounths an auricular prosthesis associated to Polyacrylamide implant, was positioned in bilateral pre-auricular area **(FIGURE 10).** Clinical and radiological follow-up demonstrated a good integration of implants and the biomaterial.

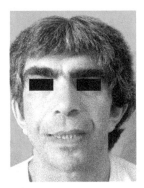

Fig. 10. Frontal view of the patient after treatment

3. Ultrasonography monitor follow-up

The ideal biomaterial should be easy to implant and to remove, and simple to be identified by a low-dose radiation and low-cost radiologic technique. Authors wanted to evaluate ultrasonography (US) as a technique in monitoring biomaterial status after operation. Ultrasonography has been shown as an excellent way to visualize clinical features and a possible pathologic process of an implanted biomaterial; it is a non-invasive, low-radiation and low-cost dose radiologic technique. Reconstruction in facial deficit diseases needs adequate biomaterial to implant and a careful patients observation, that is, both clinical and radiologic. Ultrasonography is a fundamental component of the follow up of implanted biomaterial patients. the use of synthetic materials instead of an autolog tissue is codified from years and is widely diffused. In the last years, maxillofacial surgery has adopted poliacrilamide for the soft tissue, which is already used in esthetic surgery such as "last generation filler" to overwhelm the defects of the time such as wrinkles and furrows. Such material has replaced paraffin and silicone fluid used in the 1960s, and collagen and analogs used in the 1980s.In the same years, Conley and Baker experienced some slow-resorption synthetic materials that, when inserted in the derma, overwhelmed cutaneous imperfections. The biomaterials used until that moment were all very well tolerated, but they introduced the disadvantage of being "statics" materials, concrete, and above all, temporary. In the last few years, poliacrilamide has replaced, in part, the use of these materials. it does not have these common disadvantages, such as being concrete, visible during activation of mimics muscles, and having a temporary effect. For what concerns the skeletal tissue, for years we have used autologous bone grafts and cartilaginous tissues similar to many prosthetic materials. They showed plastic phenomenon and they were easy to infections and resorption instead of esthetic and functional aspects. In these years, porous polyethylene results to be a suitable material for bony integration; it is easy to use and has great reconstructive quality and low susceptibility to infections. A general problem of the same biomaterials is a lack of visibility on conventional radiographs; they can be seen using magnetic resonance imaging or computed tomography. These investigations are not suitable for the frequent examinations, because magnetic resonance imaging is a high-priced procedure and computed tomography has a high radiation dose. During the follow-up, we encountered some difficulties for their radiotransparency; therefore, in our study, we used a noninvasive technique such as ultrasonography (US) to estimate the filling conditions and eventually to characterize an eventual pathologic process during the early phase. The aim of this study was to examine the use of ultrasound imaging in detecting the changes in biomaterials.

All patients were grouped according to different kinds of diseases: malformative pathologies (patients with hemifacial microsomia), degenerative pathologies (patients with scleroderma and with Romberg syndrome), results of skull-facial traumas, and pure aesthetic problems such as senile aclasia. They have been examined using US (in early and late postsurgical courses) with a highresolution probe (7.5-13 MHz, Astro; Esaote Biomedica, Genoa, Italy). The protocol of the study has foreseen almost 3 ultrasound controls; with a variable follow-up of 7 days to 36 months. After 7 days ofimplantation, we made the first ultrasound control.

Polyethylene porous, being a semirigid material that needs rigid interns fixture, decreased migration and stabilization problems. The polyacrylamide is introduced as a gel. If it is not well positioned, it could migrate. Integration and migration progress can be studied by US investigations, such as object examination. Initially, in both treated groups, transplant may show a light inflammatory state that will disappear in the succeeding days. Correct evaluation to appraise for the stabilization of the materials is composed of evaluating clinical and US parameters. The clinical parameters were as follows: the alteration absence of the impending fabrics, the graft, the edema absence, manque´ mobilization, or migration of the implantation. The US parameters were as follows: absence of massive harvests of liquidate, inflammatory reactions of the surrounding fabrics, and good visualization of the implantation and the surrounding tissue. With this worktop, we have been able to appraise the diverged characteristics of the biomaterial and visualize the tissues reactions. In our results porous polyetylene showed strong ecogenic features such as the bone and vanishing margins; however, the implantation (like a titanium screw) appears as a reverberated ultrasound bundle. We could evidence the stability of the biomaterials, namely, its integration, eventual nearby tissue alterations, in the early and late phases. Therefore, polyacrilamide appears anecogenous with a water-like aspect in the recent implants and corpuscolated in the older ones. Sometimes, such as in connective tissue degenerative pathologies (such as scleroderma) with an increase of the fibrotic component, we can visualize more vacuolized structures not for a lack of fibrotic integration but for the pathologic fibrotic beams. Although the implant seemed to be surrounded by a fibrotic tissue envelope, US technique can be considered an excellent way to visualize the clinical features and the pathologic process of implanted biomaterials. It represents a low-cost and low-radiation dose technique for a careful follow-up of patients affected by facial deficit disease after biomaterial implantation. The US technique shows a high radiologic sensibility in evaluating the features of the biomaterials. For its use, the limited cost and lack of investigation on harmful US is the key for studying of the biomaterials. (18)

4. Discussion

A biomaterial is defined as a several composed structure, able to interface with the biological systems in order to increase the volume, to give support or to replace a tissue. The performances of the installed materials are evaluated on the basis of bio-functionality and bio-availability therefore bio-compatibility (1).

The bio-functionality is the property of a bio-material to produce a determinate function from the physical and mechanical point of view while the bio-availability is the capacity of a bio-material to develop a determinate function during all the useful life of the plant (2). The final properties of a material depend both on the intrinsic molecular structure of the polymer and on the chemical and physiques processes to which it is exposed and can be widely manipulated intervening on the 10 operating conditions of such processes and on the polimerizzation's reaction. The immediate answers of the human body to the action of a bio-material is divided in two phases: an inflammation is initially developed because of the first defensive reaction of the organism to an foreign body; subsequently there is a restorative process of the damage. In general if the installed material is toxic, this causes the necrosys of the surrounding tissues; if it is not toxic and inert under the biological point of view a fibrous capsule around to the plant is formed (this answer is quite rare because the biomaterial is usually not completely inert); if, at last, the material is bioactive, it stimulates a precise biological answer and it is progressively supplemented with the surrounding tissue .

In most cases the material undergone some degradation form and the products of such process are released in the tissues. Such products, if they are not biologically active and they are not toxic, are removed with the normal metabolic processes, if however their concentration reaches high values they can locally accumulate and give an acute or chronic pathologies .In case, instead, of toxic products, a persistent inflammation developed; the products of the degradation processes, can stay in the releasing zone, with only local effects, or they can spread in the vascolare system and have so effects also on organs and tissues far from the releasing zone. The progress of the medical research has allowed the perfectioning and the development of new biomaterial in the reconstructive surgery, that has aesthetic licence to obtain excellent results by no much invasive surgical techniques and immediate results. An ideal bio-material presents these characteristics: absence of toxicities, anti-allergic properties, bio-compatibility (2), biofunctionality, easy to use and easy to remove. In our study the porous polyethylene and the bio bone have been analysed as substitutive of the hard tissues and the polyacrylamide and polialkylimide for the soft tissues (3). The bony reintegration is a complex and multi-factory process studied end analyzes in the time (4). At present several substitutive alternatives of the bone by autologus and eterologus bone, biomaterial are possible. Since the past what better choice was considered the allograft bone (5-6) which was useful for replacement of big bony deficits even if with difficult 11 vascularization (16-50% of fractures)(7). The bone-conductive, mechanical and immunological properties were good but correlated to the seat of the withdrawal and to the type of bone processing **(5-7-8-9)**. The Demineralized Bone Matrix can be easily extracted with 0.5 or 0.6 NHCL **(10-11)**

from the allograft bone so as to obtain an increase in the bone-inductive capacities, the loss of the immunological power and an increase in his supplementary effectiveness **(12)**. One of the limits of the Demineralized Bone Matrix is his malleable consistency which makes him suitable for the bony filling most than to the bony replacement, above all in districts of the face that is not subject to the traction of the soft tissues **(13)**. Everybody notice the porous polyethylene, used for the reconstruction of the hard tissues of the face; This material presents bio-compatibility with the human organism and easy malleability **(14)**. This includes that the surgical time are considerably reduced and the surgical procedure are not much invasive. One of the advantages of this material is his property to restore big bony deficits, contributing to give volume to the absent structures . It is fixed by screws, which contribute to the maintenance of a fixes position in the time, as it is also taken back in literature **(15)**. His utilization can be corollary of the orthognathic surgery in the restoration of the facial harmony or in the treatment of facial asymmetries from pathologies.Sharpen for this big variety of applications the porous polyethylene exists in various forms concerning the various facial places to restore: orbital margin **(15)**, nasal dorsum, mandibular ramus. The injectable filler are very commonly used for the treatment of soft alterations of the tissues of the face; these can be of different nature and composition. Are commonly divided in temporary and permanent; in our work take the permanent filler into consideration like the plyacrylamide and polialkylimide. These are also defined of the "hidrogel", for the high water content; their use has prevalently aimed at the filling of soft tissues, whit satisfactory results. Their principal properties are: elasticity, permeability and high bio-compatibility with the organism. Of these we remember the polialkylimide, injective bio-material formed by 2.5% of synthetic polymer of polyalkylimide and for 97.5% from apirogen water. Studies have been also executed in 12 vitro and they have shown a low toxicity and cutaneous sensibilization, following prolonged treatment. The polyacrilyamide **(16- 17)** is an injectable bio- material, formed by 3% of reticulate polymer of polyacrylamide and 97% from not pirogenic water. It is an hydrophilus absorbent gel, what comes infiltrated in deficit of the soft tissue. His mechanism of action is to add volume to the soft tissues, restoring the normal structures of the face. This bio material has an easy applicability also in different pathologies; important is not to use in post herpetic phase. It needs several applications for the partial absorption which can verify.

5. Results and conclusion

The study's results, based on our experience, have been more than satisfactory. The 290 treated patients, have given optimum results to a 6 year follow up. And also true that the experience acquired during the years about the porous polyethylene is previous of 12 years from the beginning of our study, that is happened in 2000. The justify for which in such work we considered the patient treated from this date, is due to the fact that before we did not have a sufficient sample of patient treated with other filling biomaterial to compare with porous polyethylene already mentioned. In aesthetics surgery the new fillers got rich of greater biological qualities and of easy utilization and their application has made possible for various infectious, malformativ or post-traumatic pathologies. This comparison has been optimum, in the perspective of a patient's better management and treatment of the deficits of the facial soft tissue: in order to choise the biomaterial as much as more suitable and-bio compatible with the host organism. To a careful evaluation of every single case, the results have not been completely uniforms; in the patients treated with polialkylimide, adverse reactions have happened: in two cases has verified a multi-capsule formations around the biomaterial , with relative capsule formation and an only case the rejection, with adverse cutaneous reaction. In particular the case of rejection has verified in subject with Hemi-facial atrophy. The phenomenon of compartimentalizzazione was verified in only two cases, it was not due to the absence of integration of the biomaterial with the organism but it was consequent to the 13 autoimmune pathology. For the subjects treated with polyacrilyamide adverse reactions have not happened; the material has shown a good integration in patients with various pathologies . This biomaterial turns out to be of easier utilization and provided with greater fluidity injection. For what concerns the patients affected with infectious pathologies like HIV, their treatment with filling biomaterial has given both recent and after 3 year distance good results; rejection events have not happened. Such patients are constantly evaluated by the infective point of view; we have experiences of biomaterial's utilization in HIV+ patients with HAART treatment which present lipodystrophy lesions . The malleability and easy application of the hydrogel, an utilization of theirs has made possible in combination to other biomateriali. The use of different biomaterials in the same patient is necessary when the deficit to fill was high and when the simple application of an only one biomaterial was turning out insufficient. The treated cases have given good results and good tolerance between the installed material. Porous polyethylene, bio-bone, Polyalkylimide, Polyacrylamide offer a good alternative in selected cases to the traditional reconstructive surgery, for the immediate results and easy way of application. A follow up of at last six years through echography and Computed Tomography exams shows their bio-compatibility, stability and inactivity, The restoring of symmetry and harmony of a face can be reached through both traditional surgery and also using biomaterials, that can substitute tissues graft or osteotomy with an easy way of application, their stability and their long lasting, not last a good esthetic result.14

6. References

Bulbulian AH. Maxillofacial prosthetics: evolution and practical application in patient rehabilitation. J Prosthet Dent 1965; 15:554Y569

Fini Hatzikiriakos G. Uno sguardo al passato, curiosita` sulle protesi nasali. Il Valsala 1985;61:61Y64

Tjellstrom A. Osteointegrated implants for replacement of absent or defective ears. Clin Plast Surg 1990;17:355Y366

Tjellstrom A, Granstrom G. One stage procedure to establish osteointegration: a zero to five years follow-up report. J Laryngol Otol 1995;109:593Y598

Schaaf NG. Maxillofacial prosthetics and the head and neck cancer patients. Cancer 1984;54:2682Y2690

Labbe´ D, Be´nateau H, Compe`re JF, et al. Implants extra-oraux: indications et contre-indications. Rev Stomatol Chir Maxillofac 2001;102:239Y242

Be´nateau H, Crasson F, Labbe´ D, et al. Implants extra-oraux et irradiation: tendances actuelles. Rev Stomatol Chir Maxillofac 2001;102:266Y269

Granstrom G, Jacobsson M, Tjellstrom A. Titanium implants in irradiated tissue: benefits from hyperbaric oxygenation. Int J Oral Maxillofac Implants 1992;7:15Y25

Markt JC, Lemon JC. Extraoral maxillofacial prosthetic rehabilitation at the M.D. Anderson Cancer Center: a survey of patient attitudes and opinions. J Prosthet Dent 2001;85:608Y613

Tjellstrom A, Granstrom G. One-stage procedure to establish osteointegration: a zero to five years follow-up report. J Laryngol Otol 1995;109:593Y598

Ramires PA, Miccoli MA, Panzarini E, Dini L, Protopapa C. In vitro and in vivo biocompatility of a Polylkymide Hydrogel for soft tissue augmentation. J.Biomed.Mater.Res.Part.2004; 72: 230-238. 2004.

Christensen LC, Breiting VB, Aasted A, Jorgensen A, Kebuladze I. Long-Term effects ofpolyacrylamide hydrogel in human breast tissue. Plast.Reconstr.Surg.2003; 11:1883-1890

Rees TD, Ashlet FL, Delgado JP. Silicone fluid injectons for facial atrophy: a 10 years study.Plast.Reconstr.Surgery. 1985;52: 118-125.

Greenwald AS,bBoden SD, goldberg VM et al.Bone graft substitutes: Facts, fictions and applications. J. Bone Joint Surg Am 2001; 83-A:S98-S103

Bauer TW, Muschler GF. Bone Graft Materials. An overview of The basic scienze. Clin Orthop. 2000; 371:10-27.

Oka Y, Ikeda M. Treatment of severe osteochondritis dissecans of the elbow using osteochondral grafts from a rib. J Bone Joint Surg Am. 2001 83-b: 738-739

Stevenson S. Biology of bone grafts. Orthoped Clin North Am. 1999; 30:543-552.

Skowronski PP, An Yh. Bone graft materials in orthopaedics.MUSC Orthopaed J. 2003;6:58-66

Betz RR. Limitation of autograft and allograft: new synthetic solutions. Orthopedics 2002;25(Suppl):S561-S560

Hollinger JO, Mark DE, Goco P, et al. A comparison of four particulate bone derivatives. Clin

Pietrzak WS, MillervSD, Kucharzyk DW, et al. Demineralized bone graftvformulations: Design, development, and a novel exemple. Proceedings of the Pittsburg BonevSymposium, Pittsburgh, PA, August 19-23,"003,557-575.

Davy DT.Biomechanical issues in bone transplantation. Orthop Clin North Am.1999;30:553-56 science and technology for the Craniomaxillofacial Surgeon.J Craniofac. Surg. 005(16);6:981-988.15

Mauriello JA, McShane R, Voglino J. Use Vicryl(Polyglactina 910) mesh implant for correcting enophtalmos A study of 16 patients. Ophthal Plast Reconstr Surg 1990; 6:247-251.

S. Ozturk, M Sengezer, S Isik et all. Long Term Outcomes of Ultra Thin Porous Polyethylene Implants used for Reconstruction of Orbital Floor Defects. J Craniofac Surg. 2005 (16) 6:973-977.

S. Buelow, D Heimburg, N. Pallua. Efficacy ad Safety of Polycrylamide Hydrogel for Facial Soft-Tissue Augmentation. Plast Reconstr Surg 2005 (15);1137-1146. KW Broder, SR Cohen. An Overview of Permanent and Semipermanent Fillers. Plast Reconstr Surg 2006 118(3 suppl.) S7_S14 Biomaterial Implantation in Facial Esthetic Diseases: Ultrasonography Monitor Follow-Up

Elena Indrizzi, MDS, Luca Maria Moricca, MD, Valentina Pellacchia, MD, Alessandra Leonardi, MD, Sara Buonaccorsi, MD, Giuseppina Fini, MDS, PhD. The Journal of Craniofacial Surgery –Vol.19, N. 4 -July 2008

Part 3

Prevention and Management of Biological Phenomena at Biomaterial/Cell Surfaces

15

Candida Biofilms on Oral Biomaterials

Philippe Courtois
Université Libre de Bruxelles, Brussels
Belgium

1. Introduction

Biological as well as inert surfaces of the oral cavity are exposed to an abundant microflora that is able to initiate the formation of biofilms. Yeasts are frequently involved, such as *Candida* (especially *albicans*), a low-level commensal of oral, gastrointestinal, and genitourinary mucosae in humans. *In vivo* and *in vitro* studies have shown *Candida* incorporation into biofilms covering different biomaterials used in the oral cavity for the manufacturing of dentures, orthodontic appliances, etc. Yeast (*Candida* genus) biofilms can then induce device-related infections mainly in the elderly and in medically-compromised patients with subsequent morbidity and occasional mortality, all bearing high social and financial costs. Generally, scientific literature does not integrate all aspects of material/tissue interfaces: mechanisms of *Candida* biofilm development and biomaterial maintenance, the welfare of patients, and prevention of candidosis. *In vitro* investigations were mainly undertaken with mono-species biofilms whereas *Candida* incorporation into biofilms on oral surfaces tends to correspond with an increase in the yeast/bacteria ratio. This illustrates the need for interdisciplinary insight. This chapter will review 1) the literature data concerning material surfaces in support of *Candida* biofilms in the oral environment, 2) the *in vitro* approaches to understanding the mechanisms of *Candida* biofilm formation on materials, 3) the interfaced manipulations in order to prevent *Candida* biofilm onset, and 4) the precautions when testing new devices *in vivo* in the oral cavity.

2. Materials as a support of *Candida* biofilm in the oral environment

Biomaterials placed in the oral environment offer new surfaces prone to biofilm formation. Rough surfaces allow more biofilms to develop than smooth ones. In contrast with free microorganisms in suspension (defined as planktonic), which are able to grow in liquid, biofilm development is theoretically divided into three stages: 1) attachment to the surface (Figure 1), 2) proliferation into a monolayer of anchoring cells, and 3) growth into several layers of budding cells (blastoconidia) with filamentous structures as hyphae or pseudohyphae (Figure 2).

Numerous studies indicated the presence of *Candida* on oral dentures (Vandenbussche & Swinne, 1984; Abu-Elteen & Abu-Elteen, 1998; Busscher et al., 2010) and other oral devices such as orthodontic appliances (Addy et al., 1982; Hägg et al., 2004). Some authors (Arendorf & Addy, 1985; Jewtuchowicz et al., 2007) demonstrated an effect of *Candida* carriage in the oral environment caused by wearing devices. Indeed, orthodontic appliances have been shown to increase *Candida* counts in the mouth (Arendorf & Addy, 1985) and in the periodontal pockets of dental device wearers with gingivitis (Jewtuchowicz et al., 2007). Inserted devices act as new reservoirs able to imbalance oral microflora. In a healthy mouth, saliva protects oral mucosa against candidosis; in contrast, dry mouth is associated with increased yeast counts and candidosis risk. *In vitro,* cigarette smoke condensates increased adhesion of *Candida albicans* on orthodontic material surfaces such as bands, brackets, elastics, and acrylic resin (Baboni et al., 2010).

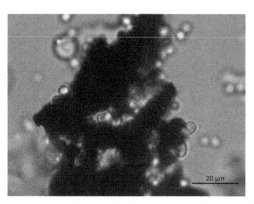

Fig. 1. *Candida albicans* suspension mixed with titanium powder directly observed on microscope in the absence of any stain procedure. Some blastoconidia are already adherent to material (attachment can be attested by MTT test after 3 washings). Magnification: x1000.

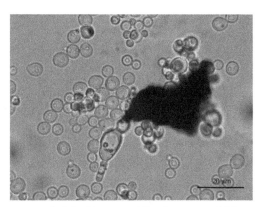

Fig. 2. Titanium grain surrounded by blastoconidia clusters and filamentous structure (pseudohyphae) after a two-week incubation. Magnification: x1000.

2.1 *Candida* on dentures

Candida albicans is often detected on methyl methacrylate polymers or acrylic resins from dentures. Biofilm formation on dentures results from complex interactions among yeast, bacteria, nutrients, and saliva or even serum proteins (Nikawa et al., 1997; Nikawa et al., 2000). *Candida* carriage on acrylic resin has been reported in the literature as varying up to more than 80% of the investigated dentures. For instance, yeasts were found in 14% of isolates from previously worn dentures in the Northeast and Southwest regions of the United States (Glass, 2010); in their conclusions, the authors pointed out frequent denture use without appropriate disinfection and biofilm formation within the pores of the material. In a previous study (Vanden Abbeele et al., 2008), authors reported *Candida* contamination of upper prosthesis in 76% of denture wearers hospitalized for long-term care in geriatric units. The most frequently isolated species were *C. albicans* (78%), *C. glabrata* (44%) and *C. tropicalis* (19%). Carriage of more than one yeast species was found in 49% of the contaminated dentures. There was a significant association between denture contamination and palatal mucosa colonization, making *ex vivo* denture decontamination mandatory, together with *in vivo* mucosa disinfection. *Candida* carriage has been observed in different types of dentures, both with and without soft liner fittings (Bulad et al., 2004; Mutluay et al., 2010).

Adhesion of *Candida* to the base materials of the dentures is associated with denture plaque (i.e. denture biofilm) and denture-related stomatitis. Even if many observations support the presence of *Candida albicans* in the biofilms on dentures, insufficient data are available to assess the etiology and to understand the pathogenesis of *Candida*-associated denture stomatitis. Review of the literature (Radford et al., 1999; Pereira-Cenci et al., 2008) does not permit settling specific and non-specific plaque hypotheses. Indeed, denture plaque comprises an ill-defined mixture of bacteria (such as *Streptococcus spp.*, *Lactobacillus spp.*, *Staphylococcus aureus*, and Gram-negative anaerobic bacteria) with *Candida spp.* also apt to cause mucosa inflammation.

2.2 *Candida* on other materials inserted in the oral cavity

Candida spp. was detected in low proportions at peri-implantitis sites and in failing implants associated with periodontopathogenic bacteria such as *Porphyromonas spp.*, *Prevotella spp.* and *Actinobacillus actinomycetemcomitans* (Alcoforado et al., 1991; Leonhardt et al., 1999; Pye et al., 2009), but ecological relationships with their surrounding and eventual pathological roles are yet to be understood. *In vitro*, *Candida albicans* may also adhere to pieces of biodegradable membranes used for periodontal tissue regeneration (Molgatini et al., 1998) and to tissue-conditioning materials for denture relining (Kulak & Kazazoglu, 1998). Additionally, presence of *Candida albicans* has been documented on obturator prostheses (whatever the material may be: silicone, polymethyl methacrylate, or titanium) in patients with maxillary defects (suffering from congenital malformation, tumors, or trauma), and on the mucosa adjacent to the prosthesis (Depprich et al., 2008; Mattos et al., 2009); these patients can present prosthesis-induced stomatitis. Finally, the use of orthodontic appliances leads to an increased carriage rate during the appliance-wearing time, with a significant fall of salivary pH and an increase of *Candida* count observed at different oral sites through various sampling techniques (Hibino et al., 2009).

2.3 *Candida* on devices used outside the oral cavity

Materials inserted in other sites can be colonized by yeasts as well, causing device-related infections (Cauda, 2009): articular prosthesis, cardiac devices (Falcone et al., 2009), catheters, vascular access devices (Brouns et al., 2006), and voice prostheses (Kania et al., 2010). These infections require prolonged antifungal therapy and often device removal.

3. Experimental approach

A better understanding of interface biology and material surface treatments requires experimental models to produce *in vitro* biofilms on supports that are easy to manipulate in the laboratory (Chandra et al., 2001) and the ability to investigate *in vivo* biofilm growth or drug susceptibility. Different studies have already described such models, mainly addressing procedures that are able to limit yeast adherence and biofilm formation. Some of these technologies were originally proposed as artificial dental plaque biofilm model systems (reviewed by Sissons, 1997), especially for plaque biology studies in caries and periodontitis research. This section will report on the experimental models producing *Candida* biofilms onto biomaterials. Table 1 summarizes some contributions from literature.

design	reference	material
in vitro models		
static culture models		
- material dived in solution	Chandra et al., 2001	acrylic, silicone
- poloxamer gel in Petri dish	Percival et al., 2007	material covered by gel
- 96-well culture microtiter plate	Peeters et al., 2008	polystyrene
- titanium powder	Ahariz & Courtois, 2010	titanium powder
continuous culture models		
- continuous flow culture system	Uppuluri et al., 2009	silicon elastomer
- Modified Robbins Device	Coenye et al., 2008	
- constant depth film fermenter	Lamfon et al., 2003	enamel, dentine, acrylic
in vivo models		
- animal models	Andes et al., 2004	central venous catheter
	Nett et al., 2010	acrylic denture
	Ricicová et al., 2010	subcutaneous catheter
- human models	Budtz-Jörgensen et al., 1981	tape or acrylic disk

Table 1. Experimental designs reported in the literature to produce *Candida*-biofilms.

Candida albicans can be grown with or without the addition of saliva on different materials used in dentistry including acrylic resins, denture-lining material, porcelain, composite, amalgam, hydroxyapatite, silicone, and polystyrene.

3.1 *In vitro* models
In vitro models consist of static cultures or continuous cultures.

3.1.1 Static culture models
Mono-species biofilms can be experimentally grown on pieces of materials currently used in dentistry, such as polymethylmethacrylate strips or silicone elastomer disks (Chandra et al., 2001). Some surfaces prepared from mucosa (epithelial cell cultures) or hard tooth tissues (enamel, dentine) were used as well. The materials are immersed in assay tubes or in multi-well culture plates containing a contaminated solution similar to saliva for a predefined incubation period. The polystyrene wall of the container itself has been used as an adherent surface for biofilm formation: 96-well plates allow management of experiments with numerous replications in various conditions (Peeters et al., 2008). Specific studies of the adhesion phase require an intermediate washing step to remove non-adherent yeast cells. Evaluation of the biofilm growth phase is generally based on microbial cell staining (crystal violet, among others), metabolic activities (tetrazolium test), biomass determination (dry weight, incorporation of radioactive tracer such as amino acids), or microscopic examination.

Poloxamer hydrogel, being liquid at low temperatures and solid at cultivation temperatures, has been proposed (Percival et al., 2007) as a culture support in the Petri dish to induce bacteria and yeast biofilm-like aggregates. The thermoreversible gelation makes the preparation and the recovery of biofilm samples easy and reproducible but still requires further confirmation for use in biofilm biology. Powder material, such as titanium powder, provides increasing support surfaces that are similar to cell culture on beads; moreover, it allows the anchored phase to be easily separated from the planktonic phase by simple sedimentation (Ahariz & Courtois, 2010). The titanium surface is not antimicrobial by itself, so it can be used as support for *Candida* biofilm. Titanium is widely employed for implant manufacturing due to its good biocompatibility and mechanical properties, but infection remains as a primary cause for failure, leading to removal. *Candida albicans* biofilms on titanium powder could offer a simple and reliable model for further investigation of new antimicrobial strategies; moreover, the model could be extendable to other microorganisms contaminating implanted materials. Making implant surfaces resistant to microbial colonization should reduce infectious complications; however, such developments need an *in vitro* model that allows the effect of surface modification and coatings on biofilm production to be studied. This aspect will be detailed in the next section.

The approach by means of static cultures simplifies the *in vivo* complexity to interactions between one single species and one single support without considering the numerous salivary compounds and abundant oral microflora in the real oral environment. Microorganisms in biofilms *in vivo* display properties different from those observed under laboratory conditions. The single-species procedures can be extended to a two-membered microbial co-culture or a characterized microbial consortium in order to reconstitute a medium for approaching oral microcosm and containing sterilized or artificial saliva. Multi-species biofilms have already been investigated on various dental materials such as enamel, amalgam, composite, and acrylic to assess the role of surface roughness (foremost in the first steps of biofilm formation) and impact of (pre)conditioning by saliva (Dezelic et al., 2009). Diffusion of drugs through biofilms, including *Candida* biofilms, can be documented by an experimental perfusion system superposing disk, filters, biofilm, and agar containing the drug under evaluation (Samaranayake et al., 2005). Perfusion of drugs in biofilms allows the putative factors that lead to biofilm antimycotic resistance to be evaluated.

3.1.2 Continuous culture models

Contrary to static cultures, continuous culture models (flow cells, Modified Robbins Device, chemostats, artificial mouths, and constant depth film fermenter) take into account oral flow and oral bathing conditions as shear forces and nutrient supplies (Bernhardt et al., 1999; Ramage et al., 2008). Indeed, oral biofilms on prosthetic materials are exposed to salivary fluxes conveying water and nutrients to the aggregated microorganisms in the saliva. For instance, liquid flow has been shown to influence the production of an extracellular matrix by *Candida albicans* biofilms *in vitro* (Hawser et al., 1998). Taking extracellular material produced in static cultures as a basal value, a gentle stirring significantly multiplied this by ~1.25, while a more intense stirring led to complete inhibition. Other data (Biswas & Chaffin, 2005) reported the absence of *Candida albicans* biofilm formation in anaerobiosis even if this yeast can grow in anaerobic environment. Yeast retention against a continuous flow of medium has been used as a marker for yeast adhesion to a surface in the same manner as the retention after liquid washes (Cannon et al., 2010). Cultures under continuous flow conditions facilitated the penetration of *Candida albicans* into silicone elastomers when compared to static conditions, especially when pure cultures were tested. The presence of *Streptococci* reduced material invasion by the yeast, thereby revealing the importance of testing materials in a biological environment (Rodger et al., 2010).

A continuous flow culture (CFC) aims at cultivating microorganisms in a continuous flow of fresh medium to mimic physiological conditions and to avoid the decline of the culture by nutrient depletion. Such instrumentation is easy to assemble in a microbiology laboratory equipped with an incubator, peristaltic pump, fresh medium flask, tubing, and chamber(s) where the material of interest, on which biofilm was already initialized, is inserted (Uppulari et al., 2009). Serial chambers allow independent samples to be produced under similar conditions (nutrients, flow rate, temperature, and incubation time). *Candida albicans* biofilms have been reported to grow more rapidly under continuous flow than in static culture (Uppulari et al., 2009).

Specific devices have been developed for continuous flow studies. The Modified Robbins Device (MRD) is a small channel-shaped chamber with different openings in which biomaterial discs could be inserted when the instrument is mounted to form the channel wall. These devices are provided with a liquid circulator for low-pressure applications. Microorganisms introduced into the fluid stream can adhere to the plugs and generate a biofilm that is easy to remove for analysis. This instrumentation was used to evaluate some maintenance protocols for oral devices (Coenye et al., 2008), as well as for other purposes.

The constant depth film fermenter (CDFF) is an instrument that is able to generate standardized biofilms on any materials that can thereafter be removed for subsequent investigations; the advantages of such a device includes controlled experimental conditions for growth that mimic the oral environment. Some authors (Lamfon et al., 2003) used the CDFF to produce *Candida* biofilms in the presence of artificial saliva on enamel, dentine, and denture acrylic disks. Their data demonstrated that the roughness of the material influences formation and development of *Candida* biofilms.

3.2. *In vivo* models

The rat catheter biofilm infection model (Andes et al., 2004; Lazzell et al., 2009; Ricicová et al., 2010) allows evaluation *in vivo* of the comportment of one microorganism that is exposed to host proteins and immune factors. Other animal models have been utilized to monitor materials and devices that are placed in the bathing conditions of the oral cavity and to mimic denture stomatitis with fungal invasion and neutrophil infiltration of the adjacent mucosa. The rodent acrylic denture model was developed for such purposes (Nett et al., 2010). In humans, abrased pieces of self-adhesive (Budtz-Jörgensen et al., 1981) or acrylic resin disk (Avon et al., 2007) were fixed on dentures to gain some information *in vivo* concerning biofilm formation in the oral cavity.

4. Surface treatment to control *Candida* biofilms

Candida in biofilms on prosthetic materials is difficult to remove. If it is absurd to eradicate a commensal of the oral environment, it is important to consider the prosthesis and the patient simultaneously because the prosthesis is a nest for *Candida* growth and a possible source of infection for the oral mucosa. Daily brushing should be encouraged in all denture-wearers; in denture stomatitis, decontamination of the denture becomes a mandatory part of treatment. Often, the family or professional caregivers must compensate for the difficulties that the elderly face due to loss of independence, dexterity, or memory. Thus, antifungals (azoles, nystatin) should be reserved for patient treatment and are less active against biofilms on dentures (see *infra*); moreover, they may cause the emergence of resistant strains. Use of antiphlogistic solutions has been highly successful; as many of them are fungicidal. However, there are no comparative studies that examine all aspects of the problem.

Rarely, if ever, do study authors take into account the views of all professionals involved in denture care, not only the opinions of the dentist who treats the patient, but also those of the prosthetist who sees the potential deleterious effects of some decontaminating procedures and the microbiologist who isolates and studies yeast *in vitro*. Microbiologists are able to determine the minimum inhibitory concentrations of antifungals on *Candida* growth in suspensions, or better (but less often), on *Candida* biofilms produced in the laboratory. Dental technicians are involved in the relationship with the materials they provide, biofilms, and the decontamination procedures. Clinicians' decisions cannot rely on evidence-based studies alone since they lack data from large-scale clinical trials (i.e., *in vivo* studies).

4.1 Anti-biofilm agents

Many molecules that are embedded in antiseptic mouthwash or in effervescent tablets are candidacidal. Sodium hypochlorite, a major component of bleach that is also produced *in vivo* by myeloperoxidase from the neutrophils, has an anti-*Candida* effect. Ozonated water with or without ultrasound reduces yeasts' adherence to the resin. The use of ultrasound reduces the concentration required for effectiveness of most antiseptics or fungicide antimycotics. The use of a microwave oven is not recommended because the conditions that suppress the yeasts are too close to those that damage some prosthesis materials. When misused, some products can damage the materials: the repeated use of chlorhexidine colors resins brown; hypochlorite at a high dose bleaches them. Hydrogen peroxide is active only at a very high concentration that is close to the mucosal toxicity level; moreover, in the presence of hydrogen peroxide, *Candida* over-expresses catalase and glutathione oxidase, which in turn reduces the concentration of hydrogen peroxide and protects the yeast cells against oxidation.

An alternative to prevent biofilm formation could involve a reduction of microorganism adherence to materials by anti-adhesive/anti-microbial coatings, with or without drug release. Indeed, the anchoring of microorganisms to surfaces such as mucosa, teeth, or biomaterial is a pivotal step in initiating biofilms into the oral cavity. Adhesion can be quantified by measurement of the microorganisms' retention after fixed incubation periods and washings or by the microorganisms' retention against a continuous flow of medium (Cannon et al., 2010). Many protocols have been proposed to limit biofilm formation on various materials used in dentistry (recently reviewed by Busscher et al., 2010): antibiotic and peptide coatings, silver and polymer-brush coatings, and quaternary ammonium couplings. *In vitro*, titanium dioxide coating inhibits *Candida* adhesion to the denture's base in acrylic resin (Arai et al., 2009). Surface protection from bacteria and yeast by chitosan coating is also worthy of further pharmacologic and clinical studies (Carlson et al., 2008). Surface treatment must solve numerous challenges before clinical implementation: the amount of bioavailable drug on the material's surface, kinetic and safety of the released compounds, interferences with the oral environment, and the quest for multifunctional effects such as biofilm control, tissue integration, and/or tissue regeneration.

4.2 New strategies based on research in *Candida*-biofilm biology

The decreased susceptibility of yeast biofilms to classical antifungal drugs encouraged scientists to explore other means to inhibit *Candida* and to limit the deleterious effects of its biofilm. Knowledge of interactions between *Candida* and oral tissues and between *Candida* and oral bacteria should present new perspectives for therapy. Microorganisms in biofilms (yeast included) are less sensitive to antimicrobial agents than free microorganisms in suspension. Authors (Thurnheer et al., 2003) have shown a decreased drug diffusion rate through polyspecies biofilms, containing *Candida* among others, proportional to the cubic root of the drug's marker molecular weight, suggesting the deviousness of the diffusion route through biofilm depth as the cause of delay in molecule penetration. Nevertheless, drug resistance could also be attributed to metabolic properties and gene expression induced by microorganisms living in the community and to the production of an extracellular matrix.

Other *in vitro* studies (reviewed by Nobile et al., 2006) suggest several molecular factors that explain biofilm development and biofilm drug resistance, such as specific biofilm phenotypes (Finkel & Mitchell, 2011), adhesins, cell to cell communication, and quorum sensing (Deveau & Hogan, 2011). The link between hyphae production and *Candida* biofilm development *in vitro* and between hyphae production and pathological conditions *in vivo* led to the investigation of the genetic regulation of hyphal morphogenesis. The rapid initiation of biofilm in the presence of new surfaces available for anchoring oriented the genetic analysis towards a gene expression distinct from that found in the planktonic state.

Quorum sensing pathways that allow microorganism colonies (including yeast) to sense their cell density involve small molecules such as farnesol and tyrosol. The former is known to promote resistance against oxidative stress and inhibit hyphal morphogenesis and biofilm formation, whereas the latter is a putative biofilm facilitator. The (over)expression and polysaccharidic matrix production of adhesins is also linked to biofilm formation. All of these biological characteristics contribute to make *Candida* biofilms "a well-designed protected environment" (Mukherjee P. et al., 2005). A better knowledge of molecular events in *Candida* biofilm formation could present new strategies to prevent oral candidiasis contracted from biomaterials inserted in the oral cavity.

4.3 New strategies based on research in exocrine biology

Antimicrobial molecules/systems derived from exocrine secretions are interesting topics of research. Studies *in vitro* have already shown the benefits of lysozyme, lactoferrin, histatin (Pusateri et al., 2009), and peroxidase systems with thiocyanate, chloride, and especially iodide. However, transferring such data to *in vivo* studies hasn't yet provided the expected results because of the immense complexity of the oral environment. Again, the publication of large clinical studies is still being awaited.

Research in peroxidase biology is an illustrative example of the multiple facets of knowledge transfer from fundamental sciences to clinical applications. In the presence of hydrogen peroxide, for example, peroxidases in exocrine secretions are able to catalyze the production of hypohalous compounds that carry an antimicrobial effect: hypoiodite *in vitro* and hypothiocyanite in saliva. Previous studies have shown that a 30-minute exposure to hypoiodite was sufficient to inhibit planktonic growth *in vitro* (Majerus & Courtois, 1992). Moreover, the development of *Candida* biofilm on material surfaces could be reduced or even suppressed by lactoperoxidase-generated hypoiodite and hypothiocyanite. This was the case not only when peroxidase and substrates system were dissolved in the liquid phase into which the material was immersed, but also when peroxidase precoated on material was activated by simple addition of the substrates to the liquid surrounding this material. Those data also demonstrated the efficiency of peroxidase systems against a *Candida* strain during a three-week incubation period and concomitantly suggested a possible interest in coating the material with peroxidase.

Other investigations demonstrated 1) that lactoperoxidase activity was not modified by coating onto titanium, 2) that lactoperoxidase incorporated in oral gel maintained its activity for at least one year, and 3) that the substrate exhaust (namely hydrogen peroxide and iodide) was the true limiting factor (Bosch et al., 2000; Ahariz et al., 2000). Previous investigations indicated an antibacterial effect on Gram-positive and Gram-negative bacteria, which suggests a non-specific inhibitory effect of hypoiodite on microbial metabolism and growth (Courtois et al., 1995). The ability to transfer this knowledge from bench to clinic is questionable. Indeed, the immunogenicity of a material surface coated with lactoperoxidase should restrict the applications of this system to *ex vivo* conditions. Besides the toxicity of oxidant products on host cells, the competition between iodide and thiocyanate is another limiting factor for *in vivo* use. Thiocyanate is not only present in several exocrine secretions (e.g., human saliva) but is also the preferential substrate of lactoperoxidase. Simultaneous incorporation of iodide and thiocyanate in the same gel decreased the beneficial effect of 2 mM iodide in the presence of increasing concentrations of thiocyanate ranging from 0.25 to 4 mM, which correspond to the normal range of this ion in saliva.

5. Precautions for testing new devices *in vivo*

Finally, the investigators must be aware of the biases frequently encountered in clinical trials that evaluate microbial contamination and colonization of oral devices and prosthesis. Recommendations and guidelines to evaluate the benefits of prophylactic anti-*Candida* procedures are similar to these advocated for any oral care product. Two important biases that must be taken into account are the influence of investigators on patients' hygiene behavior (Grimoud et al., 2005) and the galenic formulation of products lacking the active molecule. Evaluation of dentifrice efficiency for denture hygiene also needs other controls: one testing the product without brushing and one testing the mechanical brushing alone. The abrasive effect of the product must be evaluated, and the abrasiveness of saliva itself is another concern to be considered.

Quantification of the *Candida* biomass that is adherent to the device is difficult in practice. Yeast samples from the oral environment can be collected by rinsing, imprinting, or swabbing. Swabs and imprints are more suitable for gathering yeasts attached to surfaces, and swabbing is easier for clinical studies on a larger scale. Procedures to quantify yeast biomass *in vitro* are not applicable *in vivo* for epidemiological studies and hygiene purposes, particularly since denture-wearers are not always compliant. Dentures can be rinsed in saline and brushed in standard condition to harvest microbial cells. The suspension is thereafter serially diluted for counting (Panzeri et al., 2009).

Another study (Vanden Abbeele et al., 2008) documented the reliability of oral swabbing to investigate yeast carriage on denture. Sampling dentures for *Candida* is more than just a diagnostic tool: it could present an opportunity to verify the patient's compliance with hygiene advice as well as the efficiency of new topical antifungals. Yeast counts after swab culture reflect the biomass present on the oral surfaces, but this is not the number of yeast cells included in oral biofilms. Colony forming units (CFU) counted on the agar medium represented only a small part of the cells harvested by the cotton, as assessed by three successive spreadings of the same cotton that provided similar data (in the range interval of 0.1 logarithmic units). Furthermore, two successive swabs of the same oral surface yielded similar quantities of yeast cells (in the range interval of 0.5 logarithmic units).

Finally, investigators themselves can influence the hygiene behavior of the subjects under study. The study previously quoted (Vanden Abbeele et al., 2008) also analyzes the effect of the oral care program. In the absence of any hygiene advice, a second denture swabbing taken in 46 patients after an interval of one week demonstrated only minor variations, thus minimizing the hygiene-stimulation effect produced by pursuing the collection of samples. Repeated sampling (at one-week intervals) of 46 different healthy denture-wearers demonstrated yeast counts remaining in the same range (±1 logarithmic unit) for more than 85% of the denture swabs and mucosa samples. Values below the lower limit (-1 logarithmic unit) occurred in less than 15% of denture and mucosa swabs. This was attributed to behavioral changes in hygiene practice following the investigators' first visit. By contrast, a hygiene program including a placebo oral gel (tested to be inactive *in vitro*) led to a decrease of yeast carriage after two weeks.

6. Conclusion

Yeasts belonging to the *Candida* genus usually colonize the human oral cavity. *In vivo* and *in vitro* studies have shown *Candida* incorporation into biofilms covering different biomaterials and devices such as dentures. These biofilms may indicate an increased risk factor for invasive candidosis when the host immune system is compromised. Daily denture brushing is recommended to all denture (and other device) wearers in order to prevent the development of *Candida* biofilm. Family members and healthcare workers must assume this task when there is a deficiency in dexterity and/or a loss of autonomy, especially in elderly and disabled persons. In case of candidosis in denture-wearers, decontamination of dentures is mandatory. Antimycotics such as azoles or nystatin must be reserved for curative treatment of infected patients; they are less active against *Candida* biofilms on dentures and could lead to emergent resistance if applied daily to dentures in order to prevent yeast colonization. Nevertheless, few studies, if any, integrate all aspects of denture care: welfare of denture-wearers, prevention of candidosis, biomaterial defects after decontamination processing, and possible *Candida* biofilm development. Daily brushing of dentures remains the key recommendation. A better understanding of *Candida* biology in the oral environment will provide new tools to control *Candida* biofilms, the possible development of more appropriate biomaterials for dentistry (or surface improvements), and better management of biomaterial use in the oral cavity. Further investigations in this field will require cooperation among dentists, biologists, and engineers.

7. Acknowledgment

The author reports no conflicts of interest in this work. The author wishes to thank the students from the *Haute Ecole Francisco Ferrer* (Medical Biology section) and the *Université Libre de Bruxelles* (Faculty of Medicine, Dentistry section) in Brussels who enthusiastically participated in studies on biofilm biology in the frame of their final memory or doctorate.

The author also thanks Pr M. Stas, MD PhD for her helpful advice and review of the manuscript.

8. References

Abu-Elteen, K. & Abu-Elteen, R. (1998). The prevalence of *Candida albicans* populations in the mouths of complete denture wearers. *The New Microbiologica*, Vol.21, No.1, (January 1998), pp. 41-48, ISSN 1121-7138

Addy, M.; Shaw, W.; Hansford, P. & Hopkins, M. (1982). The effect of orthodontic appliances on the distribution of *Candida* and plaque in adolescents. *British journal of orthodontics*, Vol.9, No.3, (July 1982), pp. 158-163, ISSN 0301-228X

Ahariz, M.; Mouhyi, J.; Louette, P.; Van Reck, J.; Malevez, C. & Courtois, P. (2000). Adsorption of peroxidase on titanium surfaces: a pilot study. *Journal of biomedical materials research*, Vol.52, No.3, (December 2000), pp. 567-571, ISSN 0021-9304

Ahariz, M. & Courtois, P. (2010). *Candida albicans* susceptibility to lactoperoxidase-generated hypoiodite. In: *Clinical, cosmetic and investigational dentistry*, , Vol.2010, No.2, (August 2010), pp. 69-78, ISSN 1179-1353, Available from http://www.dovepress.com/candida-albicans-susceptibility-to-lactoperoxidase-generated-hypoiodit-peer-reviewed-article-CCIDEN

Ahariz, M. & Courtois, P. (2010). *Candida albicans* biofilm onto titanium : effect of peroxidase precoating. In: *Medical Devices: Evidence and Research*, Vol.2010, No.3, (August 2010), pp. 33-40, ISSN 1179-1470, Available from http://www.dovepress.com/candida-albicans-biofilm-on-titanium-effect-of-peroxidase-precoating-peer-reviewed-article-MDER

Alcoforado, G.; Rams, T.; Feik, D. & Slots, J. (1991). Microbial aspects of failing osseointegrated dental implants in humans. *Journal de parodontologie*, Vol.10, No.1, (February 1991), pp. 11-18, ISSN 0750-1838

Andes, D.; Nett, J.; Oschel, P.; Albrecht, R.; Marchillo, K. & Pitula A. (2004). Development and characterization of an *in vivo* central venous catheter *Candida albicans* biofilm model. *Infection and immunity*, Vol. 72, No.10, (October 2004), ISSN 0019-9567

Arai, T.; Ueda, T.; Sugiyama, T. & Sakurai, K. (2009). Inhibiting microbial adhesion to denture base acrylic resin by titanium dioxide coating. *Journal of oral rehabilitation*, Vol.36, No.12, (December 2009), pp. 902-908, ISSN 0305-182X

Arendorf, T. & Addy, M. (1985). Candidal carriage and plaque distribution before, during and after removable orthodontic appliance therapy. *Journal of clinical periodontology*, Vol.12, No.5, (May 1985), pp. 360-368, ISSN 0303-6979

Avon, S.; Goulet, J. & Deslauriers, N. (2007). Removable acrylic resin disk as a sampling system for the study of denture biofilms *in vivo*. *Journal of prosthetic dentistry*, Vol.97, No.1, (January 2007), pp.32-38, ISSN 0022-3913

Baboni, F.; Guariza Filho, O., Moreno, A. & Rosa, E. (2010). Influence of cigarette smoke condensate on cariogenic and candidal biofilm formation on orthodontic materials. *American journal of orthodontics and dentofacial orthopedics*, Vol.138, No.4, (October 2010), pp. 427-434, ISSN 0889-5406

Bernhardt, H.; Zimmermann, K. & Knoke, M. (1999). The continuous flow culture as an *in vitro* model in experimental mycology. *Mycoses*, Vol.42, Suppl.2, (1999), pp.29-32, ISSN 0933-7407

Biswas, S. & Chaffin, W. (2005). Anaerobic growth of *Candida albicans* does not support biofilm formation under similar conditions used for aerobic biofilm. *Current microbiology*, Vol.51, No.2, (August 2005), pp. 100-104, ISSN 0343-8651

Bosch, E.; Van Doorne, H. & De Vries S. (2000). The lactoperoxidase system: the influence of iodide and the chemical and antimicrobial stability over the period of about 18 months. *Journal of applied microbiology*, Vol.89, No.2, (August 2000), pp. 215-224, ISSN 1364-5072

Brouns, F.; Schuermans, A.; Verhaegen, J.; De Wever, I. & Stas, M. (2006). Infection assessment of totally implanted long-term venous access devices. *Journal of vascular access*, Vol.7, No.1, (January-February, 2006), pp. 24-28, ISSN 1129-7298

Budtz-Jörgensen, E.; Theilade, E.; Theilade, J. & Zander, H. (1981). Method for studying the development, structure and microflora of denture plaque. *Scandinavian journal of dental research*, Vol.89, No.2, (April 1981), pp. 149-156, ISSN 0029-845X

Bulad, K.; Taylor, R.; Verran,J. & McCord, J. (2004). Colonization and penetration of denture soft lining materials by *Candida albicans*. *Dental materials*, Vol.20, No.2, (February 2004), pp. 167-175, ISSN 0109-5641

Busscher, H.; Rinastiti, M.; Siswomihardjo, W. & van der Mei H. (2010). Biofilm formation on dental restorative and implant materials. *Journal of dental research*, Vol.89, No.7, (July 2010), pp. 657-665, ISSN 0022-0345

Cannon, R.; Lyons, K.; Chong, K. & Holmes, A. (2010). Adhesion of yeast and bacteria to oral surfaces. Methods in molecular biology, Vol.666, (2010), pp. 103-124, ISSN 0164-3745

Carlson, R.; Taffs, R.; Davison, W. & Stewart, P. (2008). Anti-biofilm properties of chitosan-coated surfaces. *Journal of biomaterials science. Polymer edition*, Vol.19, No.8, (August 2008), pp. 1035-1046, ISSN 0920-5063

Cauda, R. (2009). Candidaemia in patients with an inserted medical device. *Drugs*, Vol.69, Suppl.1, (November 2009), pp. 33-38, ISSN 0012-6667

Chandra, J.; Mukherjee, P., Leidich S.; Faddoul, F.; Hoyer, L.; Douglas, L. & Ghannoum, M. (2001). Antifungal resistance of candidal biofilms formed on denture acrylic *in vitro*. *Journal of dental research*, Vol.80, No.3, (March 2001), pp. 903-908, ISSN 0022-0345

Chandra, J.; Kuhn, D.; Mukherjee, P.; Hoyer, L.; McCormick, T. & Ghannoum M. (2001). Biofilm formation by the fungal pathogen *Candida albicans*: development, architecture, and drug resistance. *Journal of bacteriology*, Vol.183, No.18, (September 2001), pp. 5385-5394, ISSN 0021-9193

Coenye, T.; De Prijck, K., De Wever, B. & Nelis, H. (2008). Use of the modified Robbins device to study the *in vitro* biofilm removal efficacy of NitrAdine, a novel disinfecting formula for the maintenance of oral medical devices. *Journal of applied bacteriology*, Vol.105, No.3, (September 2008), pp. 733-740, ISSN 1364-5072

Courtois, P.; Vanden Abbeele, A.; Amrani, N. & Pourtois, M. (1995). *Streptococcus sanguis* survival rates in the presence of lactoperoxidase-produced OSCN- and OI-. *Medical sciences research*, Vol.23, No.3 , (1995), pp. 195-197, ISSN 0269-8951

Depprich, R.; Handschel, J.; Meyer, U. & Meissner, G. (2008). Comparison of prevalence of microorganisms on titanium and silicone/polymethyl methacrylate obturators used for rehabilitation of maxillary defects. *Journal of prosthetic dentistry*, Vol.99, No.5, (May 2008), pp. 400-405, ISSN 0022-3913

Deveau, A. & Hogan, D. (2011). Linking *quorum sensing* regulation and biofilm formation by *Candida albicans*. *Methods in molecular biology*, Vol. 692, (2011), pp. 219-233, ISSN 1064-3745

Dezelic, T.; Guggenheim, B. & Schmidlin, P. (2009). Multi-species biofilm formation on dental materials and an adhesive patch. *Oral health and preventive dentistry*, Vol.7, No.1, (March 2009), pp. 47-53, ISSN 1602-1622

Falcone, M.; Barzaghi, N.; Carosi, G.; Grossi, P.; Minoli, L.; Ravasio, V.; Rizzi, M.; Suter, F.; Utili, R.; Viscoli, C.; Venditti, M. & Italian Study on Endocarditis. (2009). *Candida* infective endocarditis: report of 15 cases from a prospective multicenter study. *Medicine*, Vol.88, No.3, (May 2009), pp. 160-168, ISSN 0025-7974

Finkel, J. & Mitchell, A. (2011). Genetic control of *Candida albicans* biofilm development. Nature reviews – microbiology, Vol.9, No.2, (February 2011), pp. 109-118, ISSN 1740-1526

Glass, R.; Conrad, R.; Bullard, J.; Goodson, L.; Metha, N.; Lech S. & Loewy, Z. (2010). Evaluation of microbial flora found in previously worn prostheses from the Northeast and Southwest regions of the United States. *Journal of prosthetic dentistry*, Vol.103, No.6, (June 2010), pp. 384-389, ISSN 0022-3913

Grimoud, A.; Lodter, J.; Marty, N.; Andrieu, S.; Bocquet, H.; Linas, M.; Rumeau, M. & Cazard, J. (2005). Improved oral hygiene and *Candida* species colonization level in geriatric patients. *Oral diseases*, Vol.11, No.3, (May 2005), pp. 163-169, ISSN 1354-523X

Hägg, U.; Kaveewatcharanont, P.; Samaranayake, Y.; Samaranayake L. (2004). The effect of fixed orthodontic appliances on the oral carriage of *Candida* species and Enterobacteriaceae. *European journal of orthodontics*, Vol.26, No.6, (December 2004), pp. 623-629, ISSN 0141-5387

Hawser, S.; Baillie, G. & Douglas, L. (1998). Production of extracellular matrix by *Candida albicans* biofilms. *Journal of medical microbiology*, Vol.47, No.3, (March 1998), pp. 253-256, ISSN 0022-2615

Hibino, K.; Wong, R.; Hägg, U. & Samaranayake, L. (2009). The effects of orthodontic appliances on *Candida* in the human mouth. *International journal of paediatric dentistry*, Vol.19, No.5, (September 2009), pp. 301-308, ISSN 0960-7439

Jewtuchowicz, V.; Brusca, M.; Mujica, M.; Gliosca, L.; Finquelievich, J.; Lovannitti, C. & Rosa, A. (2007). Subgingival distribution of yeast and their antifungal susceptibility in immunocompetent subjects with and without dental devices. *Acta odontológica latinoamericana*, Vol.20, No.1, (July 2007), pp. 17-22, ISSN 0326-4815

Kania, R.; Lamers, G.; van de Laar, N.; Dijkhuizen, M.; Lagendijk, E.; Huy, P.; Herman, P.; Hiemstra, P.; Grote, J.; Frijns, J. & Bloemberg, G. (2010). Biofilms on tracheooesophageal voice prostheses: a confocal laser scanning microscopy demonstration of mixed bacterial and yeast biofilm. *Biofouling*, Vol.26, No.5, (July 2010), pp. 519-526, ISSN 0892-7014

Kulak, Y. & Kazazoglu, E. (1998). *In vivo* and *in vitro* study of fungal presence and growth on three tissue-conditioning materials on implant supported complete denture wearers. *Journal of oral rehabilitation*, Vol.25, No.2, (February 1998), pp. 135–138, ISSN 0305-182X

Lamfon, H.; Porter S.; McCullough, M. & Pratten J. (2003). Formation of *Candida albicans* biofilms on non-shedding oral surfaces. *European journal of oral sciences*, Vol.111, No.6, (December 2003), pp. 465-471, ISSN 0909-8836

Lazzell, A.; Chaturvedi, A.; Pierce, C.; Prasad, D.; Uppuluri, P. & Lopez-Ribot, J. Treatment and prevention of *Candida albicans* biofilms with caspofungin in a novel central venous catheter murine model of candidiasis. *Journal of antimicrobial chemotherapy*, Vol.64, No.3, (September 2009), pp.567-570, ISSN 0305-7453

Leonhardt, A.; Renvert, S. & Dahlén, G. (1999). Microbial findings at failing implants. *Clinical oral implants research*, Vol.10, No.5, (October 1999), pp. 339-345, ISSN 0905-7161

Majerus, P. & Courtois, P. (1992). Susceptibility of *Candida albicans* to peroxidase-catalyzed oxidation products of thiocyanate, iodide and bromide. *Journal de biologie buccale*, Vol.20, No.4, (December 1992), pp. 241-245, ISSN 0301-3952.

Mattos, B.; Soussa, A.; Magalhães, M.; André, M.; Brito, E.; Dias, R. (2009). *Candida albicans* in patients with oronasal communication and obturator prostheses. *Brazilian dental journal*, Vol.20, No.4, (2009), pp. 336-340, ISSN 0103-6440

Molgatini, S.; González, M.; Rosa, A.; Negroni, M. (1988). Oral microbiota and implant type membranes. *Journal of oral implantology*, Vol.24, No.1, (January 1998), pp. 38–43, ISSN 0160-6972

Mukherjee, P.; Zhou, G.; Munyon, R. & Ghannoum, M. (2005). *Candida* biofilm: a well-designed protected environment. *Medical mycology*, Vol.43, No.3, (May 2005), pp. 191-208, ISSN 1369-3786

Mutluay, M.; Oğuz, S.; Ørstavik, D.; Fløystrand, F.; Doğan, A; Söderling, E. ;Närhi, T. & Olsen, I. (2010). Experiments on *in vivo* biofilm formation and *in vitro* adhesion of *Candida* species on polysiloxane liners. *Gerodontology*, Vol.27, No.4, (December 2010), pp. 283-291, ISSN 0734-0664

Nett, J.; Lepak, A.; Marchillo, K. & Andes, D. (2009). Time course global gene expression analysis of an *in vivo Candida* biofilm. *Journal of infectious diseases*, Vol.200, No.2, (July 2009), pp. 307-313, ISSN 0022-1899

Nett, J.; Marchillo, K.; Spiegel, C. & Andes, D. (2010). Development and validation of an *in vivo Candida albicans* biofilm denture model. *Infection and immunity*, Vol.78, No.9, (September 2010), pp.3650-3659, ISSN 0019-9567

Nikawa, H.; Nishimura, H.; Hamada, T.; Kumagai, H. & Samaranayake, L. (1997). Effects of dietary sugars and, saliva and serum on *Candida* biofilm formation on acrylic surfaces. *Mycopathologia*, Vol.139, No.2, (August 1997), pp. 87-91, ISSN 0301-486X

Nikawa, H.; Nishimura, H.; Makihira, S.; Hamada, T.; Sadamori, S. & Samaranayake, L. (2000). Effects of serum concentration on *Candida* biofilm formation on acrylic surfaces. *Mycoses*, Vol.43, No.3-4, (May 2000), pp. 139-143, ISSN 0933-7407

Nobile, C. & Mitchell, A. (2006). Genetics and genomics of *Candida albicans* biofilm formation. *Cellular microbiology*, Vol.8, No.9, (September 2006), pp. 1382-1391, ISSN 1462-5814

Panzeri, H.; Lara, E.; Paranhos, H.; Lovato da Silva, C.; de Souza, R.; de Souza Gugelmin, M.; Tirapelli, C.; Cruz, P. & de Andrade, I. (2009). *In vitro* and clinical evaluation of specific dentifrices for complete denture hygiene. *Gerodontology*, Vol.26, No.1, (March 2009), pp. 26-33, ISSN 0734-0664

Peeters, E.; Nelis, H. & Coenye, T. (2008). Comparison of multiple methods for quantification of microbial biofilms grown in microtiter plates. *Journal of microbiological methods*, Vol.72, No.2, (February 2008), pp. 157-165, ISSN 0167-7012

Percival, S.; Bowler, P. & Dolman, J. (2007). Antimicrobial activity of silver-containing dressings on wound microorganisms using an *in vitro* biofilm model. *International wound journal*, Vol.4, No.2, (June 2007), pp. 186-191, ISSN 1742-4801

Pereira-Cenci, T.; Del Bel Cury, A.; Crielaard, W. & Ten Cate, J. (2008). Development of *Candida*-associated denture stomatitis: new insights. *Journal of applied oral science*, Vol.16, No.2, (March-April 2008), pp. 86-94, ISSN 1678-7757

Pusateri, C.; Monaco, E. & Edgerton, M. (2009). Sensitivity of *Candida albicans* biofilm cells grown on denture acrylic to antifungal proteins and chlorhexidine. *Archives of oral biology*, Vol.54, No.6, (June 2009), pp. 588-594, ISSN 0003-9969

Pye, A.; Lockhart, D.; Dawson, M.; Murray, C. & Smith, A. (2009). A review of dental implants and infection. *Journal of hospital infection*, Vol.72, No.2, (June 2009), pp. 104-110, ISSN 0195-6701

Radford, D.; Challacombe, S. & Walter J. (1999). Denture plaque and adherence of *Candida albicans* to denture-base materials *in vivo* and *in vitro*. *Critical reviews in oral biology and medicine*, Vol.10, No.1, (February 1999), pp. 99-116, ISSN 1045-4411

Ramage, G.; Wickes, B. & López-Ribot, J. (2008). A seed and feed model for the formation of *Candida albicans* biofilms under flow conditions using an improved modified Robbins device. *Revista iboamericana de micologia*, Vol.25, No.1, (March 2008), pp. 37-40, ISSN 1130-1406

Ricicová, M.; Kucharíková, S.; Tournu, H.; Hendrix, J.; Bujdáková, H.; Van Eldere, J.; Lagrou, K. & Van Dijck, P. (2010). *Candida albicans* biofilm formation in a new *in vivo* rat model. *Microbiology*, Vol.156, No.3, (March 2010), pp. 909-919, ISSN 1350-0872

Rodger, G.; Taylor, R.; Pearson, G. & Verran, J. (2010). *In vitro* colonization of an experimental silicone by *Candida albicans*. *Journal of biomedical materials research. Part B, applied biomaterials*, Vol.92, No.1, (January 2010), pp. 226-235, ISSN 1552-4973

Samaranayake, Y.; Ye, J.; Yau, J.; Cheung, B. & Samaranayake, L. (2005). *In vitro* method to study antifungal perfusion in *Candida* biofilms. *Journal of clinical microbiology*, Vol.43, No.2, (February 2005), pp. 818-825, ISSN 0095-1137

Sissons, C. (1997). Artificial dental plaque biofilm model systems. *Advances in dental research*, Vol.11, No.1, (April 1997), pp. 110-126, ISSN 0895-9374

Thurnheer, T.; Gmür, R.; Shapiro, S. & Guggenheim, B. (2003). Mass transport of macromolecules within an *in vitro* model of supragingival plaque. *Applied and environmental microbiology*, Vol.69, No.3, (March 2003), pp. 1702-1709, ISSN 0099-2240

Uppulari, P.; Chaturvedi, A. & Lopez-Ribot, J. (2009). Design of a simple model of *Candida albicans* biofilms formed under conditions of flow: development, architecture, and drug resistance. *Mycopathologia*, Vol.168, No.3, (September 2009), pp. 101-109, ISSN 0301-486X

Vanden Abbeele, A.; de Meel, H.; Ahariz, M.; Perraudin, J.-P.; Beyer, I. & Courtois, Ph. (2008). Denture contamination by yeasts in the elderly. *Gerodontology*, Vol.25, No.4, (December 2008), pp. 222-228, ISSN 0734-0664

Vandenbussche, M. & Swinne, D. (1984). Yeasts oral carriage in denture wearers. *Mykosen*, Vol.27, No.9, (September 1984), pp. 431–435, ISSN 0027-5557

16

Biomaterials in Urology - Beyond Drug Eluting and Degradable - A Rational Approach to Ureteral Stent Design

Dirk Lange, Chelsea N. Elwood and Ben H. Chew
The Stone Centre at Vancouver General Hospital, University of British Columbia
Canada

1. Introduction

Ureteral stents are commonly used in urology to provide urinary drainage of the upper tracts, particularly following treatment of urolithiasis. Stents are commonly plagued with infections and encrustation, particularly in stone-forming patients (Denstedt and Cadieux 2009). This involves a multistep process outlined in Figure 1. The first step is formation of a conditioning film comprised of urinary proteins, ions, and crystals that are deposited at the stent surface (Tieszer, Reid et al. 1998). The conditioning film becomes an attractive surface for bacteria to adhere to and forms a biofilm which can lead to a urinary tract infection or encrustation (Wollin, Tieszer et al. 1998; Choong and Whitfield 2000; Choong, Wood et al. 2001; Shaw, Choong et al. 2005). Bacteria have been demonstrated to adhere to the stent surface in up to 90% of indwelling stents, which in 27% of cases leads to a positive urine culture (Reid, Denstedt et al. 1992).

Ideally, ureteral stent biomaterials would be able to limit or completely prevent the processes shown in Figure 1. Various attempts have been made to reduce the deposition of crystals, bacteria, and protein on stent surfaces including using low surface energy biomaterials (Tieszer, Reid et al. 1998), heparin coating (Cauda, Cauda et al. 2008), antimicrobial eluting biomaterials (Cadieux, Chew et al. 2006; Chew, Cadieux et al. 2006; Wignall, Goneau et al. 2008; Cadieux, Chew et al. 2009), diamond-like carbon coatings (Laube, Kleinen et al. 2007), polyethylene glycol and marine mussel adhesive proteins (Pechey, Elwood et al. 2009) to name just a few. Most have limited effectiveness and some have even shown increased bacterial adhesion compared to controls in the case of heparin coating (Lange, Elwood et al. 2009). While drug-eluting technology of biomaterials is both readily available and used clinically in other fields, its role in urology has been limited. Triclosan was used recently in ureteral stents. It held promise in *in vitro* (Chew, Cadieux et al. 2006) and animal infection models (Cadieux, Chew et al. 2006) but it fared poorly in those patients requiring chronic ureteral stents (Cadieux, Chew et al. 2009). The current methodology of technological implementation surrounding ureteral stent coating and design have come from trial and error or are borrowed technologies from coronary and vascular stenting. The time has come for the urological world to apply the same types of scientific discovery and development to problems specific to urinary devices rather than just applying the end product from other areas of medicine. What works in vascular stenting may not be directly applicable to the urinary environment.

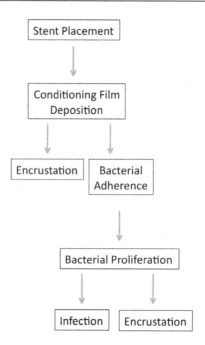

Fig. 1. The processes following stent insertion that lead to bacterial adhesion, infection, and encrustation. Following stent insertion, urinary components deposit on the surface, forming anchor points for both bacteria as well as ions/minerals. Subsequent bacterial colonization and proliferation leads to the formation of a resistant biofilm that leads to subsequent infection. In addition the interaction of ions/minerals with conditioning film components and bacterial induced crystallization will facilitate encrustation of the device. Inhibition of the common step of conditioning film deposition is a potential step in preventing patient symptoms associated with indwelling ureteral stents.

2. Problems arising from stents

Stents have been associated with increased morbidity causing infection,(Riedl, Plas et al. 1999) encrustation,(Paick, Park et al. 2003) hematuria,(Damiano, Oliva et al. 2002; Joshi, Newns et al. 2003; Joshi, Chitale et al. 2005) and discomfort (Joshi, Stainthorpe et al. 2001; Joshi, Okeke et al. 2002). The Ureteral Stent Symptom Questionnaire (USSQ) looks at different facets of life including sexual function which has been shown to be negatively affected by the presence of a stent (Sighinolfi, Micali et al. 2007). In fact, ureteroscopes, intracorporeal lithotriptors, and ureteroscopic techniques have improved to the point that the major morbidity of ureteroscopy has become the stent left *in situ* postoperatively. Studies evaluating differences in postoperative complications and stone-free rates in stented compared to non-stented patients have shown that stents are not a routine requirement following uncomplicated ureteroscopy (Hosking, McColm et al. 1999; Borboroglu, Amling et al. 2001; Denstedt, Wollin et al. 2001). The uncommon, but most severe problem arising from stents is the "forgotten stent" that is left in place for several months to years. These stents become encrusted and create difficulty for both patient and urologist, especially since their removal involves multiple surgical procedures (Rana and Sabooh 2007) and may result in loss of the renal unit or potentially even death (Singh, Srinivastava et al. 2005).

3. Current stent biomaterials

The synthetic polymer, polyethylene, was previously used in stent construction, but was abandoned due to its stiffness, brittleness, and tendency to fragment. Blends of polyethylene and other polymers, such as polyurethane, have been shown to resist encrustation (Gorman, Tunney et al. 1998; Gomha, Sheir et al. 2004). Silicone is currently the most biocompatible stent material as it is the most resistant to biofilm formation, infection and encrustation (Watterson, Cadieux et al. 2003), and is one of the most lubricious materials available (Jones, Garvin et al. 2004); however, its softness and elasticity make it difficult to handle, particularly through tortuous or tight ureters. In addition, the low tensile strength of silicone makes it susceptible to extrinsic compression. The development of new stent materials aimed to meld the flexible and elastic properties of silicone with the rigidity of polyethylene which resulted in the development of polyurethane, the most common class of polymer currently used in stents. Polyurethane, however, is a stiff material that causes patient discomfort and significant ureteral ulceration and erosion have been reported in an animal model (Marx, Bettmann et al. 1988). New proprietary materials and combinations are softer, more comfortable, and easier to maneuver within the urinary tract. Examples of commonly used materials in stents include Percuflex® (Boston Scientific Corporation, Natick, MA), Silitek® (Surgitek, Medical Engineering Corporation, Racine, WI, USA), C-Flex® (Consolidated Polymer Technologies, Clearwater, FL, USA), Tecoflex® (Thermedics, Wilmington, MA, USA), and ethylene-vinyl-acetate (from the polyefin family of which polyethylene is a member). They have been designed to provide rigidity to facilitate handling by the surgeon and to provide adequate drainage while being soft enough to limit patient discomfort.

4. New materials

New materials include metal stents that are designed to keep the ureter open despite extrinsic ureteral compression secondary to lymphadenopathy due to malignancy. Ureteric obstruction may result in decreased renal function, pain, or infection requiring urinary diversion (Chitale, Scott-Barrett et al. 2002; Allen, Longhorn et al. 2010). As these stents must remain in place for long periods of time, they require frequent exchanges because they are susceptible to infection and encrustation with increased indwelling time. The goal in this patient population is to develop a stent that maintains ureteral patency during extrinsic compression, is soft to minimize discomfort, and is resistant to encrustation and infection.

5. Metal ureteral stents

Metal ureteral stents were introduced by Pauer in 1992 (Pauer and Lugmayr 1992) and have been utilized in the treatment of malignant ureteric obstruction (Kulkarni and Bellamy 2001; Liatsikos, Karnabatidis et al. 2009; Masood, Papatsoris et al. 2010; Papatsoris and Buchholz 2010; Sountoulides, Kaplan et al. 2010), ureteral strictures (Daskalopoulos, Hatzidakis et al. 2001; Papatsoris and Buchholz 2010), and ureteropelvic junction obstruction (Barbalias, Liatsikos et al. 2002; Masood, Papatsoris et al. 2010; Benson, Taylor et al. 2011). Current problems associated with metallic stents include biofilm formation, infection, migration, and tissue hyperplasia leading to luminal obstruction (Barbalias, Liatsikos et al. 2002; Wah, Irving et al. 2007; Liatsikos, Karnabatidis et al. 2009; Sountoulides, Kaplan et al. 2010).

Metal stents were originally used for the relief of end-stage malignant disease, where the ureteral stricture ws either directly caused by the tumor or indirectly via pressure of a tumor on the ureter. Pauer and Lugmayr used metallic ureteral stents in 1996 (Pauer and Lugmayr 1996) to treat 54 malignant stenoses of the ureter in 40 patients via the implantation of a self-expandable permanent endoluminal stent (SPES), the Wallstent™. During a follow-up period of 10.5 months, 51 ureters maintained adequate patency. Of these, 51% needed no further intervention, while 49% needed intervention to re-establish patency. In comparison insufficiency was noted at a mean 4.3 weeks in control patients with an indwelling Double-J catheter. One of the drawbacks of metal stents is, however, that they induce local urothelial hyperplasia, with ingrowth of tissue through the struts that may result in recurrent obstruction during long term placement. Recently, a nickel-cobalt-chromium-molybdenum-alloy stent double pig-tailed stent (Resonance™ Stent, Cook Urological, Spencer, IN, USA) has been developed to provide long-term urinary drainage in patients with malignant ureteric strictures (Blaschko, Deane et al. 2007; Wah, Irving et al. 2007; Lopez-Huertas, Polcari et al. 2010; Wang, Lee et al. 2010). The tight winding of this metal stent helps to maintain stent flexibility and movement, while resisting in-growth of tissue. In addition to this, the movement of the stent causes opening of the coils, allowing the fluid to access the lumen. In a study of 15 patients, this metal stent provided adequate long-term (up to 12 months) urinary drainage in patients with malignant ureteric obstruction without significant bulky pelvic disease. These stents were also found to have minimal encrustation (Wah, Irving et al. 2007). An *in vitro* study has shown that this metal stent provides its best drainage when the ureter is tightly compressed onto its outer surface which is likely due to the result of increased flow between the coils of the metal stent. It was this feature that makes the Resonance™ stent useful in patients in which the ureter is obstructed due to malignancy (Blaschko, Deane et al. 2007; Liatsikos, Karnabatidis et al. 2009; Masood, Papatsoris et al. 2010; Sountoulides, Kaplan et al. 2010; Wang, Lee et al. 2010; Benson, Taylor et al. 2011).

6. Biodegradable stents

Despite the fact that the biocompatible materials and stent designs have improved over the years, they have one key disadvantage, which is the fact that they have to be removed via a separate procedure unless the retention suture is left on the stent. Avoiding a secondary procedure for ureteral stent removal would decrease patient morbidity and make this technology attractive. More recent research has focused on the design of stents that do not need to be removed and are biodegradable. The design of a biodegradable stent must take into consideration the biocompatibility properties of the material, as well as the degree of expansion and degradation rates, and most importantly it must be able to fulfill the basic requirement of a stent in that it must be able to guarantee urinary flow from the renal pelvis through the ureter and into the bladder for the desired period of time. Degradable materials also must retain their properties after sterilization and be able to withstand a long shelf-life before use. One of the challenges involved in designing a biodegradable stent is the control of the rate and direction of degradation. Schlick and Planz (Schlick and Planz 1997; Schlick and Planz 1998) designed a stent composed of plastic, the degradation of which was dependent upon the urine pH. In vitro experiments with artificial urine showed that the stents were stable at urine pH less than 7.0 for at least 30 days, while they dissolved completely within 48 hours at pH greater or equal to 7.0. The principle behind this stent is that it would remain stable at physiological urine pH of 5-6, but can be triggered to dissolve by medically altering urinary pH. Although very promising, this technology remains at an experimental stage and awaits animal trials. An additional factor that may need to be taken into consideration is the influence of encrustation protein deposition as it can form a platform for bacterial adherence and infection. The influence of encrustation and protein deposition must also be considered as it can form a platform for bacterial adherence, all of which may influence urine pH. Uropathogens in general are known to increase urine pH, and may have an effect *in vivo*. In addition to this, medically increasing the urinary pH may introduce an additional risk for infection as more alkaline pH favours bacterial survival and some increased stone formation (calcium phosphate and struvite stones). In addition, encrustation of the stent may also prevent May also prevent its exposure to the altered pH environment, environment and the altered pH, thus limiting its rate of decomposition.

A spiral stent (Spirastent®, Urosurge Medical, Coralville, IA) is a polyurethane stent with metal helical ridges designed to prevent kinking and compression in chronically obstructed patients. In *in vitro* studies, this stent increased flow and theoretically increased the space between stent and ureter to facilitate passage of stone fragments.(Stoller, Schwartz et al. 2000) The spiral design has been incorporated into biodegradable materials for urethral stents.(Isotalo, Talja et al. 2002; Laaksovirta, Isotalo et al. 2002) Laaksovirta et al used a self-reinforced poly-L-lactic and poly-L-glycolic acid (SR-PLGA) copolymer spiral urethral stent (SpiroFlow stent, Bionx Implants Ltd, Tampere, Finland) following prostatic laser coagulation.(Laaksovirta, Isotalo et al. 2002) This stent degraded in 6–8 weeks and resisted encrustation at 4 weeks in artificial urine. After 8 weeks, the SpiroFlow® stent was significantly less encrusted than the metal urethral stents Prostakath® (Engineers and Doctors A/S, Copenhagen, Denmark) and Memokath® 028 (Engineers and Doctors A/S). SR-PLGA is the most commonly utilized material for prostatic stents, but it has also been developed as a ureteral stent and may be incorporated into new degradable, encrustation-resistant ureteral stents in the future.(Olweny, Landman et al. 2002)

An alginate based ureteral stent was designed to stay in place for at least 48 hours after uncomplicated ureteroscopy.(Lingeman, Preminger et al. 2003; Lingeman, Schulsinger et al. 2003) Of 87 patients, 80.5% of patients retained their stents greater than 48 hours. Seventeen patients had early stent passage (earlier than 48 hours), but did not require any supplemental procedures to insert another stent. Although these results were promising, several patients (14) still had stent fragments remaining at 30 days, while 3 patients had stent fragments remaining after 90 days. All three of these patients underwent shockwave lithotripsy and two went on to have endoscopic ureteroscopy to remove the fragments from the kidney. Because of the lack of all stents to degrade by 3 months, this stent is no longer commercially available. (Lingeman, Preminger et al. 2003)

The authors are currently involved in developing a new biodegradable stent (Poly-Med Inc., Anderson, SC) that dissolves within 1 to 4 weeks in a porcine model. The animals stented with degradable stents displayed less histologic inflammation than animals stented with control polyurethane biostable Double J stents (Chew, Lange et al. 2010). Weekly intravenous pyelograms displayed less hydronephrosis in the degradable stent group. All stents degraded by 4 weeks and degradation began after 1 week in a very controlled fashion and no animal had a distal obstructed ureter Due to retained stent pieces. Properties such as stent softness from these biodegradable stents may improve patient comfort. Clinical studies will be necessary to determine if biodegradable stents are more comfortable.

7. Stent coatings

One of the most common stent coatings is hydrogel, which consists of hydrophilic polymers that absorb water and increase lubricity and elasticity.(Marmieri, Pettenati et al. 1996; John, Rajpurkar et al. 2007) These properties facilitate stent placement, making the device rigid and easily maneuverable in its dry state, but once exposed to urine, the hydrogel begins to absorb and trap water in its polyanionic structure, causing it to soften and theoretically increase patient comfort. Data on encrustation and infection are less convincing, as hydrogel has been shown to both reduce (Gorman, Tunney et al. 1998) and increase encrustation and biofilm formation (Desgrandchamps, Moulinier et al. 1997). Hydrogels have been used in an attempt to soak and retain antibiotics but an *in vitro* study did not show increased efficacy of bacterial killing compared to non-antibiotic soaked hydrogel coated stents (John, Rajpurkar et al. 2007).

Glycosaminoglycan (GAG), a normal constituent of urine, is a natural inhibitor of crystallization. Other novel stent coatings include pentosan polysulfate (Zupkas, Parsons et al. 2000) (a member of the Glycosaminoglycan family a normal constituent of urine and a natural inhibitor of crystallization), phosphorylcholine (Stickler, Evans et al. 2002) (a constituent of human erythrocytes that mimics a natural lipid membrane), and polyvinyl pyrrolidone (Tunney and Gorman 2002) (a hydrophilic coating, similar to hydrogel, that absorbs water).

Attempts to reduce encrustation have included other stent coatings, such as the bacterial enzyme, oxalate decarboxylase, which has been shown to decrease encrustation in silicone discs placed in rabbit bladders.(Watterson, Cadieux et al. 2003) A novel coating of mPEG-DOPA$_3$, a natural constituent produced by mussels that produces strong adhesive properties, also has the ability to avoid biofouling in the environment. The polyethyelene (PEG) component provides the antifouling property while the DOPA$_3$ provides the adherence that PEG lacks on its own. Adherence of these combined compounds on silicone disks has resulted in a strong ability to resist bacterial adherence and growth in vitro.(Ko 2007) Further development of this type of coating was studied *in vivo* using a rabbit *E. coli* cystitis model (Pechey, Elwood et al. 2009). This study showed that the anti-adhesive coating was successful at inhibiting bacterial adhesion and decreased the incidence of infection, however it was unable to prevent non-bacterial mediated encrustation.

Plasma deposited diamond like carbon coatings have been used to coat stents in an attempt to prevent encrustation (Laube, Kleinen et al. 2007). *In vitro* experiments have shown a 30% decrease in encrustation of these stents in artificial urine compared to the non-coated controls. Ongoing clinical trials appear to indicate a further enhancement of these results *in vivo*, however a mechanism for this needs to be elucidated. Encrustation of ureteral stents remains one of the most common problems associated with ureteral stenting and more research will need to be done for an optimal stent design which resists the deposition of bacteria, minerals and proteins.

In vascular medicine, the anticoagulant heparin has been shown to inhibit bacterial attachment to venuous catheters (Ruggieri, Hanno et al. 1987; Appelgren, Ransjo et al. 1996), which has been attributed to its highly negative charge. Similarly, effects of heparin have also been observed for ureteral stents. Riedl *et al.* (Riedl, Witkowski et al. 2002) used heaprin-coated and uncoated polyurethane ureteral stents and inserted them into obstructed ureters with indwelling times between 2 and 6 weeks. Electron microscopy showed that the uncoated control stents were covered with amorphous anorganic deposits and bacterial biofilms as early as 2 weeks following stent insertion, while the heparin-coated stents remeined unaffected by encrustation following 6 weeks of indwelling time. Cauda *et al.* (Cauda, Cauda et al. 2008) performed a long term study involving patients with bilateral ureteral obstructions treated via the insertion of a heparin-coated stent into one ureter, and an uncoated control stent into the other ureter. Overall, the uncoated control stents were found to be encrusted with amorphous, crystalline inorganic deposits and bacterial biofilm as early as 1 month post-insertion, while the heparin-coated stents remained visibly free of encrustation as long as 10 months post-insertion. Biofilm encrustation was evident only on the external surface of the coated stent after 1 year of being in place. Heparin coated ureteral stents (Radiance Stent, Cook Urological) were tested in an *in vitro* model of infected urine and did not display any reduction in bacterial adherence compared to control stents (Lange, Elwood et al. 2009). These preliminary results are somewhat promising, but clinical trials involving a larger number of patients are needed to ensure that heparin coating of stents is effective across a broader patient range.

8. Drug eluting stents

The most serious complications of long term stenting involve infection triggered by bacterial adherance and biofilm formation on the surfaces of stents as well as patient discomfort due to stent placement. Much research has gone into the prevention of infection, and the most promising results have come from drug eluting stents. Triclosan is an antimicrobial used in many products including soap, surgical scrub, toothpaste, and mouthwash. It inhibits the highly conserved bacterial enoyl-ACP reductase, which is responsible for fatty acid synthesis and cell growth. Cadieux *et al.* reported that, compared to control stents, triclosan-loaded stents implanted in rabbit bladders infected with *Proteus mirabilis* were associated with significantly fewer urinary tract infections.(Cadieux, Chew et al. 2006) Chew *et al.* have shown that bacterial adherence to triclosan eluting stents is markedly reduced compared to regular stents.(Chew, Cadieux et al. 2006) These studies indicate that human clinical trials involving these stents are warranted.

Ureteral stents may also be loaded with pharmaceuticals to aid patient comfort, and to prevent encrustation. Irritative and painful stent symptoms have traditionally been managed with oral medications such as anticholinergics and analgesics, or even by stent removal. Drug-eluting stents release a medication that acts locally on the bladder to decrease irritation and pain. In an attempt to determine which medication might improve stent-related symptoms, Beiko *et al.* intravesically instilled 3 different medications into the bladder of 40 patients who received a ureteral stent at the time of shockwave lithotripsy.(Beiko, Watterson et al. 2004) Intravesical ketorolac significantly reduced flank pain scores following stent insertion compared to lidocaine or oxybutynin following SWL. A ketorolac-eluting ureteral stent was designed and shown to produce the highest levels of ketorolac in the ureteral tissues in an porcine model (Chew, Davoudi et al. 2010). The levels of ketorolac in the ureter were 11 fold of that found in the serum thereby reducing potential systemic side effects while delivering medication directly to the target area. The stent was biocompatible and systemic levels of ketorolac were negligible. A double-blinded prospective randomized controlled trial comparing ketorolac-eluting ureteral stents to controls showed no difference in pain scores except in young males who had less symptoms with the ketorolac eluting stent (Krambeck, Walsh et al. 2010).

Liatsikos *et al* have tested paclitaxel eluting metal stents in the pig ureter to examine the tissue effects and stricture formation.(Liatsikos, Karnabatidis et al. 2007) Paclitaxel eluting stents produced less ureteral inflammation and hyperplasia of the surrounding tissue compared to the bare metal stents. Ureteral patency was lost in the control stents and maintained by the Paclitaxel eluting stents. These studies were carried out over a 21 day period and require further validation via long term animal trials.

Stent encrustation worsens with increased indwelling time and concurrent infection with urease-producing organisms. Oxalate is normally broken down in the gastrointestinal tract by the enzyme oxalate decarboxylase, which is found in a commensal organism *Oxalobacter formigenes*. Oxalate that escapes degradation and fecal excretion is absorbed into the bloodstream and filtered in the kidneys where, under certain conditions, it can combine with calcium to form calcium oxalate stones. Watterson *et al.* coated silicone disks with oxalate decarboxylase and implanted these into rabbit bladders.(Watterson, Cadieux et al. 2003) After 30 days, the oxalate decarboxylase-coated disks were significantly less encrusted than control disks. Coating ureteral stents with such an enzyme could theoretically prevent encrustation as the stent would elute an enzyme to degrade urinary oxalate.

9. Identifying potential targets in stent design

When considering the design of new indwelling ureteral devices such as stents or catheters, the sequential steps triggering a given side effect should be taken into consideration, however this has been complicated by the complexity of mechanisms involved. Rational drug design hypothesizes that the alteration of a biological target has therapeutic value and forms the basis for the invention of new medications predicated on the identification and knowledge of a specific biological target. The first step involves turning to basic science and considering the molecular and biochemical pathways involved in the condition to identify specific targets for drug design. Once a target has been identified, its molecular structure is determined and a suitable drug that will alter it in a favorable manner is designed. Usually the target is a key molecule in a metabolic or signaling pathway specific to a disease condition or pathology (Mandal, Moudgil et al. 2009).

We believe that the same principals can also be applied to the design of ureteral stents, as the current stent designs have failed to live up to their expectations in the complex environment of the urinary tract. Given the fact that the mechanisms causing stent symptoms are unknown makes it difficult to identify a key target in the context of rational drug design to relieve patient symptoms. The identification of such a target in the urinary tract would be beneficial, as it will allow for the reduction or elimination of stent symptoms by targeting a single mechanism. However, in order for that to become a possibility, key steps in the mechanisms surrounding stent-related symptoms need to be identified to allow for their inhibition.

Although identifying a single receptor or enzyme target in the development of stent encrustation and infection is unlikely, a more solid understanding of the mechanism involved in this process is required. Several processes occur following stent insertion and the cumulative effect can result in stent associated symptoms suffered by the patient. It has been well documented that a urinary conditioning film deposits on the stent surface shortly following device insertion that consists of urinary components (Tieszer, Reid et al. 1998).

These components are believed to facilitate bacterial adhesion leading to bacterial colonization, proliferation, and biofilm formation with subsequent infection. Once a biofilm has formed, this environment facilitates recurrent infection and eradication of bacteria is difficult. Bacteria embedded within the biofilm change to a low metabolic state and undergo a low replication rate, thus rendering antibiotics (which are most effective against bacteria in high metabolic states and undergoing replication) ineffective. In many cases, embedded bacteria are also protected since antibiotics cannot penetrate the biofilm and the protecting exopolysaccharide layer excreted by the bacteria onto its surface. Thirdly, bacteria can upregulate resistance genes once inside the biofilm (Lewis 2005).

Aside from biofilm formation and infection, another symptom associated with patient morbidity caused by indwelling ureteral stents is device encrustation. Stent encrustation can be idiopathic and caused by calcium oxalate crystals. In other instances, stent encrustation can be attributed to the presence of urease producing bacteria, which break down urinary ammonia into ammonium (thus effectively taking a hydrogen ion), which results in a rise in urinary pH and crystallization of magnesium, ammonium and phosphate ions. These crystals then adhere to the surface of the stent via the interaction with components of the conditioning film. The conditioning film on the stent surface is considered to be a great contributor to bacterial associated encrustation because it facilitates bacterial adhesion and crystal adhesion to the stent surface. In addition to this, the conditioning film has also been implicated in idiopathic encrustation (in the absence of bacteria) of the stent with calcium oxalate crystals. As such, certain conditioning film components have been proposed to be able to bind minerals from the urine, forming a nidus for crystal growth and device encrustation. To date, we have identified 3 potential targets to interrupt the sequence of events involved in the evolution of stent encrustation and infection: 1) preventing conditioning film formation, 2) preventing initial adherence and encrustation and 3) inhibition of further bacterial proliferation.

10. Current stent biomaterial design

Over the years, attempts have been made at preventing stent associated symptoms by targeting either bacterial adhesion and encrustation or inhibition of bacterial proliferation. Drug eluting technology to prevent bacterial adhesion has previously been used in a triclosan-eluting ureteral stent. Triclosan is an antimicrobial found in over 800 commercially available products such as soaps, hand scrubs, and toothpaste. This stent proved to be successful at eliminating bacterial loads *in vitro* (Chew, Cadieux et al. 2006) as well as a *Proteus mirabilis* urinary tract infection in a rabbit model (Cadieux, Chew et al. 2006), but did not show any significant differences in long term clinical trials (Cadieux, Chew et al. 2009). Similarly a heparin- coated stent was designed to prevent bacterial adhesion given the material's highly negative charge. This stent was shown to decrease encrustation in patients (Hildebrandt, Sayyad et al. 2001; Riedl, Witkowski et al. 2002; Cauda, Cauda et al. 2008), however was unable to prevent bacterial adhesion (Lange, Elwood et al. 2009). The use of diamond-like amorphous carbon as a coating on stents is a new technology that has shown some promise in terms of inhibiting encrustation (Laube, Bradenahl et al. 2006; Laube, Kleinen et al. 2007), however experiments aimed at determining its ability to inhibit bacterial adhesion is lacking. One of the drawbacks of these new technologies is the fact that they are susceptible to blockage by the deposition of the urinary conditioning film, which covers any coating and blocks elution of drugs from the stent, rendering it ineffective and promoting bacterial adhesion and encrustation via mechanisms discussed above.

11. Seeking novel targets in the pursuit of rational stent design.

Given the fact that the urinary conditioning film has been directly implicated in causing both bacterial adhesion and associated/non-associated encrustation, it becomes important to switch our focus from designing a biomaterial that inhibits direct bacterial and ion/mineral deposition to one that inhibits conditioning film components. It is important to focus on understanding this biological target further and the first step is to identify components of the conditioning film. Santin et al have previously identified human serum albumin as well as Tamm-Horsfall Protein (THP) as major conditioning film components found on four stents removed from patients (Santin, Motta et al. 1999). More recently, Canales et al have studied the conditioning films of stents removed from 27 patients, identifying hemoglobin alpha and beta chain, albumin, calgranulin B, fibrinogen beta chain, vitronectin, annexin A1, calgranulin A, fibrinogen gamma chain, and THP as the ten most common adherent components (Canales, Higgins et al. 2009). In addition, this group also hypothesized that the presence of histones likely contribute to stent encrustation given their unique net positive charge. Despite the fact that these papers have contributed to a large extent to the identification of conditioning film components, it still needs to be determined whether urinary conditioning films differ between stent types or patients, as the molecules targeted in stent design should be "universal" and need to be common between patients and stent types.

Our group has recently compared the composition of conditioning films found on certain stents from Boston Scientific (Polaris) to those on Bard stents (Inlay) after they have been removed from patients (Lange et al, unpublished data). Both of these stents differ in their biomaterials, as the Polaris stent is made of an olefinic copolymer, while the InLay stent is made of polyurethane. To date, there does not appear to be a significant difference in the conditioning film composition from patients with the same stent type or between the two different stent types, indicating that conditioning film deposition is not affected by different stent biomaterials or patients. Similar results have also been obtained by Tieszer et al, who showed via X-ray photoelectron spectroscopy that the elemental composition of conditioning film components was unaffected by stent biomaterial or patient characteristics (Tieszer, Reid et al. 1998). Our study found that the fifteen most common proteins include cytokeratins, serum albumin, hemoglobin subunits alpha and beta, THP, fibrinogen gamma chain, protein S100A9, vitronectin and apolipoprotein. Interestingly, the majority of the fifteen most commonly found proteins are binding sites for bacteria and thus facilitate bacterial adhesion and biofilm formation. In addition to this, the presence of calcium binding proteins such as the S100 proteins or THP may act as a nidus for encrustation. We found that significantly less Polaris stents contained THP and fibrinogen gamma chain compared to the InLay stent, eliminating these two proteins as potential targets. Overall these results validate specific conditioning film components as targets for future stent biomaterial design as they appear to play a role in stent associated infection and encrustation. Further analysis will have to be performed to determine whether commonalities exist between the physical characteristics of these components and whether they can be targeted to inhibit their deposition.

Our current experiments are aimed at studying the temporal deposition of urinary components onto the surface of stent pieces, as some proteins such as serum albumin are known to bind to other proteins rather than the surfaces themselves. In the context of urinary component deposition, it is possible that certain proteins with a higher affinity to the stent surface form a base layer to which other proteins such as serum albumin attach. Such a mechanism of deposition would be favorable for the purpose of rational stent design, as the proteins forming the base layer would make excellent targets for adhesion prevention. If temporal aspects can be determined in addition to the various layers of proteins, potential targets could be identified to prevent the initiating events of encrustation and infection.

12. Conclusions

We propose that the principles used by our colleagues in pharmaceutical research in the pursuit of rational drug design can be transferred to the design of novel ureteral stent biomaterials: 1) To understand the potential targets in ureteral stent encrustation and infection and 2) Develop biomaterials to limit these processes. Current research to date has focused on the prevention of bacterial adherence and encrustation; however, we propose that research interests should shift to the initial primary steps in conditioning film formation, thus perhaps preventing the whole cascade of events from occurring. Once these processes have been more clearly defined, the pursuit of highly specific engineered biomaterials can be started. Identification of specific targets would help direct the development of new materials and hopefully succeed where previous work has failed.

13. References

Allen, D. J., S. E. Longhorn, et al. (2010). "Percutaneous urinary drainage and ureteric stenting in malignant disease." Clin Oncol (R Coll Radiol) 22(9): 733-739.

Appelgren, P., U. Ransjo, et al. (1996). "Surface heparinization of central venous catheters reduces microbial colonization in vitro and in vivo: results from a prospective, randomized trial." Crit Care Med 24(9): 1482-1489.

Barbalias, G. A., E. N. Liatsikos, et al. (2002). "Ureteropelvic junction obstruction: an innovative approach combining metallic stenting and virtual endoscopy." J Urol 168(6): 2383-2386; discussion 2386.

Barbalias, G. A., E. N. Liatsikos, et al. (2002). "Externally coated ureteral metallic stents: an unfavorable clinical experience." Eur Urol 42(3): 276-280.

Beiko, D. T., J. D. Watterson, et al. (2004). "Double-blind randomized controlled trial assessing the safety and efficacy of intravesical agents for ureteral stent symptoms after extracorporeal shockwave lithotripsy." J Endourol 18(8): 723-730.

Benson, A. D., E. R. Taylor, et al. (2011). "Metal ureteral stent for benign and malignant ureteral obstruction." J Urol 185(6): 2217-2222.

Blaschko, S. D., L. A. Deane, et al. (2007). "In-vivo evaluation of flow characteristics of novel metal ureteral stent." J Endourol 21(7): 780-783.

Borboroglu, P. G., C. L. Amling, et al. (2001). "Ureteral stenting after ureteroscopy for distal ureteral calculi: a multi-institutional prospective randomized controlled study assessing pain, outcomes and complications." J Urol 166(5): 1651-1657.

Borin, J. F., O. Melamud, et al. (2006). "Initial experience with full-length metal stent to relieve malignant ureteral obstruction." J Endourol 20(5): 300-304.

Cadieux, P. A., B. H. Chew, et al. (2006). "Triclosan loaded ureteral stents decrease proteus mirabilis 296 infection in a rabbit urinary tract infection model." J Urol 175(6): 2331-2335.

Cadieux, P. A., B. H. Chew, et al. (2009). "Use of triclosan-eluting ureteral stents in patients with long-term stents." J Endourol 23(7): 1187-1194.

Canales, B. K., L. Higgins, et al. (2009). "Presence of five conditioning film proteins are highly associated with early stent encrustation." J Endourol 23(9): 1437-1442.

Cauda, F., V. Cauda, et al. (2008). "Heparin coating on ureteral Double J stents prevents encrustations: an in vivo case study." J Endourol 22(3): 465-472.

Chew, B. H., P. A. Cadieux, et al. (2006). "In-vitro activity of triclosan-eluting ureteral stents against common bacterial uropathogens." J Endourol 20(11): 949-958.

Chew, B. H., H. Davoudi, et al. (2010). "An in vivo porcine evaluation of the safety, bioavailability, and tissue penetration of a ketorolac drug-eluting ureteral stent designed to improve comfort." J Endourol 24(6): 1023-1029.

Chew, B. H., D. Lange, et al. (2010). "Next generation biodegradable ureteral stent in a yucatan pig model." J Urol 183(2): 765-771.

Chitale, S. V., S. Scott-Barrett, et al. (2002). "The management of ureteric obstruction secondary to malignant pelvic disease." Clin Radiol 57(12): 1118-1121.

Choong, S. and H. Whitfield (2000). "Biofilms and their role in infections in urology." BJU Int 86(8): 935-941.

Choong, S., S. Wood, et al. (2001). "Catheter associated urinary tract infection and encrustation." Int J Antimicrob Agents 17(4): 305-310.

Damiano, R., A. Oliva, et al. (2002). "Early and late complications of double pigtail ureteral stent." Urol Int 69(2): 136-140.

Daskalopoulos, G., A. Hatzidakis, et al. (2001). "Intraureteral metallic endoprosthesis in the treatment of ureteral strictures." Eur J Radiol 39(3): 194-200.

Denstedt, J. D. and P. A. Cadieux (2009). "Eliminating biofilm from ureteral stents: the Holy Grail." Curr Opin Urol 19(2): 205-210.

Denstedt, J. D., T. A. Wollin, et al. (2001). "A prospective randomized controlled trial comparing nonstented versus stented ureteroscopic lithotripsy." J Urol 165(5): 1419-1422.

Desgrandchamps, F., F. Moulinier, et al. (1997). "An in vitro comparison of urease-induced encrustation of JJ stents in human urine." Br J Urol 79(1): 24-27.

Gomha, M. A., K. Z. Sheir, et al. (2004). "Can we improve the prediction of stone-free status after extracorporeal shock wave lithotripsy for ureteral stones? A neural network or a statistical model?" J Urol 172(1): 175-179.

Gorman, S. P., M. M. Tunney, et al. (1998). "Characterization and assessment of a novel poly(ethylene oxide)/polyurethane composite hydrogel (Aquavene) as a ureteral stent biomaterial." J Biomed Mater Res 39(4): 642-649.

Hildebrandt, P., M. Sayyad, et al. (2001). "Prevention of surface encrustation of urological implants by coating with inhibitors." Biomaterials 22(5): 503-507.

Hosking, D. H., S. E. McColm, et al. (1999). "Is stenting following ureteroscopy for removal of distal ureteral calculi necessary?" J Urol 161(1): 48-50.

Isotalo, T., M. Talja, et al. (2002). "A bioabsorbable self-expandable, self-reinforced poly-L-lactic acid urethral stent for recurrent urethral strictures: long-term results." J Endourol 16(10): 759-762.

John, T., A. Rajpurkar, et al. (2007). "Antibiotic pretreatment of hydrogel ureteral stent." J Endourol 21(10): 1211-1216.

Jones, D. S., C. P. Garvin, et al. (2004). "Relationship between biomedical catheter surface properties and lubricity as determined using textural analysis and multiple regression analysis." Biomaterials 25(7-8): 1421-1428.

Joshi, H. B., S. V. Chitale, et al. (2005). "A prospective randomized single-blind comparison of ureteral stents composed of firm and soft polymer." J Urol 174(6): 2303-2306.

Joshi, H. B., N. Newns, et al. (2003). "Ureteral stent symptom questionnaire: development and validation of a multidimensional quality of life measure." J Urol 169(3): 1060-1064.

Joshi, H. B., A. Okeke, et al. (2002). "Characterization of urinary symptoms in patients with ureteral stents." Urology 59(4): 511-516.

Joshi, H. B., A. Stainthorpe, et al. (2001). "Indwelling ureteral stents: evaluation of quality of life to aid outcome analysis." J Endourol 15(2): 151-154.

Ko, R. C., PA; Dalsin, JL; Lee, BP; Elwood, CN, Razvi, H (2007). "Novel Uropathogen-Resistant Coatings Inspired by Marine Mussels." Journal of Endourology 21(Supplement No. 1): A5.

Krambeck, A. E., R. S. Walsh, et al. (2010). "A novel drug eluting ureteral stent: a prospective, randomized, multicenter clinical trial to evaluate the safety and effectiveness of a ketorolac loaded ureteral stent." J Urol 183(3): 1037-1042.

Kulkarni, R. and E. Bellamy (2001). "Nickel-titanium shape memory alloy Memokath 051 ureteral stent for managing long-term ureteral obstruction: 4-year experience." J Urol 166(5): 1750-1754.

Laaksovirta, S., T. Isotalo, et al. (2002). "Interstitial laser coagulation and biodegradable self-expandable, self-reinforced poly-L-lactic and poly-L-glycolic copolymer spiral stent in the treatment of benign prostatic enlargement." J Endourol 16(5): 311-315.

Lange, D., C. N. Elwood, et al. (2009). "Uropathogen interaction with the surface of urological stents using different surface properties." J Urol 182(3): 1194-1200.

Laube, N., J. Bradenahl, et al. (2006). "[Plasma-deposited carbon coating on urological indwelling catheters: Preventing formation of encrustations and consecutive complications]." Urologe A 45(9): 1163-1164, 1166-1169.

Laube, N., L. Kleinen, et al. (2007). "Diamond-like carbon coatings on ureteral stents--a new strategy for decreasing the formation of crystalline bacterial biofilms?" J Urol 177(5): 1923-1927.

Lewis, K. (2005). "Persister cells and the riddle of biofilm survival." Biochemistry (Mosc) 70(2): 267-274.

Li, X., Z. He, et al. (2007). "Long-term results of permanent metallic stent implantation in the treatment of benign upper urinary tract occlusion." Int J Urol 14(8): 693-698.

Liatsikos, E. N., G. C. Kagadis, et al. (2007). "Application of self-expandable metal stents for ureteroileal anastomotic strictures: long-term results." J Urol 178(1): 169-173.

Liatsikos, E. N., D. Karnabatidis, et al. (2007). "Application of paclitaxel-eluting metal mesh stents within the pig ureter: an experimental study." Eur Urol 51(1): 217-223.

Liatsikos, E. N., D. Karnabatidis, et al. (2009). "Ureteral metal stents: 10-year experience with malignant ureteral obstruction treatment." J Urol 182(6): 2613-2617.

Lingeman, J. E., G. M. Preminger, et al. (2003). "Use of a temporary ureteral drainage stent after uncomplicated ureteroscopy: results from a phase II clinical trial." J Urol 169(5): 1682-1688.

Lingeman, J. E., D. A. Schulsinger, et al. (2003). "Phase I trial of a temporary ureteral drainage stent." J Endourol 17(3): 169-171.

Lopez-Huertas, H. L., A. J. Polcari, et al. (2010). "Metallic ureteral stents: a cost-effective method of managing benign upper tract obstruction." J Endourol 24(3): 483-485.

Mandal, S., M. Moudgil, et al. (2009). "Rational drug design." Eur J Pharmacol 625(1-3): 90-100.

Marmieri, G., M. Pettenati, et al. (1996). "Evaluation of slipperiness of catheter surfaces." J Biomed Mater Res 33(1): 29-33.

Marx, M., M. A. Bettmann, et al. (1988). "The effects of various indwelling ureteral catheter materials on the normal canine ureter." J Urol 139(1): 180-185.

Masood, J., A. Papatsoris, et al. (2010). "Dual expansion nickel-titanium alloy metal ureteric stent: novel use of a metallic stent to bridge the ureter in the minimally invasive management of complex ureteric and pelviureteric junction strictures." Urol Int 84(4): 477-478.

Minghetti, P., F. Cilurzo, et al. (2009). "Sculptured drug-eluting stent for the on-site delivery of tacrolimus." Eur J Pharm Biopharm 73(3): 331-336.

Olweny, E. O., J. Landman, et al. (2002). "Evaluation of the use of a biodegradable ureteral stent after retrograde endopyelotomy in a porcine model." J Urol 167(5): 2198-2202.

Paick, S. H., H. K. Park, et al. (2003). "Characteristics of bacterial colonization and urinary tract infection after indwelling of double-J ureteral stent." Urology 62(2): 214-217.

Papatsoris, A. G. and N. Buchholz (2010). "A novel thermo-expandable ureteral metal stent for the minimally invasive management of ureteral strictures." J Endourol 24(3): 487-491.

Pauer, W. and H. Lugmayr (1992). "Metallic Wallstents: a new therapy for extrinsic ureteral obstruction." J Urol 148(2 Pt 1): 281-284.

Pauer, W. and H. Lugmayr (1996). "[Self-expanding permanent endoluminal stents in the ureter. 5 years results and critical evaluation]." Urologe A 35(6): 485-489.

Pechey, A., C. N. Elwood, et al. (2009). "Anti-adhesive coating and clearance of device associated uropathogenic Escherichia coli cystitis." J Urol 182(4): 1628-1636.

Rana, A. M. and A. Sabooh (2007). "Management strategies and results for severely encrusted retained ureteral stents." J Endourol 21(6): 628-632.

Reid, G., J. D. Denstedt, et al. (1992). "Microbial adhesion and biofilm formation on ureteral stents in vitro and in vivo." J Urol 148(5): 1592-1594.

Riedl, C. R., E. Plas, et al. (1999). "Bacterial colonization of ureteral stents." Eur Urol 36(1): 53-59.

Riedl, C. R., M. Witkowski, et al. (2002). "Heparin coating reduces encrustation of ureteral stents: a preliminary report." Int J Antimicrob Agents 19(6): 507-510.

Ruggieri, M. R., P. M. Hanno, et al. (1987). "Reduction of bacterial adherence to catheter surface with heparin." J Urol 138(2): 423-426.

Santin, M., A. Motta, et al. (1999). "Effect of the urine conditioning film on ureteral stent encrustation and characterization of its protein composition." Biomaterials 20(13): 1245-1251.

Schlick, R. W. and K. Planz (1997). "Potentially useful materials for biodegradable ureteric stents." Br J Urol 80(6): 908-910.

Schlick, R. W. and K. Planz (1998). "In vitro results with special plastics for biodegradable endoureteral stents." J Endourol 12(5): 451-455.

Shaw, G. L., S. K. Choong, et al. (2005). "Encrustation of biomaterials in the urinary tract." Urol Res 33(1): 17-22.

Sighinolfi, M. C., S. Micali, et al. (2007). "Indwelling ureteral stents and sexual health: a prospective, multivariate analysis." J Urol 178(1): 229-231.

Singh, V., A. Srinivastava, et al. (2005). "Can the complicated forgotten indwelling ureteric stents be lethal?" Int Urol Nephrol 37(3): 541-546.

Sountoulides, P., A. Kaplan, et al. (2010). "Current status of metal stents for managing malignant ureteric obstruction." BJU Int.

Stickler, D. J., A. Evans, et al. (2002). "Strategies for the control of catheter encrustation." Int J Antimicrob Agents 19(6): 499-506.

Stoller, M. L., B. F. Schwartz, et al. (2000). "An in vitro assessment of the flow characteristics of spiral-ridged and smooth-walled JJ ureteric stents." BJU Int 85(6): 628-631.

Tieszer, C., G. Reid, et al. (1998). "Conditioning film deposition on ureteral stents after implantation." J Urol 160(3 Pt 1): 876-881.

Tieszer, C., G. Reid, et al. (1998). "XPS and SEM detection of surface changes on 64 ureteral stents after human usage." J Biomed Mater Res 43(3): 321-330.

Tunney, M. M. and S. P. Gorman (2002). "Evaluation of a poly(vinyl pyrollidone)-coated biomaterial for urological use." Biomaterials 23(23): 4601-4608.

Wah, T. M., H. C. Irving, et al. (2007). "Initial experience with the resonance metallic stent for antegrade ureteric stenting." Cardiovasc Intervent Radiol 30(4): 705-710.

Wang, H. J., T. Y. Lee, et al. (2010). "Application of resonance metallic stents for ureteral obstruction." BJU Int.

Watterson, J. D., P. A. Cadieux, et al. (2003). "Oxalate-degrading enzymes from Oxalobacter formigenes: a novel device coating to reduce urinary tract biomaterial-related encrustation." J Endourol 17(5): 269-274.

Watterson, J. D., P. A. Cadieux, et al. (2003). "Swarming of Proteus mirabilis over ureteral stents: a comparative assessment." J Endourol 17(7): 523-527.

Wignall, G. R., L. W. Goneau, et al. (2008). "The effects of triclosan on uropathogen susceptibility to clinically relevant antibiotics." J Endourol 22(10): 2349-2356.

Wollin, T. A., C. Tieszer, et al. (1998). "Bacterial biofilm formation, encrustation, and antibiotic adsorption to ureteral stents indwelling in humans." J Endourol 12(2): 101-111.

Zupkas, P., C. L. Parsons, et al. (2000). "Pentosanpolysulfate coating of silicone reduces encrustation." J Endourol 14(6): 483-488.

Permissions

The contributors of this book come from diverse backgrounds, making this book a truly international effort. This book will bring forth new frontiers with its revolutionizing research information and detailed analysis of the nascent developments around the world.

We would like to thank Prof. Rosario Pignatello, for lending his expertise to make the book truly unique. He has played a crucial role in the development of this book. Without his invaluable contribution this book wouldn't have been possible. He has made vital efforts to compile up to date information on the varied aspects of this subject to make this book a valuable addition to the collection of many professionals and students.

This book was conceptualized with the vision of imparting up-to-date information and advanced data in this field. To ensure the same, a matchless editorial board was set up. Every individual on the board went through rigorous rounds of assessment to prove their worth. After which they invested a large part of their time researching and compiling the most relevant data for our readers. Conferences and sessions were held from time to time between the editorial board and the contributing authors to present the data in the most comprehensible form. The editorial team has worked tirelessly to provide valuable and valid information to help people across the globe.

Every chapter published in this book has been scrutinized by our experts. Their significance has been extensively debated. The topics covered herein carry significant findings which will fuel the growth of the discipline. They may even be implemented as practical applications or may be referred to as a beginning point for another development. Chapters in this book were first published by InTech; hereby published with permission under the Creative Commons Attribution License or equivalent.

The editorial board has been involved in producing this book since its inception. They have spent rigorous hours researching and exploring the diverse topics which have resulted in the successful publishing of this book. They have passed on their knowledge of decades through this book. To expedite this challenging task, the publisher supported the team at every step. A small team of assistant editors was also appointed to further simplify the editing procedure and attain best results for the readers.

Our editorial team has been hand-picked from every corner of the world. Their multi-ethnicity adds dynamic inputs to the discussions which result in innovative outcomes. These outcomes are then further discussed with the researchers and contributors who give their valuable feedback and opinion regarding the same. The feedback is then collaborated with the researches and they are edited in a comprehensive manner to aid the understanding of the subject.

Apart from the editorial board, the designing team has also invested a significant amount of their time in understanding the subject and creating the most relevant covers. They scrutinized every image to scout for the most suitable representation of the subject and create an appropriate cover for the book.

The publishing team has been involved in this book since its early stages. They were actively engaged in every process, be it collecting the data, connecting with the contributors or procuring relevant information. The team has been an ardent support to the editorial, designing and production team. Their endless efforts to recruit the best for this project, has resulted in the accomplishment of this book. They are a veteran in the field of academics and their pool of knowledge is as vast as their experience in printing. Their expertise and guidance has proved useful at every step. Their uncompromising quality standards have made this book an exceptional effort. Their encouragement from time to time has been an inspiration for everyone.

The publisher and the editorial board hope that this book will prove to be a valuable piece of knowledge for researchers, students, practitioners and scholars across the globe.

List of Contributors

Miroslav Petrtyl, Jaroslav Lisal and Jana Danesova
Laboratory of Biomechanics and Biomaterial Engineering, Faculty of Civ. Engineering, Czech Technical University in Prague, Czech Republic

Nobuhiro Nagai
Division of Clinical Cell Therapy, Center for Advanced Medical Research and Development, ART, Tohoku University, Graduate School of Medicine, Japan
Division of Biotechnology and Macromolecular Chemistry, Graduate School of Engineering, Hokkaido University, Japan

Ryosuke Kubota, Ryohei Okahashi and Masanobu Munekata
Division of Biotechnology and Macromolecular Chemistry, Graduate School of Engineering, Hokkaido University, Japan

Gabriel Katana
Physics Department, Pwani University College, P.O. Box 195 Kilifi, Kenya

Wycliffe Kipnusu
Institute of Experimental Physics I, University of Leipzig, 04103 Leipzig, Germany

Yong Y. Peng, Veronica Glattauer, Jacinta F. White, Kate M. Nairn, Jerome A. Werkmeister and John A.M. Ramshaw
CSIRO Materials Science and Engineering, Bayview Avenue, Clayton, VIC 3169, Australia

Timothy D. Skewes
CSIRO Marine and Atmospheric Research, Middle Street, Cleveland, QLD 4163, Australia

Andrew N. McDevitt and Christopher M. Elvin
CSIRO Livestock Industries, Carmody Road, St Lucia, QLD 4067, Australia

Lloyd D. Graham
CSIRO Food and Nutritional Sciences, Julius Ave, North Ryde, NSW 2113, Australia

Marta Corno, Fabio Chiatti, Alfonso Pedone and Piero Ugliengo
Dipartimento di Chimica I.F.M. and NIS, Università di Torino, Torino, Italy
Dipartimento di Chimica, Università di Modena & Reggio Emilia, Modena, Italy

Tsang-Hai Huang
Institute of Physical Education, Health and Leisure Studies, National Cheng Kung University, Tainan, Taiwan

Ming-Yao Chang
Department of Biomedical Engineering, National Cheng Kung University, Tainan, Taiwan

Kung-Tung Chen
College of Humanities, Social and Natural Sciences, Minghsin University of Science and Technology, Hsinchu, Taiwan

Sandy S. Hsieh
Graduate Institute of Exercise and Sport Science, National Taiwan Normal University, Taipei, Taiwan

Rong-Sen Yang
Department of Orthopaedics, National Taiwan University & Hospital, Taipei, Taiwan

Eun Mi Rhim
The Catholic University of Korea, St. Paul's Hospital, Dept. of Conservative Dentistry, Seoul, Korea

Sungyoon Huh
Shingu University, Dept. of Dental Hygiene, Seongnam, Korea

Duck Su Kim
Kyung Hee University, Dental Hospital, Dept. of Conservative Dentistry, Seoul, Korea

Sun-Young Kim
Kyung Hee University, Dept. of Conservative Dentistry, Seoul, Korea

Su-Jin Ahn
Kyung Hee University, Dental Hospital at Gandong, Dept. of Biomaterials & Prosthodontics, Korea

Kyung Lhi Kang
Kyung Hee University, Dental Hospital at Gandong, Dept. of Periodontology, Korea

Sang Hyuk Park
Kyung Hee University, Dept. of Conservative Dentistry, Seoul, Korea
Kyung Hee University Dental Hospital at Gandong, Dept. of Conservative Dentistry, Korea

Osvaldo H. Campanella and Bhavesh Patel
Agricultural and Biological Engineering Department and Whistler Carbohydrate Research Center, Purdue University, West Lafayette, IN, USA

Hartono Sumali
Sandia National Laboratories, Albuquerque, NM, USA

Behic Mert
Department of Food Engineering, Middle East Technical University, Ankara, Turkey

Foued Ben Ayed
Laboratory of Industrial Chemistry, National School of Engineering, Box 1173, 3038 Sfax, Tunisia

Mitsugu Todo
Research Institute for Applied Mechanics, Kyushu University, Japan

Tetsuo Takayama
Graduate School of Science and Engineering, Yamagata University, Japan

Lu Yan, Zhao Xia and Cai Lihui
Department of Otorhinolaryngology, Huashan Hospital, Fudan University, Shanghai, China

Shao Zhengzhong and Cao Zhengbing
Department of Macromolecule Science, Fudan University, Shanghai, China

Serpil Ünver Saraydin
Cumhuriyet University, Faculty of Medicine, Department of Histology and Embryology Sivas, Turkey

Dursun Saraydin
Cumhuriyet University, Faculty of Science, Department of Chemistry Sivas, Turkey

Marga C. Lensen
DWI e.V. and Institute of Technical and Macromolecular Chemistry, RWTH Aachen, Germany
Technische Universität Berlin, Institut für Chemie, Nanostrukturierte Biomaterialien, Germany

Vera A. Schulte and Mar Diez
DWI e.V. and Institute of Technical and Macromolecular Chemistry, RWTH Aachen, Germany

G. Fini, L.M. Moricca, A. Leonardi, S. Buonaccorsi and V. Pellacchia
La Sapienza/ Roma, Italy

Philippe Courtois
Université Libre de Bruxelles, Brussels, Belgium

Dirk Lange, Chelsea N. Elwood and Ben H. Chew
The Stone Centre at Vancouver General Hospital, University of British Columbia, Canada

Printed in the USA
CPSIA information can be obtained
at www.ICGtesting.com
JSHW011502221024
72173JS00005B/1171

9 781632 382290